# ORIGINS

## THE QUEST FOR OUR COSMIC ROOTS

Science journalist Tom Yulsman is a professor at the University of Colorado at Boulder, where he is a collaborator in the NASA-sponsored Center for Astrobiology. Former Editor-in-Chief of *Earth* magazine, Yulsman has written for the *New York Times*, *Washington Post*, *Discover*, *Astronomy* magazine, and many other publications. He lives in Niwot, Colorado.

# ORIGINS

## THE QUEST FOR OUR COSMIC ROOTS

TOM YULSMAN

IoP

Institute of Physics Publishing
Bristol and Philadelphia

*British Library Cataloguing-in-Publication Data*
A catalogue record for this book is available from the British Library.

ISBN 0 7503 0765 X

*Library of Congress Cataloging-in-Publication Data are available*

Commissioning Editor: Tom Spicer
Production Editor: Simon Laurenson
Production Control: Sarah Plenty
Cover Design: Frédérique Swist
Marketing: Nicola Newey and Verity Cooke

Published by Institute of Physics Publishing, wholly owned by The Institute of Physics, London

Institute of Physics, Dirac House, Temple Back, Bristol BS1 6BE, UK

US Office: Institute of Physics Publishing, The Public Ledger Building, Suite 929, 150 South Independence Mall West, Philadelphia, PA 19106, USA

Typeset by Academic + Technical Typesetting, Bristol
Index by Indexing Specialists (UK) Ltd, Hove, East Sussex
Printed in the UK by MPG Books Ltd, Bodmin, Cornwall

*For Sylvia, Sam, and Anna (who got her ice cream sundae)*

# CONTENTS

# ACKNOWLEDGMENTS

I began research for *Origins* in the summer of 1996, although the roots of the project go back much further. To complete a book of this scope, I had to interview many people. To all of them—and there are far more than can be listed here—I owe a great debt of gratitude.

For reviewing parts or all of the manuscript, I am especially grateful to Ian Morison, Sheila Bullock, Laura Danly, Andrew Hamilton, Ray Jayawardhana, Tim Hawkins, Wayne Hu, and my editor, Tom Spicer. They saved me from many blunders. Any that remain are entirely my responsibility.

For spending so much time to mentor me and explain the science, I am exceedingly grateful to Alan Boss, Sandra Faber, Stephen Mojzsis, Joel Primack, Scott Sandford, Frank Shu, and Neil Turok. Most especially, I thank John Bally for the enormous help he provided—and for letting me tag along during his observing run at the Canada–France–Hawaii Telescope.

For all of the support and encouragement he provided, I offer deeply-felt thanks to Bruce Jakosky. For the helpful commiseration we often shared, thank you David Baron. For reminding me of the prizes waiting at the end of the project (in the icy waters of Rocky Mountain streams), thank you Bill Travis. And for helping keep my spirits up, over caf and decaf, my gratitude goes to David Slayden.

Many people in the School of Journalism & Mass Communication and others at the University of Colorado helped make it possible for me to invest the time needed to research and write this book, including Del Brinkman and James White. Particular thanks go to my dear colleagues in the Center for Environmental Journalism: Len Ackland, Wendy Redal and Doña Olivier. Without their help and understanding, I never would have been able to complete this project.

Finally, I am deeply indebted to Sam Yulsman for the musical accompaniment that often kept me going deep into the evening, to Anna Rose Yulsman for her bright smile, infectious laugh, and constant encouragement, and to Sylvia Fibich—for more reasons than I could ever enumerate.

# PREFACE
## *Horseshoe Canyon, Utah*

*To see a World in a Grain of Sand*
*And a Heaven in a Wild Flower*
*Hold Infinity in the palm of your hand*
*And Eternity in an hour*

— William Blake[1]

South of Interstate 70 in central Utah, the San Rafael desert rolls toward the horizon, a desolate expanse of barely vegetated hills and sand dunes. For 100 square miles, the only landmarks in an otherwise featureless landscape are chocolate-colored buttes. With crenellated summits, they look like the ruins of ancient castles.

Glimmering to the south and east, the snow-covered Manti La Sal mountains crouch low on the horizon. And to the west, the crustal fold of the San Rafael Swell shoulders up from the plains, its slopes faced with giant ramps of upturned sandstone. These distant signs of physical relief serve only to heighten the impression of expansiveness, providing a sense of scale to a landscape that is difficult to read.

Snaking into the void is a wash-boarded ribbon of dirt—a rough, backcountry road that seems to lead nowhere. But looks are deceiving. Follow this road to its end and you will come to a geological and archeological treasure.

It comes up unexpectedly. One minute the desert brush and grasses seem to roll on forever. The next minute, the land drops away. Here, Barrier Creek flowing northeast toward the Green River has cut a 700-foot-deep gash into the nap of the Earth, creating what is known today as Horseshoe Canyon.

Like many such places in the Four Corners region of the United States, Horseshoe Canyon is an ideal place to contemplate a question that is the main motivation behind this book. In a cosmos that dwarfs us, what is our place?

In canyon walls and mesa tops, the gloriously exposed stone of the Four Corners region reveals stories of Earth's past—of mountain ranges rising and then eroding, and of inland seas, and oceans of Saharan sand, invading and retreating. The signs of this rich history are so vivid they offer a connection to cosmic scales of time and space that otherwise span too many orders of magnitude to be meaningfully comprehensible.

For making that connection, Horseshoe Canyon is particularly special. Four thousand years ago, hunter-gatherers of the Barrier Creek culture appear to have used its sweeping rock amphitheaters to seek their own answers to cosmic questions.

As Four Corners' chasms go, Horseshoe Canyon is not particularly deep or wide. But what it may lack in monumental grandeur is more than outweighed by its beauty and isolation. From the rim, cottonwood trees are clearly visible in the creek's sandy wash, their heart-shaped leaves shimmering in a breeze. It takes about an hour to reach their shade along a trail that switches back and forth beneath beige cliffs of Navaho Sandstone. From there, it's an easy two-mile walk to the Great Gallery, a broadly curving canyon wall that forms a natural amphitheater. Here, along a section of rock more than 300 feet across, the Barrier Creek Indians inscribed a chorus line of ghostlike figures, some as tall as nine feet. These "anthropomorphs," as scientists call them, look something like humans. But they are clearly supernatural beings.

Their faces, set directly on their torsos with no necks, are unsettling. Some have no facial features at all, just a blank visage. Others have eyes but no mouths—blank, hollow eyes staring into eternity. The bodies of these painted figures are tapered: wide at the shoulders, narrowing toward the feet. Except there are no feet. Nor are there hands or arms or legs, for that matter. In fact, the figures look like they are wrapped tightly in cloth, as if they were mummies.

Anthropologists have long speculated on the cultural and religious significance of these figures. One intriguing hypothesis is that the natural amphitheater provided the stage for an ancient form of performance art in which the painting of the figures, as well as music and dance, were part of an elaborate exercise in ancestor worship. Of course, in the absence of hard evidence, it's tempting to project contemporary cultural preoccupations and modern artistic ideas onto these ancient people. But I can't help but feel that these hunter gatherers were attracted to Horseshoe Canyon because they found it a good place to act out their connections to a world beyond their everyday existence—and in so doing fix their place in the cosmos.

In the four millennia since, we humans have learned quite a bit more about our cosmic connections. We've traced the history of the universe back to the tiniest fraction of a second after its Ultimate Beginning. We've devised powerful theories to explain how matter came from nothing and coalesced into stars and galaxies. We've described the formation of our own solar system and deciphered the very earliest history of our own planet. And we've begun to understand the origin of life.

Scientists have also done a remarkable job taking the measure of existence. They will tell you, for example, that the Planck length, the smallest length that makes any sense in the framework of the laws of physics, is a millionth of a billionth of a billionth of a billionth of a centimeter.

But do those words, ''millionth'' and ''billionth'' really mean anything to creatures our size? Do you really *feel* just how tiny that is?

In *The Elegant Universe*, physicist Brian Greene tries to make the tiniest thing that can be more palpable for readers. ''To get a sense of scale,'' he writes, ''if we were to magnify an atom to the size of the known universe, the Planck length would barely expand to the height of an average tree.''[2]

Astonishing, for sure. But do you have a *feel* for the size of the known universe, let alone how much stuff it contains? Astronomers tell us that the edge of the known universe is 14 billion light years away, give or take a couple of billion. That's the distance a photon of light, zooming at 186 000 miles a second, can travel in 14 billion years. Suffice it to say that a photon traveling from that cosmic horizon to a telescope here on Earth must traverse many trillions of trillions of miles of space.

Along the way, the peripatetic photon will trace a path through an exceedingly narrow slice of the cosmos. To get an idea of how many galaxies that slice contains, consider a single image from the Hubble Space Telescope's Deep Field Camera.

Such an image covers a tiny area of sky: a spot equivalent to the head of a pin held at arm's length. Yet within that head of a pin, hundreds of galaxies, each with many billions of stars, typically are visible.

So in this unimaginably vast realm, in which the scales of time and space span so many orders of magnitude, where do we fit? Do we occupy some special position?

Four thousand years after the hunter gatherers of Barrier Creek abandoned their performance space in Horseshoe Canyon, a Roman Catholic cardinal, Nicholas of Cusa, advanced the first truly modern answer to the question of our place in the cosmos. His medieval contemporaries believed that space was hierarchical. Earth was at the center of the cosmos. Next came concentric celestial spheres containing the Sun, Moon, planets and stars. And outside of this physical space—in fact, outside of space and time itself—was the realm of God. In a stroke of daring genius, Cusa did away with this cosmic Rube Goldberg contraption. The universe, he said, is the same everywhere. *No* place is the center. No place has a unique perspective.[3]

Cusa's insight anticipated a bedrock assumption of modern cosmology. In fact, he was so far ahead of his time, his ideas went nowhere. It took another 100 years for the medieval conception of Earth at the center of spheres within celestial spheres to begin to crumble.

Although he didn't abandon this medieval architecture of the cosmos, Nicholas Copernicus nevertheless ignited a revolution in human thought by asserting that the Sun, not the Earth, was at the center of things. As was the case with Cusa's cosmos, on the Copernican map, our planet occupied no special place.

Thanks to the astronomers who followed in Cusa's and Copernicus's footsteps, today we think we know our physical place in the cosmos. We live on a planet that circles an average star in a run-of-the-mill galaxy among many in an ordinary supercluster that's just like countless others strewn throughout the cosmos. In parallel, the Darwinian revolution helped us fix our position within the realm of life. We now know that the *E. coli* in our guts evolved by the same process as we did. Just as important, *E. coli* and *Homo sapiens* represent two branches among very many on a bushy tree of life—and our branch occupies no more special a position than does the bacteria's.

If you leave it at that, as many scientists do, it is tempting to conclude that there is nothing special about us at all. On cosmic scales, our species can be seen as a mere spark flying up from a campfire: bright, yes, but infinitesimal, fleeting and ultimately of no consequence. But Horseshoe Canyon tells a different story. There, ironically, I don't feel so small and inconsequential.

I can't get my arms around a photon zooming at the speed of light across thousands of trillions of miles of space. But in many canyons of the Colorado Plateau, I *can* get my arms around a slab of rippled sandstone. And no high-tech equipment or degree in astrophysics is needed to understand the story the ripple marks tell. They speak of tides sloshing gently in and out of a shallow bay, ruffling the sandy bottom. By running my hands over the ripples, I can make a palpable connection to a shoreline environment that existed millions of years ago.

And in Horseshoe Canyon I can also clamber up a slope of layered Navaho Sandstone. The cross-hatched pattern of the rock is vivid evidence of an ancient sea of sand dunes that shifted across this area 200 million years ago. Buried under subsequent layers of sediment, the dunes eventually petrified, retaining their distinctive layering while turning to stone. Today, the sandstone slowly erodes, shedding ancient dune sand into the canyon bottom. There it builds up in thin, cross-hatched layers, mirroring in miniature the towering cliffs above.

I like to scoop up handfuls of this sand and contemplate the story of each grain. In so doing, I've found in Horseshoe Canyon what I believe the Barrier Creek people found there 4000 years ago: a connection to the cosmos. With the sand in my hand, I can picture a star's nuclear furnace forging hydrogen and helium, elements created in the big bang, into silicon and oxygen. I can envision the star exploding as a supernova, spewing these and other elements, including those essential for life, into interstellar space. And I can picture them in a swirling disk of dust and gas surrounding our forming Sun, and coming together into a rocky planet that would later be called Earth. I can also imagine the silicon and oxygen combining into silicon dioxide, and this compound—sand, really—finding its way into dunes drifting across an ancient North America.

How is it that we humans, unlike any other species, as far as we know, can make such remarkable connections? Consider this. According to cosmologist Joel Primack of the University of California, Santa Cruz, 60 orders of magnitude separate the size of the very tiniest thing that makes sense and the very biggest thing we know about, the universe. It turns out that we humans are more or less mid-way in size between these two extremes. And that is pretty much ideal for intelligence. If our brains were considerably larger, Primack says, nerve impulses would take too long to travel from one place to another to support the very rapid processing of information. And if our brains were considerably smaller, they wouldn't have the necessary complexity.[4]

From our vantage point in the center of the cosmic space scale, and with the intelligence that this may have made possible, we are ideally placed to understand the story of the cosmos and our place in it—the story that the rest of this book will tell.

And all that within a simple grain of sand.

# PART 1

# PROLOGUE

## A Universe From Nothing?

*Here the reason flies at once to imparticularity — to a particle —*
*to one particle — a particle of one kind — of one character —*
*of one nature — of one size — of one form — a particle, therefore,*
*'without form and void' — a particle positively a particle at all*
*points — a particle absolutely unique, individual, undivided . . .*
— Edgar Allen Poe[1]

Neil Turok thinks he may know how the universe began.

That's easy, you say. The universe began in the big bang.

But there's a problem with the big bang, says Turok, a mathematical physicist at Cambridge University. At its very heart is an inexplicable enigma.

In the Beginning, according to the standard big bang theory, there was the singularity, a mind-bogglingly small point of Brobdingnagian density and temperature that somehow — no one knows how — put the BANG in the big bang. The singularity is what you'd see if you ran the classic movie of cosmic evolution in reverse to the opening frame. Every person, every bacterium, every mountain, rock and grain of sand, all the planets and all *their* life, all the stars and galaxies, and all the intergalactic gas and dust, in short, every bit of matter and energy in the universe, would zoom backward and converge in an infinitely hot, infinitely dense — and singularly inexplicable — speck.

And here, all the laws by which scientists explain nature would break down.

The modern big bang theory does much more than describe an explosion at the beginning of time. It is a precise account of the fiery birth of the cosmos and the synthesis of the very first elements from a primordial soup of elementary particles. Combined

3

with even newer theories, it also explains the origin of galaxies and bigger structures seen in the universe today.

As a theory of cosmic evolution, the big bang has stood up to many experimental and observational tests. But the theory rests on the deep enigma of the singularity, so it really tells only part of the story of the birth and evolution of the universe. And for scientists, an explanation of creation built on the inexplicable is just not satisfying.

"The problem we have is that every particle in the universe originated in the singularity," Turok, says. "That's unacceptable because there are no laws of physics that tell you how they came out of it."[2]

And so the search is on within cosmology—the science dealing with the origin, structure, and evolution of the universe—for theories of what might have happened *before* the big bang, theories that at the very least will sidestep the worst aspects of the singularity.

"Many cosmologists are now asking the deeper questions of how the universe *really* began," says Alan Guth, a physicist at the Massachusetts Institute of Technology and father of a paradigm-shifting, before-the-big-bang theory known as cosmic inflation.[3]

According to the theory, the newborn universe hyper-inflated in a burst of exponential growth, blowing up from a patch smaller than a proton to something that may have encompassed everything in the universe we now see—and all that in much less than a billionth of a trillionth of a second. This astonishing growth also ironed tiny wrinkles into the very fabric of space and time itself, slight imperfections that would later grow into galaxies. And as inflation ended, pent up energy and matter exploded in every corner of the cosmos, giving rise to the primordial fireball known as the big bang.*

---

* A word about terminology. As it has traditionally been used, "the big bang" includes the bursting forth of the universe from a singularity. But as Turok, Guth and others have pointed out, the theory says nothing about how this happened. So I have followed the lead of a number of scientists, including Guth, in using "big bang" more narrowly. In this discussion, the big bang refers only to events the theory itself describes, beginning with an expanding fireball of radiation and fundamental particles. So in this book, "before the big bang" includes the ultimate origin of the universe and events that may have ignited the big bang's primordial fireball.

"Inflation is an attempt not to explain the ultimate origin of the universe, but to provide a theory of the *bang*, a theory of what set the universe into expansion and what created the matter we see today," Guth says.[4]

Today, many scientists accept cosmic inflation as the gold standard of cosmology. But Turok is something of a contrarian. I first met him at a conference of scientists and theologians convened to discuss "cosmic questions." Tall and lanky, sporting a goatee, he showed up in a blue suit and yellow shirt. And it seemed evident that in science, as well as in fashion, he was unafraid to go his own way. At the conference, attended by Alan Guth and other champions of inflation, Turok expressed his reservations about the theory, arguing that it is built on a quicksand of assumption and the enigma of the singularity. And he described a theory that he and Stephen Hawking, his colleague at Cambridge, had devised to help put inflation on a firmer footing—and also explain how the universe *really* began.

What kind of person would have the audacity to address such issues? While some astronomers and cosmologists trace their fascination with cosmic questions to their first look through a telescope, Turok doesn't even know the constellations in the night sky, though he has long been fascinated with nature. He traces his first awareness of nature to an early childhood memory: playing with soapy water in a sink.

The year was 1961, and Neil was just three years old. His father, Ben Turok—a white member of the Umkhonto we Sizwe, the military wing of the African National Congress—had been arrested for planting a fire bomb in a Johannesburg post office. ("Thank God it didn't go off," Turok says with obvious relief even today.) And his mother, Mary, had been arrested for putting up posters protesting against the segregationist apartheid policies of the white South African government. While his parents served out their prison terms, Neil went to live with his grandmother.

When I asked him if he recalled his first awareness of nature, Neil recounted two vivid memories from this time. "I recall playing in the sink with soap bubbles, and also carrying a clock everywhere I went and constantly asking grown ups to teach me how to tell time." Several decades later, as a mathematical physicist at Cambridge, Turok would find himself exploring the mathematical possibilities of inflating bubbles as part of a theory for the origin of time itself.

Turok's mother was released from prison after six months; his father after three years. The Turok family fled the country at that point, heading first to Kenya and then Tanzania. East Africa "was a Garden of Eden for nature," Neil says. "I would spend hours outside every day looking under stones and playing with ant lions—these bugs that catch ants."

As he grew a bit older, and with inspiration and guidance from old Victorian books on the natural history of East African insects, Neil began collecting and characterizing insects. "Through these strange and beautiful volumes, you could see people who had dedicated their lives to understanding some facet of nature in a very detailed and precise manner," he says. Those words, *detail* and *precise*, characterized his own work. "In some cases, I had to dissect the genitals of an insect to classify it because often that was the only distinguishing feature," he says. No mere passing fancy, his efforts produced a world-class collection that now resides in the Oxford Natural History Museum.

Turok's passion for detail and precision carried over into another interest that blossomed in East Africa: mathematics. It was sparked partly by his father, who worked as a land surveyor in Tanzania. "He made maps with me at home and taught me the Pythagorean theorem." At the age of 7 or 8, Neil says he was also lucky to have an inspiring teacher in primary school. The Scottish expatriate, whose twin sister also taught at the school, "got me doing projects on my own. She had me surveying the school, learning trigonometry, and doing lots of experiments. She resonated with me."[5]

Turok next moved with his family to England. He went on to Cambridge University to pursue a degree in natural science, concentrating at first in biology. Moving beyond the detailed work of classification that had enthralled him in East Africa, he hoped to find deep insights about nature by somehow combining his fascination with mathematics and his passion for biology. "I always wanted to apply math to the real world," he says. "I always wanted to use it to solve problems related to nature."

But he soon grew disillusioned as he came to see biology as, in his words, "a series of historical accidents." He realized that if he continued to pursue biology as his life's work, it would resist his efforts to boil it down to some underlying fundamental principle.

"All the different forms of life are different for different reasons," Turok notes. And in ecology, in which "things are so interconnected it's hard to make meaningful, simple generalizations," he found similar limitations.

Turok decided to take a radically different direction, pursuing a Ph.D. in theoretical physics. "Biology was still interesting to me," he recalls. "But I realized that it wasn't fundamental enough for my tastes."

Turok's advisor put him to work on a problem involving the evolution of the infant universe. Turok couldn't have known it then, but his work on the problem would mark the first steps in a journey that would ultimately lead to his collaboration with Hawking. Along the way, he would make a name for himself in the rarefied world of cosmology.

In his search for the ultimate origin of the universe, Turok has worked on two very different theories in recent years. The first, which was built on earlier work by Hawking, showed how the cosmos could have simply conjured itself from a state of non-existence.[*]

In Turok and Hawking's theory, the universe sprouts naturally in a burst of cosmic inflation from a primordial object Turok describes as, well, a "wrinkled pea." (Never mind that in his recent book, *The Universe in a Nutshell*, Hawking describes it as a kind of nut. The real precision is in the mathematics, not in the helpful analogies scientists use to explain it to us.) As inflation ends, matter and energy burst forth in the massive fireball of the big bang.

Of course, this didn't really describe the Ultimate Beginning, for where did the "pea" come from? Turok's answer was that it was the inevitable result of the laws of physics. So, given those laws, run the movie of cosmic evolution backward and you'd find the pea, Turok and Hawking claimed. In fact, in their theory, the laws of physics seem to imply that something like this creationary object simply should have *been*.

---

[*] For those of you familiar with Hawking's work, this theory is based on Hawking's well-publicized "no-boundary" proposal.

So the universe sprouts from a wrinkled pea, leading first to hyper-inflationary growth and then a big bang of energy and matter—and all this is set in motion by *the hand of physics*? Strange as it might seem, Turok and Hawking's theory is perfectly scientific in that it puts itself at risk—it makes concrete predictions about what astronomers might find if the theory is really correct. If astronomers fail to find the predicted features, the theory will fall.

Some scientists say the theory has already failed the test. In its earliest form, Turok and Hawking's primordial pea produced a universe that looked profoundly empty.

"Even if all their assumptions are right, you get an empty universe," comments Andrei Linde, a physicist at Stanford University and a leading theorist of cosmic inflation. "There were so many unsettled issues around this, about which I think they were simply wrong from the beginning."[6]

Recognizing the theory's limitations, Turok and Hawking kept working at it to bring it into closer accord with what astronomers were observing through their telescopes: a dazzling collection of galaxies and galaxy clusters arranged in luminous filaments and threads stretching across billions of light years. And as of mid-2002, Hawking was still working on the theory, which he describes in some detail in his most recent book, *The Universe in a Nutshell*.[7] But in 2001, Turok decided to hedge his bets. His motivation, however, remains the same. He's suspicious of inflation, and he is deeply interested in finding a way to avoid the infinite densities and temperatures of the singularity. This time around he has some new partners, including Princeton's Paul Steinhardt, a pioneer of inflation who seems to be hedging his own theoretical bets.

Their new theory is built on ideas collectively known as *M-theory*. Turok, Steinhardt and their colleagues have called it the *ekpyrotic universe model*, after *ekpyrosis*, a Greek word meaning *out of fire*. According to the theory, our universe exists within a brane: a four-dimensional, sheet-like object floating in a five-dimensional space. Our brane world is shadowed by another, a parallel universe in its own brane. According to the theory, the gap between the two is tinier than an atom. But nothing can cross it—with one crucial exception: *gravity*.

Fourteen billion years ago, the branes were empty. Then gravity drew the branes toward each other. Crashing together, they vibrated like cymbals and erupted in bursts of primordial energy—two big bang fireballs in two parallel brane worlds. Each fireball would go on to develop as a universe in its own way. Ours developed life and us. And the other brane? Since nothing but gravity crosses the gap, we can't see what happened there. But Turok, Steinhardt and their colleagues do seem certain of one thing: matter would not have been crushed to infinite density at the crash site, causing the laws of nature to break down.

In one respect, the colliding branes theory is the opposite of the primordial pea. The latter is an attempt to describe the Ultimate Beginning—how everything came from nothing in one act of creation. But with the ekpyrotic model, a beginning is not necessary, because the collision between branes could be a repeating process, infinitely into the past and the future.

So which is it? A pea? Or the branes? Or maybe something else entirely? And did inflation *really* happen—was that the event that put the BANG in the big bang? In the first section of this book, we'll explore how scientists are attempting to answer questions like these. Suffice it to say here that observations of the universe do not yet provide enough evidence to allow scientists to rule any of these out for sure, although the pea, still championed by Hawking (okay, the ''nut''), and the ekpyrotic model, are controversial. Meanwhile, as of mid-2002, inflation had passed some crucial first tests, and most cosmologists accept it as a cornerstone of modern cosmology.

In the coming years, scientists should begin to get clearer answers. And this is certain to have a profound impact both scientifically and philosophically. After all, scientists are now trying to find a scientific answer to a question that once was the exclusive province of religion: *why is there something rather than nothing?* As Alan Guth told me, ''It seems rather amazing that such far-reaching questions can even be debated in scientific terms.''[8]

And if cosmic inflation, primordial peas and colliding branes sound too strange to be believable, consider this. The more science has scrutinized nature on fundamental levels, the more it has strained human imagination. So suspend disbelief. Embrace the strangeness. You may just draw closer to a more comforting face of nature: simplicity. The physicist John

Wheeler, a colleague of Einstein and one of the greatest thinkers of the twentieth century, saw this clearly. Writing in *Geons, Black Holes & Quantum Foam*, he says, "We will first understand how simple the universe is when we recognize how strange it is."[9]

* * *

So now, in Part 1 of *Origins*, we embark on an exploration through many scales of space and time, from the universe considered whole, to the very tiniest realms where up and down and past and future have no meaning. The journey starts with two background chapters, the first on Einstein's general theory of relativity, which describes the universe writ large, and the second on quantum mechanics, the science of the very tiny. If you already are comfortable with these subject areas, you might skim chapters 1 and 2 and move on quickly to chapter 3, *Hot Big Bang*. This chapter describes the history of the theory and how it accounts for the origin of the first elements. It also describes attempts to recreate the big bang in a laboratory so it can be studied up close and in something approaching real time. Chapter 4, *Inflating the Universe*, explores the theory of cosmic inflation. It's followed by *The Cosmic Cobweb*, which describes how inflation could have been responsible for the characteristic features of the universe we see all around us. Is there any reason to believe that inflation really happened? I pick up that thread of the story in chapter 6, *Paradigm Shift*. And finally, to close Part 1 of the book, we return in chapter 7, *Before Inflation*, to Neil Turok, Stephen Hawking, and theories attempting to explain the Ultimate Beginning.

In Part 2 of *Origins*, we'll explore the origin of stars, planets and life. Part 2 starts with a story of a meteorite fall in Portales, New Mexico. Meteorites are fossil evidence of the events that led to the formation of our world. So starting with an account of this meteorite crash is a good introduction to the chapters that follow. Chapter 8, *Interstellar Ecology*, explores the cosmic environments in which stars and planets form. Chapter 9, *A Star is Born*, describes how a tenuous interstellar knot of dust and gas collapses to form a star and the beginnings of planets. In chapter 10, *Solar Systems*, I review how our current perspective on the universe evolved, from the pre-Copernican conception of an Earth-centered cosmos to the discovery of other planets circling alien suns. Chapter 11, *Pebbles to Planets*, completes the picture of solar system formation with a discussion of how tiny grains of dust come together to form planets. And chapter 12, *To Life*, completes the story of our cosmic origins with a discussion of how our planet came to nurture life — and give rise to a species that is now searching for its cosmic origins.

I hope you get as much out of reading these stories as I've had researching and writing them.

# 1

# SPACETIME

## *The Cosmic Stage*

*Biology in its widest sense determines the laws of physics.*
— Max Tegmark[1]

Just what are space and time?

In the early 1900s, Albert Einstein saw that they were not what they seemed. Today, his answer forms the foundation for all of modern cosmology, including Hawking and Turok's theory of cosmic origins. So that's where we'll turn now.

Everyday experience suggests space and time are immovable, unchanging and absolute, experienced the same way by all observers under all conditions. From what we see around us, space also seems to be wholly separate from time. Our senses tell us that space is the stage on which the drama of the cosmos is acted out. And time marks the drama's duration from curtain rise to curtain fall.

If this is your view, you're in good company. It was also the view of Sir Isaac Newton, the man who made modern physics possible with his laws of motion and gravity. Writing in his monumental book *Mathematical Principles of Natural Philosophy*, or *Principia*, Newton made this observation about time: "Absolute, true, and mathematical time, of itself, and from its own nature, flows equably without relation to anything external, and by another name is called duration..." About space, he said this: "Absolute space, in its own nature, without relation to anything external, remains always similar and immovable."[2]

Newton's conception of space as absolute and unchanging was manifested in his revolutionary theory of gravitation. A little over 300 years after Newton proposed it in the *Principia*, this mathematical description of gravity would allow people to land on the moon and return safely to Earth. As part of his

12

theory, Newton declared that everything with mass exerted and felt an attractive force. In naming this force, he applied the ancient Latin word *gravitas*, meaning "heaviness" or "weight." And in quantifying the force between objects, he related it to two things: their masses and the distance separating them. The greater the masses, the stronger the force; the greater the distance, the weaker the force.*

With Newton's breakthrough, the planets' movements around the Sun finally were understood to be a consequence of a universal force operating in the same way throughout the cosmos—a force that also causes high tide to swell into bays, snowflakes to flutter from clouds, and avalanches to rocket down mountainsides. (Not to mention apocryphal apples to fall on scientists' heads). Using Newton's equations, astronomers could very precisely predict the paths of the planets around the Sun, and the tracks of the moons that circled them. And thanks to Newton's insight, NASA scientists could calculate how much thrust would be needed to fling space probes free of Earth's gravity and on precise trajectories towards those far away worlds.

Newton once said, "If I have seen farther than other men, it is because I was standing on the shoulders of giants." Standing on his shoulders today, we can see the rings of Saturn, the volcanoes of Io, and the sea-ice of Europa, close up.

But as exquisitely successful as Newton's equations have been in describing what gravity does, they remain silent on just what gravity is. As Newton put it toward the end of his *Principia*, "Hitherto we have explained the phenomena of the heavens and of our sea by the power of gravity, but have not yet assigned a cause of this power." He went on to explain that he had no basis on which to propose a cause: ". . . I have not been able to discover the cause of those properties of gravity from phenomena, and I frame no hypotheses."

In Newton's equations, gravity mysteriously exerts its influence *across* space. (Physicists call this "action at a distance.") It would take Albert Einstein to show that gravity actually was a phenomenon *of* space.

---

* According to Newton's equations, the gravitational attraction between two objects is proportional to the product of their masses and inversely proportional to the square of the distance between them.

Einstein made this astounding claim in 1916 with his general theory of relativity. To understand what he meant, and to see how he came to his conclusion, we must turn first to his special theory of relativity, proposed 11 years earlier.

Einstein first got to thinking about issues of space and time as a 16-year-old. His motivation was a deeply weird aspect of nature that had emerged in 1865 from the work of James Clerk Maxwell. The physicist had combined what was known about electricity and magnetism into one mathematical framework called electro-magnetism. Maxwell's equations showed that all electromagnetic disturbances — whether visible light or invisible radiation — move at a *fixed* speed: 186 000 miles per second through empty space.

To understand why this is so weird, imagine that you are a football player going out for a pass from the rocket-armed Brett Farve of the Green Bay Packers. (My apologies to readers outside the United States!) If you're racing downfield when Farve hits you with a bomb, the ball will fall gently into your arms. The reason? You are running away from the ball as it approaches you. As a result, it's speed relative to you when you finally catch it is lower than the speed you could measure if you caught it while you were standing still. By contrast, if you were unlucky enough to be running *toward* Farve as he threw the ball, it would burrow into your gut like a bullet. The reason? The ball's speed relative to you would be higher than if you were merely standing still.

But if Farve threw you a magical football made of pure light, Maxwell showed that it would make no difference whether you were running away from or toward him. If you could measure the speed of the ball, it would approach at the same exact speed either way, because electromagnetic disturbances like balls made of pure light always travel at the same speed.

Strange as it may sound, in Einstein's hands the constancy of light speed revealed by Maxwell's equations led to a revolutionary change in humanity's conception of space and time. No longer absolute and unchanging, they became relative and shifting. To get a feel for how Einstein was led to his conclusions, imagine that you are traveling on a train, the Thought Experiment Limited. You have taken note of the fact that this is not a typical train. For one thing, the windows in your car are two-way mirrors; you can't see out, but observers at track-side can see in. The train is barreling along, but the tracks are perfectly smooth and straight—you feel no bumps, vibrations or jostling. You're also heading in a perfectly straight line at a constant velocity. If you were speeding up, slowing down or changing direction, you'd feel it. But that's not the case. There are no cues whatsoever to tip you off that you're moving. So for all you know, you could be standing still.

Now imagine that you've decided to have a catch with another passenger. For fun, you've brought along one of Brett Farve's nifty footballs made of pure light. You also just happen to have an incredibly sophisticated atomic stopwatch to time how long it takes the ball to fly between you and your friend, a distance of 10 feet. It counts off time in terms of the number of vibrations of an atom. You need this high-tech hardware because, after all, it's a photon football—traveling at the speed of light, it will cover the distance in a tiny fraction of a billionth of a second.

So you toss the ball to your fellow passenger, and for sake of argument, let's say it takes 10 "ticks" of atomic vibration on your fancy stopwatch to cover the 10 feet separating the two of you.

At the same time, a person outside of the train observes your game of catch as the train approaches her. From her point of view, the ball travels the 10 feet separating you from the passenger *plus* the 100 feet covered by the train in the time it takes for the ball to fly. So from this observer's perspective, the ball has traveled 110 feet, a greater distance than you are aware of. Using one of those fancy atomic stopwatches she just happened to bring along, she also clocks the toss at precisely 10 atomic ticks—the same time of flight that you recorded.

Later on, you bump into the observer at the station—it turns out that she's an acquaintance. She asks you whether you always play catch with photon footballs on trains. Then you get to talking about it in earnest. She asks you how long it took for the ball to fly. You reply that the ball covered the 10 feet in 10 atomic ticks.

"Just a minute," your friend says. "The ball didn't travel 10 feet. It covered 110 feet—the distance the train covered plus the distance between you and the passenger."

You pause and then realize that how far the ball traveled depended on the observer's point of view. "From your perspective, you are absolutely right," you say. "But I had no way to measure the train's movement, and in fact, I had no basis whatsoever to conclude that it was moving. So from my perspective, the ball traveled only 10 feet."

Your friend looks troubled, and after some deep thought she says there's something fishy. "According to my stopwatch, it took 10 atomic ticks for the ball to travel 110 feet. And according to your stopwatch, it took ten atomic ticks for the ball to travel *100 fewer feet*. What's going on here?"

"Well, that's easy," you say. "From your perspective, the photon ball was traveling *faster* than from my perspective. That's because you saw the train's forward motion and that speed must be added to that of the ball as seen from within the car. Meanwhile, all I saw was the ball moving in a stationary car. So from my perspective, it was moving more slowly."

But as soon as you say this, you recognize your mistake. *Light always travels at the same speed*. So the ball traveled at the same speed, regardless of who was observing it.

So how could the ball as observed by your friend travel farther in the same amount of time without traveling faster? Your friend ponders this question for a minute, and then she says she has the answer. "The only way this could have worked is if time *slowed down* for you on the moving train—in other words, if one tick of your atomic clock's second hand took longer than one tick on my atomic clock."

Time slows down on moving trains? Strangely enough, according to Einstein's theory of special relativity, the answer is yes. Relative to stationary clocks, moving clocks actually tick slower. And this is not just a phenomenon of clocks. According to the theory, moving humans age more slowly relative to stationary humans because they experience time more slowly. But don't cash in all your frequent flyer miles just yet. Spending lots of time zipping about in airplanes might seem like a relativistic anti-aging elixir, but it's not. First, time on the airplanes would seem to progress at a normal pace. True, when you came home to see friends and family they would have aged more. But at the speeds of every day existence, including airplane speeds, the slowing of time is imperceptible. Your friends would notice not a wrinkle or gray hair less than what they might have expected. The effect only becomes noticeable at relativistic speeds, that is, velocities equal to a substantial fraction of the speed of light. (Where's the Enterprise when you need it?)

So according to special relativity, observers experience time in different ways, depending on how they are moving. Einstein also showed that the same is true of space. The result of any measurement an observer makes of space will vary depending on her state of motion. So Einstein showed that Newton was wrong. Space and time are not absolute.

But that wasn't the end of it. As part of special relativity, Einstein discovered another mind-bending quality of nature: energy and mass are two forms of the same stuff. And he was able to calculate the exchange rate, given in his famous formula: *energy equals mass times the speed of light squared*, or $E = mc^2$. One of the implications of this simple equation would lead Einstein to a new theory of gravity entailing a complete reconceptualization of space.

What does that equation say? For one thing, it says that if you impart more energy to something—for sake of argument, let's say the nucleus of a gold atom—it will gain mass. It has to if energy and mass are different manifestations of the same stuff. (So if you're trying to lose weight, forget that idea of traveling around in airplanes...) Now imagine that the gold nucleus is being flung around and around in a particle accelerator at ever increasing speed. What would happen?

In fact, such an experiment has been underway at the Brookhaven National Laboratory on Long Island. (More about this in chapter 3.) At Brookhaven's Relativistic Heavy Ion Collider, or "RHIC," gold nuclei are being accelerated to 99.999% of light speed in two beams going in opposite directions around a circular racetrack. At this relativistic speed, they actually gain mass. But the goal is not to test Einstein's theory, which has been well corroborated. Instead, RHIC scientists are smashing the relativistically bulked up nuclei together in hopes that the constituent protons and neutrons will melt, releasing the subatomic quarks and gluons that reside within.

So why not bulk up the nuclei even further by accelerating them even closer to light speed? Wouldn't the researchers get an even bigger bang for the buck? It's true that $E = mc^2$ means unequivocally that the more energy of motion imparted to the nuclei, the more massive they become. (Again, the equation shows how energy and mass are basically the same thing.) On the other hand, the more massive the nuclei become, the more energy it takes to push them still faster. And it turns out that to push them past the speed of light would take an infinite amount of energy. Since infinite energy is not available, this is impossible. The gold nuclei can be accelerated to blinding speed, but in practice only to just shy of 186 000 miles per second. (Photons, the particles comprising light, can travel at light speed because they have no mass.)

So not only is the speed of light constant, as Maxwell showed. It's also the ultimate speed limit. *Nothing* can travel faster than 186 000 miles per second, not a photon, a gold nucleus, a planet, or a speeding spaceship, or even simple information. For Einstein, this was both a problem and an opportunity — a problem because an absolute cosmic speed limit was fundamentally at odds with Newton's theory of gravity, and an opportunity because now Einstein could devise a deeper, more successful theory.

According to Newton's equations of universal gravitation, the attraction between two objects with mass happens *instantaneously*. But instantaneous transmission of anything across space — instantaneous action at a distance — violates Einstein's cosmic speed limit. In order for gravitational attraction to be felt instantly between the Sun and Earth, for example, the force would have to be transmitted at faster than light speed, and we've just seen that this is impossible.

Einstein solved the problem by freeing his imagination from the three-dimensional prison of everyday experience and embracing the theoretical possibilities of a fourth dimension. In doing so, he made use of a kind of geometry invented in the mid-1800s by the German mathematician George Friedrich Bernhard Riemann.

For 2100 years before Riemann's invention, geometry had consisted of the three-dimensional framework described by Euclid, the Greek teacher and mathematician. In Euclidean geometry, space is said to be "flat." To picture what this means, imagine that you are an ant — a very special two-dimensional ant living in the two-dimensional world of a beautiful National Geographic map. (Just to be clear, this is a standard projection in which longitude and latitude run in straight lines perpendicular to each other.) As far as almost all National Geoland ant mathematicians are concerned, there's only one kind of geometry:

18

Euclidean. And Euclidean geometry states that two parallel lines extended indefinitely remain parallel. But a heretical ant mathematician has had the audacity to suggest that National Geoland's two-dimensional space could be curved—that Euclidean geometry is not all there is. In such a non-Euclidean world, parallel lines extended indefinitely would *not* remain parallel. They would converge or diverge.

So one day, you and a friend decide to undertake a scientific expedition to test whether the space of National Geoland actually is curved. To conduct the experiment, you and your friend will separately head out in precisely the same direction and keep track of what happens. If space is flat, you'll remain perfectly parallel. If not, your tracks should diverge or converge, bringing you closer or farther apart.

Starting 10 miles apart, you march exactly north. Along the way, you take regular measurements of the distance between you and your friend, and you confirm that you are indeed walking in parallel, remaining at the same exact distance from each other for the entire journey. And so you conclude that your two-dimensional space is, well, as flat as a map.

In a lecture at the University of Göttingen, Riemann was the heretical ant, proposing a non-Euclidean four-dimensional geometry in which space could be curved. To picture this, imagine that National Geoland now is the surface of a globe. Starting at the equator about 100 miles apart, you and your friend once again head precisely north. If this were the old flat map, you'd never meet. But on the curved surface of the globe, you do meet—at the North Pole, of course. Similarly, in the ''elliptic'' four-dimensional geometry proposed by Riemann, there is no line that can be said to be parallel to a given line.

In a leap of geometric imagination, Einstein lifted Reimannian geometry from the realm of mathematical abstraction and applied it to the universe. The result was his general theory of relativity, in which he wove space and time together mathematically into a four-dimensional fabric consisting of three dimensions of space and one dimension of time. After general relativity, space and time would no longer be considered separate entities; they would be known instead as *spacetime*. Moreover, space would no longer be seen as just the stage for the cosmic drama. It became a *player*. Combined with time, it responds to the other players, to things with mass, such as galaxies, stars, planets, people, even snowflakes. The result is dramatic tension on a cosmic scale: gravity. And that tension causes the other players in the drama to respond—to move, it turns out, in accordance with Newton's equations.

To develop these ideas, let's make use of another thought experiment. First, imagine the taut, flat surface of a trampoline. Now picture rolling a tennis ball across the surface. What do you observe? Of course, it rolls from one side of the trampoline to another. In this scenario, the trampoline is a stand-in for spacetime, and the tennis ball is a spaceship traversing a profoundly empty quarter of intergalactic space. With no stars or galaxies nearby to gravitationally divert the spaceship on its journey, it continues in a straight line. Space is flat.

Let's put a star into the picture to see what happens to space. To do this, imagine climbing onto the trampoline and standing in the center. You are the star. And what has happened to the surface of the trampoline, our simulated spacetime? Your weight has distorted it, creating a depression where you stand. Obviously, a spaceship traveling across this surface will be affected by the warping. To demonstrate this, imagine that a friend rolls the tennis ball across the surface—not directly toward you, but just a little off center. Close to the edge of the trampoline, away from where you're standing, the taut surface is barely warped, so the ball rolls straight and true. But as it gets closer to you, the warping becomes more pronounced. So the ball's trajectory across the trampoline begins to bend toward you.

Now imagine that the trampoline is completely invisible. And a neighbor has come over to check out the strange goings on. Your neighbor doesn't know about the invisible trampoline, so she is stunned to see you suspended in mid-air, exerting no discernible effect on anything around you. Except the ball, of course. From the neighbor's perspective, you seem to draw the ball toward you with some sort of invisible force. That invisible force is analogous to gravity. And it's not a force in the conventional sense—it's not like some sort of invisible tractor beam or magnetism. It's actually the warping of the invisible surface of the trampoline. In a similar way, according to Einstein, the Sun warps the invisible four-dimensional fabric of spacetime

20

around itself, explaining why the Sun gravitationally attracts objects that come within its sphere of influence.

This gravitational attraction, as Newton pointed out, is what keeps planets in orbit around the Sun. But how do we simulate that on the trampoline? Well, you can imagine that if your friend gave the ball a push with a gentle enough touch, it would not have enough momentum to dip down into the well and back out. Instead, it would roll around the depression you make in the trampoline. And if your friend could roll it with precisely the right amount of gentle force, it would go 'round and 'round, establishing an orbit around you.

Well, because of friction, that orbit wouldn't last long. Even if your friend could give the ball just the right amount of momentum, friction would cause the ball to slow down and spiral in to the center. But hey, this is a trampoline, so what do you expect? In empty space, where there is no friction, it works just fine.

Finally, let's extend the thought experiment just one more step to see how the idea of mass warping space actually solves the problem of instantaneous gravitational action at a distance. Recall that this problem is what motivated Einstein to devise his new conception of spacetime and gravity.

So you're standing in the middle of the trampoline and you decide to take a jump. When you land, does the new depression appear instantaneously everywhere on the trampoline? Your friend is standing to the side, and she reports that the depression seemed to expand outward at a very quick rate, but certainly not instantly. In a similar fashion, if a large star were to appear suddenly in space—or more realistically, if one suddenly disappeared, as in the case of supernova —spacetime would not be disturbed everywhere instantly. According to Einstein's calculations, it would emanate *at the speed of light*.

John Archibald Wheeler, one of the most influential physicists of the twentieth century, sums up general relativity in this breathtakingly succinct way: ''Spacetime tells matter how to move; matter tells spacetime how to curve.''[3]

It's important to keep in mind that thought experiments, such as a ball rolling across a trampoline to simulate curve space, are imperfect. But they help three-dimensional creatures like us grasp the concept of curved, four-dimensional space. In the end, though, even with thought experiments, it's doubtful that many of us really get it at the gut level. For this reason, it's easy to dismiss it, as well as other theories that defy common sense. And in Einstein's day, many scientists did just that. As Helge Kragh, an historian of science at the University of Oslo in Norway, writes, "To the majority of astronomers, and of course to most laypersons, Einstein's reconceptualization of the universe was unknown, irrelevant, unintelligible, or objectionable."[4]

The late physics Nobelist Richard Feynman frequently advised his students not to reject ideas just because they defied common sense. He argued that whether a theory was strange or objectionable was irrelevant. What was important, he said, was how it stood up to *testing*. Speaking to students at UCLA, Feynman put it this way:

> *"I'm going to describe to you how Nature is — and if you don't like it, that's going to get in the way of your understanding it. It's a problem that physicists have learned to deal with. They've learned to realize that whether they like a theory or they don't like a theory is* not *the essential question. Rather, it is whether or not the theory gives predictions that agree with experiment."*[5]

Feynman was speaking about the strange quantum theory of light and matter, which he helped pioneer and which we will get to shortly. But his point pertains equally well here.

What do experiments have to say about Einstein's radical conception of spacetime? If Einstein was right that gravity is the warping of spacetime by mass, then here's a prediction: light beams should bend as they pass by massive objects, just as our tennis ball curved on the trampoline. (Remember that energy and mass are equivalent and, since light is a form of energy, it should respond to gravity, just as mass does.) To test that very prediction, the famed English astronomer Arthur Eddington led a scientific expedition to Príncipe, an island off the western coast of Africa. During a total solar eclipse viewed from the island, the scientific party photographed background stars whose positions in the sky were very close to the blacked-out Sun. (Without the eclipse, the Sun's light would swamp that from the stars, making them invisible.) Analysis of the photographs revealed that the stars' locations in the sky were slightly shifted from their normal positions. This meant that the light from those stars had bent slightly as it slalomed on the curved space around the Sun. This corroboration of Einstein's theory was heralded in newspaper headlines as a scientific triumph. And over the years, the theory has withstood many other tests. As Stephen Hawking has written, "General relativity is a beautiful theory that agrees with every observation that has been made."[6]

Today, observational cosmologists routinely take advantage of the bending of light rays by massive objects to probe the deepest reaches of the universe. The phenomenon is called gravitational lensing, because it can magnify the light from background objects in much the same way that an ordinary lens does.

The lens in, say, a telescope, bends light rays coming from a distant object so that they converge at a focal point. In this way, the lens concentrates photons, thereby magnifying the brightness of a dim object so it can be seen. In gravitational lensing, the lens consists of a massive object—a cluster of galaxies, for example. By warping the fabric of spacetime around itself, the cluster causes light rays from objects behind it to bend and converge. If the alignment is just right, light from a background object will be bent so that some of the rays converge toward Earth. This allows astronomers to use a telescope to gather more photons from the object than would otherwise be possible.

In October of 2001, astronomers announced that they had focused the Hubble Space Telescope on a gravitational lens 2 billion to 3 billion light-years from Earth. The lens is a giant cluster of galaxies called Abell 2218. Using Hubble to peer through it, astronomer Richard Ellis of the California Institute of Technology and his colleagues captured an image of a faint cloud of stars estimated to be 13.4 billion light-years from Earth. In other words, it has taken the light from these stars 13.4 billion years to reach us. And the image, as captured by Ellis and his colleagues, reveals what the stars looked like all those many eons ago. Since the universe is believed to be about 14 billion years old, give or take, these stars date to when the universe was very young. After traveling all that distance, the light from the star cloud is incredibly dim—so much so that without the gravitational lens, Hubble would never have detected it.

In one stroke, Einstein revolutionized our understanding of space and time and provided a means for observing the very farthest reaches of the universe, allowing us to learn something of our cosmic origins.

As a new conception of gravity materialized from his equations of general relativity, Einstein began to consider how it might apply universally. After all, snowflakes, stars, galaxies and other objects are not just individual, disconnected players. They form an ensemble of energy and matter. And Einstein wondered how the gravitational tension created by the ensemble might shape the universe overall. As he and then others labored to answer this question, a new cosmology began to emerge, one that led eventually to an idea as radical and transforming as relativity itself: *the big bang*.

It's natural to think that new knowledge of this sort is gained through a straightforward process in which one discovery leads logically and inexorably to the next, gradually expanding our understanding of how nature works. But scientific reality is not so tidy. Scientists often make discoveries and then fail to understand the implications. Other times, they are fully aware of the implications but reject them out of hand because they rattle the cage of conventional wisdom. And in mathematical physics, in which conclusions emerge from the solution to equations, it's not unheard of for practitioners to write off compelling conclusions as having no significance in the physical world, however interesting they may be mathematically. In the application of general relativity to the universe, all of these things happened. So while it is certainly true that Einstein's equations were a kind of road map showing the route to the big bang, only in retrospect does the path now seem obvious—as you're about to see.

Einstein presented his theory to the Prussian Academy of Sciences on November 25, 1915.[7] A little more than a year later, he published another paper, *Cosmological Considerations on the General Theory of Relativity*. As part of his research, Einstein made a simple assumption that the universe was homogenous, meaning that matter was distributed evenly throughout, and isotropic, meaning that it looked the same from every direction. With this as a given, he worked through the equations of general relativity to construct a simple mathematical representation— scientists call it a ''model''—of a theoretical universe. By solving the equations in this way, he hoped to learn what general relativity might reveal about the universe overall. His hope, of course, was that this model universe would look just like our own, which would mean that his theory offered an accurate description not just of gravity but of the entire cosmos.

But the model told him something he wasn't prepared to hear, something that did not conform to what Einstein and other scientists thought they knew. According to the model, the universe could not stand still. The gravitational field from all the matter in the universe should ultimately pull everything together, leading to a cosmic collapse. This result bothered him because he wanted his work to conform as much as possible to what observations of the universe were revealing. And in Einstein's day, there was no observational evidence suggesting that the universe was anything but static. The prevailing view was that the universe had always existed and never changed. So he altered his equations to stabilize the model universe, inserting a new mathematical term called the *cosmological constant*. It represented a repulsive force that perfectly balanced the attractive force of gravity. In this way, Einstein was able to make his model static again.

The cosmological constant fits naturally within Einstein's theory of the universe. But it was ad hoc, and ultimately, many scientists were not very satisfied with it. Even Einstein, who clung to the concept for a time, eventually was forced by new observations of the universe to drop it. Oddly enough, however, belief in the weird anti-gravity never died altogether. And today, it's back with a vengeance. But now we're getting ahead of the story...

According to Einstein's solution of the equations of general relativity, the warping of spacetime overall by the cosmic ensemble of matter and energy forms a *closed* geometry. To visualize this configuration of curved space, let's revisit the two-dimensional surface of a globe.

In two dimensions, you and your fellow ants live in a closed world—one which is finite in extent yet has no bounds. It is finite because the surface area of a sphere is limited. And it is unbounded because there is no place on the surface of a sphere where the surface "begins" or "ends." (Remember, because you are two-dimensional creatures living on a two-dimensional surface, you are unaware that your world is just the skin of a three-dimensional globe.) Head north, south, east, west or any other direction in such a world and you will eventually return to where you began. In an analogous way, Einstein's model universe was closed in the three dimensions of space and the one dimension of time. It was finite and unbounded.

Working with Einstein's equations, the Dutch astronomer Willem de Sitter found a somewhat different solution in 1917. The result was a second model of a static, closed universe. But unlike Einstein's universe, de Sitter's was utterly devoid of matter. The universe we live in obviously is not so lonely. But as de Sitter pointed out, observations at the time suggested that the density of matter in our universe was pretty low. So maybe as an approximation, his model wasn't far from the mark.[8]

There was one other aspect of de Sitter's model universe that bears mentioning. He found that if it were to contain matter, the spectrum of electromagnetic radiation from distant stars or glowing blobs of gas would be shifted from their normal positions toward the red end of the spectrum.[9] What does this mean?

A rainbow is a kind of spectrum—a series of colored lines representing the different wavelengths of electromagnetic radiation. And with a simple prism, one can fan out the different wavelengths of light comprising the glow from, say, a simple light bulb.

The shifting of light toward the red end of the electromagnetic spectrum, like that found by de Sitter in his model universe, can be caused by the Doppler effect. You've probably experienced the effect if you've ever heard a train's whistle as it raced by you. As the engine receded, the sound waves were stretched, deepening the tone of the whistle. Similarly, if a light source is racing away from you, the light waves stretch, causing an overall reddening, or a *redshift*, of the light.

So based on the redshifts of light sources "placed" in his model universe, de Sitter could have concluded that stars and gas blobs were moving apart. And from this, he could have reasoned that the very fabric of spacetime was *expanding*, carrying the embedded stars and gas blobs with it. But he did not reach these conclusions. De Sitter regarded the apparent movement of light sources that his model allowed as "spurious," a kind of artifact resulting from relativistic effects.[10]

At the time, de Sitter's solution was seen as important because it showed that when general relativity was applied on a cosmic scale, *it permitted more than one kind of universe*. Einstein is said to have been disappointed about this outcome because he had hoped there would be a unique solution to his equations — just one possible kind of universe: his own!

Five years later, a Russian meteorologist named Alexander Friedmann discovered that even more cosmic possibilities lay hidden in Einstein's equations of general relativity. Abandoning the assumption that the universe must be static, he jettisoned Einstein's cosmological constant. The result, described in two papers in 1922 and 1924, was an entire *class* of solutions.

Among these possible universes is one with a positive curvature, meaning it is closed, as in the Einstein and de Sitter models. Even without the cosmological constant, though, this universe automatically *expands*. Moreover, in such a universe, Friedmann showed that the density of matter would exceed a certain threshold, now called the "critical density." And this would result in enough of a gravitational field overall to bring the expansion to a halt and ultimately lead to a collapse.

Another possibility revealed by Friedmann's calculations is a universe with an "open" geometry. In this case, the density of matter is low, so the curvature of spacetime is negative. Cosmologists offer an admittedly imperfect analogy for such a curvature: the surface of a saddle. Such a surface curves away from itself. A cosmic geometry with this kind of curvature provides for *infinite* space. Moreover, such a universe would have less than the critical density of matter required to stop expansion. So this universe would expand forever.

According to Alan Guth, the discoverer of cosmic inflation and author of *The Inflationary Universe*, a third kind of universe was implicit in Friedmann's solutions (although Friedmann didn't discuss it in his papers).[11] This universe is balanced perfectly between collapse and eternal expansion, and the overall geometry is flat — a *Euclidean* geometry. Masses do warp spacetime here and there. But their collective effect does not curve the sheet in on itself, as with a closed universe, or away from itself, as with an open universe. The mass density in such a universe precisely equals the critical value, so gravity is not strong enough to cause a collapse. This universe keeps expanding like the open universe. But the speed of the expansion continually slows, forever approaching but never quite reaching zero.

Friedmann's work, although unheralded at the time, is now a cornerstone of modern cosmology. Today we know that the universe we live in is expanding as a consequence of the big bang. And the three basic geometries Friedmann's solutions revealed are considered to be the three basic possibilities for our own universe. Over the past two decades, scientists have been working hard to determine which one—open, closed or perfectly flat—describes the world we live in.

Think about Friedmann's expanding universes. Over time, they get bigger and bigger. One eventually stops expanding and then collapses. The others keep expanding. All of this implies that the universe *evolves*. And just as you can follow the evolution forward to imagine the fate of the universe, you can follow it backward to see how it has changed in the past. If you do this—if you rewind the cosmic videotape in your mind's eye—you might be led back to a point in time when the entire universe was crushed into a very small volume. That's not an inevitable conclusion, because you could always suppose that the universe began expanding from some pre-existing static universe, like a static Einstein or de Sitter universe. But it is a *possible* conclusion. And, in fact, one of the model universes that emerged from Friedmann's solutions to Einstein's equations was a universe that began from a singularity. "Here we have, for the first time, the idea of an expanding universe originating in a singularity—a big bang universe," the University of Oslo's Helge Kragh writes.[12] Moreover, Friedmann calculated the age of the universe. His answer? Tens of billions of years. Remarkably, the Russian meteorologist was in the ballpark.

But lest we conclude that Friedmann was a visionary, Kragh also points out that he never developed his idea of the universe beginning from a singularity. Solving Einstein's equations of general relativity was no more than an interesting mathematical exercise for Friedman.[13]

Friedmann's papers were published in the *Zeitschrift für Physik*, one of the leading physics journals at the time. Responding to the work publicly, Einstein rejected it as flawed. In May of 1923, he conceded that Friedmann's solutions, while mathematically correct, had no significance in the real world. Friedmann felt vindicated, and for a while, that's where the issue came to rest. Beyond forcing Einstein to respond, Friedmann's work had little immediate impact, perhaps because it lacked a connection to observations.[14]

In the summer of 1925, two years after Einstein validated Friedmann's calculations, if not their potential to describe reality, the Russian meteorologist rode in a balloon to high altitude, setting a Soviet record. Two months later, at the age of 37, he died. By some accounts, the cause was typhoid. But a student of Friedmann's, George Gamow—who would go on to propose the modern theory of the big bang—reported that Friedmann died of pneumonia contracted during his ride to the icy heights of the atmosphere.

So far, we've been talking exclusively about the work of theoreticians. But what were observers discovering about the universe?

In 1842, the third Lord Rosse oversaw the casting of a 72-inch mirror for what would reign as the largest telescope in the world until 1917. Installed in Ireland, it became known as the "Leviathan of Parsonstown." In 1848, Rosse focused the telescope on *nebulae*, glowing, enigmatic smudges that had long been observed in the night sky. In more than a dozen cases, he observed an intriguing feature in these dim objects: spiral arms. He and his observing colleagues (including his son) concluded that these *spiral nebulae* were collections of stars. In years to come, astronomers would debate whether they were other galaxies like our own Milky Way, or simply big clouds of gas and dust sitting inside our galaxy.

A century before that question was even asked—indeed, before Rosse even discovered spiral nebulae—the philosopher Immanuel Kant had foreseen the answer. In his 1755 work, *Universal Natural History and Theory of the Heavens*, Kant proposed that a vast archipelago of "island universes"—giant agglomerations of stars just like the Milky Way—stretched off to infinity.

So were Kant and Rosse correct? Even with the Leviathan, the technology of the day was not up to answering that question definitively. And by the end of the century, most astronomers came around to the view that the Milky Way galaxy was the sum total of the universe, with the spiral nebulae residing within it, not outside of it. Beyond the Milky Way lay a cosmic void, how expansive no one knew.

But by 1912, an astronomer working at the Lowell Observatory in Flagstaff, Arizona, began to amass evidence that these spiral nebulae might in fact lie outside our galaxy. The astronomer, Vesto Slipher, was analyzing the spectrum of electromagnetic radiation from spiral nebulae. The measurements were difficult because the nebulae were excruciatingly dim. As a result, a single measurement could take as long as a week. But Slipher persisted, and he soon found that the Andromeda nebula was *blueshifted*. This is the opposite of redshifting. It happens when a source of light is moving *toward* an observer, compressing the waves of electromagnetic radiation. Based on the degree of blueshift he observed, Slipher argued that Andromeda was rushing toward us at the extraordinary speed of 300 kilometers per second.[15]

With further work, Slipher found that many other spiral nebulae were moving at high speed, and other observers corroborated his findings. These objects were seen to be moving much faster than anything firmly known to reside in our own galaxy. So some astronomers began to wonder whether the spiral nebulae actually lay outside our own galaxy after all.

By 1925, Slipher had found spectral shifts for 45 nebulae. Of these, 41 actually were redshifts, indicating that the light was stretching and that therefore the objects were flying away from us — exactly what one would expect if the universe itself were expanding. But how far away were these objects? Slipher could not say. As a result, astronomers continued debating the precise significance of his discoveries.

By the 1920s, the American astronomer Edwin Hubble was also observing spiral nebulae. And he had an advantage over Slipher: the 100-inch reflecting telescope atop Mt. Wilson overlooking Los Angeles. With a more powerful scope, Hubble was able to gather much more light from the distant glowing smudges. And within the galaxies, he identified actual detail more diagnostic than spiral arms: *individual stars*. Among them was a kind of star called a Cepheid variable. Cepheids are unique in that they brighten and darken at regular intervals. And this proved to be the key to determining the distance to Andromeda.

When seen from a distance, any source of light seems dimmer than it would appear if it were viewed from close up.

And it turns out that brightness falls off with distance according to a simple relationship that works for candles and stars alike. Brightness falls as the square of the distance. So a candle held at one arm's length from the eyes will seem four times dimmer than one held at half an arm's length. (It is twice the distance, and two squared is four.) Theoretically, then, if you know the actual brightness of a particular star—the brightness you would perceive if it were very close to the solar system—you can use the relationship between dimming and distance to determine how far away it is. But what was the actual brightness (astronomers call it the "absolute magnitude") of the Cepheid variable stars Hubble had seen? Luckily, astronomers in Hubble's day had figured out the actual brightness of nearby Cepheids in our galaxy, and this proved to be very helpful.

It turned out that some Cepheid variable stars pulsated slowly, others quickly, and still others at in-between rates. A careful analysis of such stars showed that their period of pulsation was precisely related to their absolute brightness. The slower the pulsation, the brighter the absolute brightness. By studying nearby Cepheids at a known distance from Earth, astronomers were able to pin down this relationship so well that it became possible to determine a Cepheid's actual brightness simply by observing its period of pulsation. And that's the technique Hubble used to determine the actual brightness of the Cepheid variable stars he had spied in the Andromeda spiral nebula. With that information in hand, he compared the absolute brightness of the various stars to their apparent brightness, as seen through the telescope, to determine how much they were dimmed. Finally, he used the equation relating dimming to distance to calculate how far away the Cepheids—and therefore Andromeda—were from Earth.

Hubble's answer was *900 000 light-years away*.[16] This was far beyond even the most expansive boundaries proposed for the Milky Way. (It turns out that he was off by quite a bit. Modern measurements place Andromeda at two million light-years away.)

Hubble soon discovered Cepheid variable stars in other spiral nebulae, and these too were very distant. On January 1, 1925, Hubble's findings were announced at the American Astronomical Society meeting. (Hubble didn't attend the meeting;

someone else read his paper.) And with this, all doubt was removed. Our galaxy is but one of many in the universe.

So by the mid-1920s, Edwin Hubble and Vesto Slipher had expanded humanity's cosmic horizon. But was the cosmos itself expanding? That question still remained unanswered.

While some astronomers were making spectral measurements, Georges Édouard Lemaître, was taking a different tack. The Catholic priest and astronomer from Belgium was working out his own solutions to Einstein's equations of general relativity. Like Friedmann, Lemaître dispensed with the cosmological constant—the anti-gravity Einstein had inserted into his equations to keep his model universe from collapsing. But then he went even further, adding a realistic detail to the equations: pressure exerted by radiation in the universe.

Like Friedmann's models, Lemaître's expanded. But the Belgian priest-astronomer believed his model had real, physical significance. Connecting it to the astronomical observations of the day, he said the model described a universe with spiral nebulae that were redshifted because of cosmic expansion.[17]

With the addition of Lemaître's model, the theoretical view of the cosmos deepened, just as Hubble's discovery had deepened the observational view. Now there was Friedmann's class of universe models, and Lemaître's very similar model, both *predicting* cosmic expansion. Moreover, Lemaître's described redshifting, as did de Sitter's earlier model. Simultaneously, there were observations that actually *showed* redshifting. Given the weight of all that evidence, you could be forgiven for thinking that scientists quickly traded in the static universe for a dynamic one. But they did no such thing. Lemaître had published his ideas in an obscure journal, so not many scientists at the time heard of them. Moreover, Friedmann's work continued to be regarded as just a mathematical curiosity. So for all intent and purposes, Slipher and Hubble's discoveries were observations without an adequate theoretical explanation. Moreover, redshifted galaxies did not necessarily equate with an expanding universe. More evidence was needed.

And Hubble was about to provide it.

He continued to observe spiral nebulae, using Cepheid variable stars to determine their distances. By 1929, he had determined the distance to 24 redshifted galaxies. Thanks mostly to the work of Slipher, the *degree* of redshift, a measure of how fast a galaxy is receding from us, was known for each of them. Hubble then plotted the distance of the galaxies against their velocities. The result was a linear relationship, such that the more distant the galaxy, the faster it was racing away. Ever since, this has been known as Hubble's law. And it was exactly what one would have expected if the universe was expanding.

"Edwin Hubble changed our understanding of the cosmos," astronomer and author Donald Goldsmith writes. "He did what scientists do naturally, taking two or more sets of data that describe a group of objects in different ways and searching for correlations that may reveal an underlying truth."[18]

But how can astronomers be sure of that underlying truth? In Hubble's day, it would take a combination of the galaxy observations and a reconsideration of the dynamic theoretical models to close the case, as we'll see shortly. But on a deeper level, how can astronomers be sure that their limited observations of the cosmos provide reliable insight into the true nature of the universe as a whole?

Hubble sampled only a very tiny fraction of the countless billions of galaxies in the universe. But in the decades since, astronomers have surveyed many more. Nothing they've found has led to the repeal of Hubble's Law, and there is no scientific doubt that the universe is expanding. Even so, the total of surveyed galaxies still is a relative handful. It's always possible, therefore, that we are missing something. As Goldsmith points out, it's possible that while the galaxies in the regions of the universe we've observed are indeed receding from each other, galaxies in other, heretofore unseen regions actually are *converging*. Observations have consistently told a different story, but this possibility, however remote, can't be completely ruled out.

In fact, this issue applies to all of science. One can never have complete confidence in a universal conclusion drawn from particular observations.

Of course, that hasn't stopped scientists from believing that galaxies are receding because the universe is expanding. Or more fundamentally, that the laws of physics are the same throughout the universe. They have faith that what they observe in our neck of the woods applies to more remote, inaccessible regions of the cosmos as well, an idea known as the *cosmological principle*. Because we're not God—we don't stand outside of space and time to see the cosmos whole and complete—there is no way to know for sure that the cosmological principle is right. But there are good reasons to have faith in it. For one, it makes space science worth doing. (Why bother spending all that time, effort and money observing the universe if what you learn can't be trusted?) Moreover, repeated observations of the universe do indeed support the idea that our neighborhood is no different than any other. Finally, there's a philosophical reason to believe in the cosmological principle: explanations of nature should be as simple as possible. This idea was championed back in fourteenth century by William of Ockham, a Franciscan philosopher, theologian and writer. He wielded it so sharply that it came to be known as *Ockham's razor*. Applying it in a cosmological context, one should assume that the universe is as simple as possible. So when we are confronted with a limited but convincing set of data showing that galaxies are receding from each other, we should accept it to be true of the universe as a whole because that's *simpler* than supposing galaxies behave every which way depending on their location.

If you accept the cosmological principle, then Hubble's discovery means every galaxy in the universe is receding from every other galaxy with a speed directly proportional to the separation between them. Observers in another galaxy would see the Milky Way and all other galaxies receding from them. And from our vantage point, we see the same phenomenon. (Actually, because of gravity from large concentrations of matter here and there, some galaxies actually *converge*. That explains why Vesto Slipher saw Andromeda and a few other spiral nebulae as being blueshifted. But this is the exception to the rule.)

It would be natural to interpret Hubble's observations to mean that galaxies were racing away from us *through* space. But general relativity shows something different. To picture it, imagine drawing tiny dots on a deflated balloon. The surface of the balloon will represent spacetime, the dots galaxies. Now imagine inflating the balloon. The dots remain fixed to the surface—they do not move away from each other on the surface of the balloon. Instead, the *space* between the dots expands as the balloon is inflated, creating a greater and greater surface area. The same is true for the universe. The fabric of spacetime is expanding, taking the embedded galaxies along for the ride. And as this happens, the volume of the universe gets larger and larger. This is true even if the universe is not closed geometrically like a balloon but either open or flat as well.

But the balloon analogy also can be misleading, because it suggests that the universe is expanding into some higher-dimensional space. According to general relativity, however, the universe is spacetime and all its constituents, and that's all there is. There is nothing—no space, no time, no existence, no reality—outside of it. So to use the analogy of the balloon correctly, envision its surface as the sum total of spacetime.

Hubble's observation of a linear relationship between the distance of a galaxy and its speed of recession was enormously powerful. But it is only half of the picture. The other half is the current expansion *rate*. It is expressed as a mathematical relationship between distance and the speed of recession. To calculate it, Hubble continued his observations, determining the distance of many galaxies and comparing them to their velocities. Based on this work, he proposed that for every megaparsec of separation between a galaxy and Earth, the speed of recession increases by 530 kilometers per second. (A megaparsec is equal to 3.26 million light years.) This relationship is called the Hubble constant, and it appears in cosmological equations simply as "H."

The number was based on Hubble's distance estimates, which turned out to be inaccurate. So his value for the constant was off. In the decades since, astronomers have been working hard to pin down a more accurate number for Hubble's constant. Much depends on this effort, since the constant sets the size of the observable universe. Other things also depend on the expansion rate, such as the formation of galaxies early in the universe.

It's fitting that more precise measurements of the constant have come recently from the instrument that bears Edwin Hubble's name: the Hubble Space Telescope. In a project led by Wendy Freedman of the Carnegie Observatories, HST was used to observe Cepheid variable stars in 18 galaxies at various distances. Freedman and her colleagues then used these data with other measurements to estimate the constant. Their finding, announced in 2001: about 70 kilometers per second per megaparsec.[19] If that number is correct, it means a galaxy at a distance of 1 megaparsec is receding from us at 70 kilometers per second. Another galaxy at 10 megaparsecs is moving at 700 kilometers per second. And so on.

Other researchers have calculated similar values. So it seems that observational cosmologists are zeroing in on one of the fundamental parameters of the universe, its expansion rate, and thereby helping to complete work begun by Hubble decades ago.

Given his remarkable contributions, it's tempting to regard Hubble as the "discoverer" of cosmic expansion. Indeed, we like to believe that nature reveals herself to individual discoverers who receive this wisdom through dogged determination or blinding flashes of insight. This may humanize and simplify the narrative of science, but new knowledge rarely comes from just one person at one discrete moment in time. After all, Vesto Slipher had found redshifted spiral nebulae. And Friedmann and Lemaître had proposed models of an expanding universe. In fact, Einstein himself had "discovered" the possibility of a dynamic universe in his original application of general relativity to the universe as a whole. So who *really* discovered cosmic expansion and when?

But it is fair to say that before Hubble's observation of receding galaxies, cosmological theory was akin to a blueprint for a beautiful and daring building that no one dared complete. The reason? There were several design variations and no one was sure which one would stand the test of time. With Hubble's observations, scientists felt confident that they could move forward with construction, eliminating static parts of the design and focusing instead on dynamic ones.

For Willem de Sitter, who like Einstein had proposed a static model of the universe, Hubble's finding suggested that the focus should shift to expanding models. Einstein agreed, finally giving up his cosmological constant and famously calling it his biggest blunder. In 1930, de Sitter, Einstein and other scientists dusted off Lemaître and Friedmann's expanding universes and used them as the foundation for further work.

So as Helge Kragh puts it, "With the 'discovery' of the works of Lemaître and Friedmann in 1930, cosmology experienced a paradigmatic shift. It was only [then] that Hubble's discovery was transformed to become, i.e., interpreted to be, a discovery of the expanding universe."[20]

The acceptance of cosmic expansion naturally raised a simple question: how did the expansion *begin*? As you may remember, in one of Friedmann's models the universe began as an infinitely dense and hot dot—a singularity. But an expanding universe didn't have to begin that way. In fact, it didn't have to have a beginning at all. Our universe could have evolved from an eternally existing static universe that became unstable and began to expand.

In 1930, Lemaître began to think about this issue. And borrowing from the new science of quantum physics, he proposed that in the beginning, all the energy and matter of the universe were packed into a unique "quantum"—an unstable, primeval superatom. Developing this idea mathematically in 1931, he came up with the first big bang theory of the origin of the universe.

In the beginning, Lemaître proposed, the primeval superatom exploded, shattering into pieces with the mass of stars and giving rise to radiation. At first, expansion proceeded at an ever-increasing clip; then in a second phase, it slowed to a more sedate and steady pace, allowing for the formation of stars and galaxies; finally, in a third phase, the one we find ourselves in today, the expansion accelerated again.

Lemaître's theory of an exploding superatom at the beginning of space and time is not really the same as the modern big bang. Today's theory is much more detailed, explaining not only cosmic expansion but also the origin of elements in the nuclear furnace of the big bang fireball. But the Belgian priest and physicist did correctly intuit the general outlines of the modern theory: the universe began with a great explosion, spraying out radiation and the seeds of stars. Moreover, Lemaître realized that general relativity alone would not be enough to explain the origin of the universe. Quantum physics would be needed too.

So before we turn to the modern theory of the big bang, and then on to even deeper insights about our cosmic origins, we turn to the quantum physics revolution.

# 2

## COSMIC CASINO

### *The Realm of the Quantum*

*...the way we have to describe nature is generally incomprehensible to us.*

— Richard Feynman[1]

*The universe is not only queerer than we suppose, but queerer than we can suppose.*

— John B. S. Haldane[2]

*The surprise is not that our theories are flawed, but that they work at all.*

— John Gribbin[3]

For understanding the universe writ large—the fabric of spacetime, the nature of gravity, and the overall shape of the cosmos—Einstein's general theory of relativity is what you need. It is the fundamental science of the monumentally huge. But for understanding what the universe contains—molecules, atoms, subatomic particles, photons—you need quantum mechanics, the fundamental science of the infinitesimally tiny.

And therein lies a problem. The two branches of science explain reality in different ways. General relativity describes a reality of *smooth* spacetime and classical predictability with no fundamental role for chance. Quantum mechanics, on the other hand, describes a reality of *quanta*, discrete little lumps of energy and matter. And chance is everything; probabilities prevail. Quantum theory says that on microscopic scales you can never calculate exactly where something is, where it's heading, and how much energy it has. In the quantum world of the very tiny, nature is *fuzzy*.

So for a complete and unified description of how the universe and all its contents came to be, scientists have been struggling to find a theory that covers all the scales of the cosmos, from big to small, in a seamless and consistent way. Such a theory would explain what happened when spacetime and all the matter and energy in the universe were smooshed together to infinitesimally small scales. To accomplish this goal, many scientists have been trying to develop a theory that would bring the smooth spacetime of general relativity into the grainy fold of quantum physics—a quantum theory of gravity. Neil Turok and Stephen Hawking's pea instanton represents such an attempt. Another approach, taken by Turok and Paul Steinhardt in their colliding branes description of the origin of our universe, is to apply string theory to the universe as a whole.

The metaphors used to describe these and similar attempts at bringing relativity and quantum mechanics together—the universe sprouting from a primordial pea or exploding into existence when branes collide—are quite seductive. But the theories have so far proved inconclusive. On the other hand, insights drawn from the separate disciplines of general relativity and quantum mechanics have been cobbled together to form what some scientists are calling a standard model of the universe. The model, which includes the hot big bang theory and cosmic inflation, may not yet fulfill the ultimate dream of unifying relativity with quantum mechanics in one elegant and all-encompassing equation. But starting at less than a trillionth of a second after time itself began, it describes almost all of cosmic evolution: from the origin of fundamental particles and forces through the formation of stars, galaxies and planets, and even beyond to the ultimate fate of the universe.

\*   \*   \*

So to fully appreciate the modern story of cosmic evolution, as described by the standard model of the universe, we need to journey down the quantum road. And to get started on our trip, consider something deceptively mundane: flipping on a light switch when you walk in the house.

Why do electrons cause the bulb to emit light? In discovering the physical principles needed to answer that simple question, scientists launched the quantum revolution.

When you flip on the switch, negatively charged electrons flow through the wire, generating a magnetic field as they move. That's interesting, because magnetism is what prompts those electrons to move in the first place. Within the electric generators at the power plant, coils of wire spin through a magnetic field, causing electrons in the wire to dance.

Recognizing that electricity and magnetism were intimately related, James Clerk Maxwell united the two phenomena under one theoretical tent, as we saw earlier in the discussion of Einstein's development of the special and general theories of relativity. But it's helpful to understand in more detail exactly what electromagnetism is, and for this you need to know something about fields. To help visualize what a field is, let's consider magnetism first. You can visualize the field as the lines of force between a magnet and a piece of metal. The lines are like contracting rubber bands, pulling the metal inward toward the magnet. And at any point along one of these lines, one can calculate the strength of the attraction.

Like magnetism, gravity also is a field. As discussed earlier, anything with mass warps spacetime, and the degree of warping depends on the mass. In this way, there is a gravitational field between the Sun and the Earth—a field of ''attractiveness,'' if you like, that varies from place to place depending on the position between the two bodies.

Likewise, you can think of an electric field as lines of attraction or repulsion between charged particles. Like charges repel, of course; opposite charges attract. And what a charged particle ''feels'' when it is in the presence of other charged particles depends on where it is within the field generated by all those lines of attraction and repulsion.

In fact, the entire universe is filled with electric fields; variations in the fields from place to place are due to the presence of charged particles and are responsible for electrical phenomena. Similarly, the entire universe is filled with a gravitational field; variations in the field—the warping of spacetime—are due to the presence of mass, and these variations create the gravitational phenomena we see all around us.

Now, back to electrons and magnetism. An electric field surrounds negatively charged electrons, and when they are accelerated through a wire, the field changes. This disturbance, in turn, creates a magnetic field around the wire, which then feeds back to alter the electric field. In Maxwell's day, electrons had not been discovered. Even so, in his theory of electromagnetism Maxwell was able to show that these constantly changing and reinforcing fields of electricity and magnetism zoom off together into space at the fixed speed of 186 000 miles per second. If that speed sounds familiar it's because this is the speed of light — and nothing other than light goes that fast. So these undulations in the electric and magnetic fields were actually invisible light waves, or, more scientifically, electromagnetic radiation in an invisible portion of the spectrum.

Experiments by Thomas Young confirmed this wave-like character of light. He directed a beam of light through two narrowly separated parallel slits in a screen and recorded the pattern that appeared on a second screen. What he saw was an *interference* pattern consisting of alternating bands of light and dark — a phenomenon characteristic of waves.

To understand why, let's first explore some basic features of waves. Imagine a train of ocean waves heading for a beach. They consist of wave crests, where water is mounded up, and wave troughs, the low spots between the crests. The *wavelength* is the distance between two adjacent crests. And the *frequency* is the number of wave crests hitting the beach during a standard interval — let's say a minute. As any self-respecting body surfer can tell you, the distance between wave crests may at times be quite long, and therefore the frequency of waves hitting the beach low. But when conditions are choppy, the wavelength is short and therefore the frequency is quite high.

Of course there's one other aspect of ocean waves that a body surfer must deal with: the energy of the waves. Low, languid rollers don't offer much in the way of a thrilling ride. What you really want are wave crests that rise high — waves with a high *amplitude.* These carry enough energy to shoot a body surfer toward shore.

Now if light comes in waves, it should behave something like waves in water. And that's what Young tested with his double slit experiment. To get through the two slits, light waves must split up. As they emerge, the waves should spread out, like the expanding pattern of ripples produced when you toss a pebble into a pond. As they spread, the waves should intersect, interfering with each other both constructively and destructively. In those places where wave crests just happen to coincide, they should build on each other, producing higher intensity light. In places where crests overlap troughs, they should cancel each other. And the resulting pattern of light on the detector screen is not too difficult to imagine: bright bands wherever waves build on each other, and dark bands wherever they cancel — precisely the interference pattern Young produced with his double-slit experiment.

In 1897, the English physicist J. J. Thomson filled in the picture even more. In experiments, he discovered that the negative charges responsible for the emission of light waves were tiny point-like objects: electrons.

So whenever electrons are accelerated, whether in a wire or not, waves of electromagnetic radiation are produced. When you throw the light switch in your house, these waves radiate from the wires. You don't see them because they are in invisible parts of the electromagnetic spectrum.[4] (When your car radio goes fuzzy as you pass under power lines, you are experiencing interference from the invisible electromagnetic radiation emanating from the wires.)

But what about the visible glow from a light bulb? In the bulb, electrons move from the circuitry of your house into a tiny filament that actually resists their free passage. The filament is resistant because it has few available free electrons that can continue the propagation of the electrical current. So as the electrons enter the light bulb, some collide with electrons bound tightly to the atoms in the filament. Since these cannot flow, they jiggle anxiously instead. If you've ever touched a lit bulb, you've experienced the result of this hyperactivity: heat. In fact, when electricity flows through a bulb's filament, the temperature can exceed 1000 degrees. Of course the heat is not the point. Visible light is. But it's produced for the same reason. The electrons are being *accelerated* as they jiggle, and so electromagnetic waves zoom off, including invisible *infra-red waves*, which we feel as heat, and visible yellowish light waves (with a shorter wavelength and higher frequency).

Like a light bulb filament, a chunk of charcoal in a barbecue grill and a cast-iron poker left in a blazing fireplace both glow because heating causes electrons to jiggle and emit electromagnetic waves. Now you might imagine that since these objects are made from very different kinds of material, the kind of radiation each emits would differ. But as long as they are heated to exactly the same temperature, the objects emit radiation with the same distribution of frequencies. And in each case, the highest intensity light—the light with the highest energy content—consists of the same frequencies. Another way of saying this is that at a given temperature, there is a peak in the spectrum of light coming from the object, with some frequencies contributing the most intensity. Crank up the temperature and the peak shifts to higher frequencies and their corresponding shorter wavelengths. That's why a briquette, a poker, and a filament will all glow orange when heated to the same, modestly hot temperature, then yellow when heated further, and then white when they reach the same very high temperature. Each temperature produces a peak in intensity at a characteristic frequency and wavelength.

Physicists have a name for the kind of radiation that depends on the temperature alone: *blackbody radiation*. And it's very common. In addition to the glow from charcoal and fireplace pokers, the light emitted at the surface of the Sun is essentially blackbody radiation. So is the afterglow of the big bang, the cosmic microwave background radiation.

Today, the peak in the spectrum of blackbody radiation is understood to be the result of the quantum nature of energy. But back in the nineteenth century, quantum mechanics had not yet been developed. So scientists fell back on classical equations of thermodynamics to describe blackbody radiation. And this caused trouble, because the equations showed something absurd. The intensity, or energy content, of the light in blackbody radiation kept rising as higher and higher frequencies (or shorter wavelengths) of light were considered. Why was this absurd? Because there was no peak in intensity—the energy increased without limit, on up to *infinity*. Infinite amounts of energy are never available, so it was obvious that something was terribly wrong.

Physicists puzzled over the problem for quite some time. Then, on December 14, 1900, Max Planck told participants in a meeting of the German Physical Society that he had solved the problem. Energy is not continuous and infinitely divisible, he proposed. It is emitted only in discrete, indivisible, little lumps, or *quanta*. Each frequency of light, Planck said, is associated with a

43

minimum, indivisible denomination of energy. And the higher the frequency, the higher is the minimum energy denomination.

When the equations describing blackbody radiation take this into account, infinite energy goes away. And a peak in the spectrum appears that matches exactly the results of experiments.

To get electrons to jiggle and emit lightwaves, you have to add energy, which raises the temperature. When you do this, electrons vibrate at different rates. Some vibrate feebly and produce low-frequency (longer wavelength) light waves. Now according to Planck's quantum scheme, such low frequency light waves come in minimum-energy lumps that are very small. (Think of them as the pennies of the quantum monetary scheme.) That means it's easy to produce a large number of these low-frequency, low-energy waves. Meanwhile, some electrons vibrate quite vigorously, producing high frequency (shorter wavelength) light waves. These come only in very big minimum energy lumps. (Think of them as the hundred dollar bills of the quantum monetary system.) So only a very few can be produced.

The upshot is that on average, a blackbody will emit far more light consisting of lower-frequency, lower-energy lumps than higher frequency, higher-energy lumps. In this way, the amount of energy consisting of high-energy lumps is limited. This produces a peak in the spectrum and thereby eliminates the infinite energy problem. When the temperature is raised, more energy is available, so somewhat more higher-energy radiation lumps can be produced. This simply shifts the peak to higher frequencies.

Finally, let's bring this back to the color of a glowing chunk of charcoal in a barbecue grill. Since modest heating means fewer high-energy radiation lumps can be produced, the peak sits at a lower, *redder* frequency. Because heating to a very high temperature allows a greater number of high-energy lumps to be produced, the peak then shifts to a higher, *whiter* frequency. And that's why really high temperatures make things white hot.

How are Planck's energy lumps actually calculated? In his formulation, the basic denomination of an energy lump of any kind of light—red, blue, visible, invisible—is determined by two quantities. The first is the frequency of the light, and the second is a fundamental constant of nature, which Planck worked out and thus bears his name today. In his equation, *energy equals Planck's constant times the frequency of the light.* Mathematically, this is expressed as $E = h\nu$, where $E$ is the minimum, indivisible energy lump, $h$ is the constant and $\nu$ is the frequency. As the equation shows, the higher the frequency of light, the bigger is $h\nu$, or the minimum energy lump for that frequency.

For Planck's work on this new quantum theory, he was awarded the Nobel prize in 1918.

Now you may be wondering what those $h\nu$ energy lumps actually are. Well, so did Planck. He thought they had something to do with the emitting atoms, not some fundamental quality of light itself. But he didn't really know. And in the absence of a physical explanation, the only justification for the lumps was that they solved the blackbody radiation problem.

In 1905, Einstein took the quantum idea to a deeper level, proposing that Planck's energy lumps actually *were* a fundamental quality of light itself. To explain a puzzling phenomenon called the photoelectric effect, Einstein argued that a beam of electromagnetic waves actually consisted of billions of *particles*, each with energy equal to $h\nu$.[5] Each of these particles, later dubbed *photons*, represents the minimum lump of energy for a given frequency of light. For this work, Einstein was awarded the Nobel Prize in physics in 1921. (Odd as it may seem, he did not win the prize for his relativity work.)

At that point, physicists were left in an uncomfortable situation. Electromagnetic radiation was now described in two different ways. Sometimes it made sense to describe it with Maxwell's wave equations. Other times, it made sense to think of light as being made of photons.

It worked. But it wasn't pretty. So in some fashion, the two descriptions had to be brought together in one framework.

The Frenchman Louis de Broglie accomplished the feat in 1924—and as a doctoral student, no less. He took note of the fact that Einstein had related the energy of an *object* to its mass, with the equation $E = mc^2$. Now bear in mind that a photon actually has no mass. (If it did, it couldn't fly at the speed of light.) But a photon is like an object in that it does carry energy of motion, or momentum. De Broglie also noted how Planck had related the energy of a *wave* to its frequency, with the equation $E = h\nu$. One equation deals with the energy of objects like photons. And the other deals with the energy of electromagnetic waves. If the two could somehow be brought together, the result would be a more unified description of light.

De Broglie did just that, finding that the wavelength of light is equal to Planck's constant divided by the momentum of the light's photons. But he went even further, arguing that if this equation works for a photon — if the equations are perfectly happy with a photon acting like a particle and like a wave at the same time — then why not for an electron? Or *any* particle for that matter?

His doctoral committee was skeptical. So his advisor sent de Broglie's proposal to Einstein, who was duly impressed. De Broglie was awarded his doctorate — and a Nobel Prize in 1929, after researchers confirmed that individual electrons do indeed have wavelengths. And through this new manifestation of quantum theory, humanity gained a new example of how strange the universe can be.

And it was only going to get worse.

By this point, you're probably wondering whether any of this has a physical basis in the real world. After all, just because something works in an equation doesn't mean that it actually works in nature. So to convince you that de Broglie's Nobel Prize was justified, consider a version of the two-slit experiment I described earlier. In this case, a beam of electrons is fired at a screen with two small parallel slits. The diameter of the beam is wide enough so that electrons can go through either of the slits. And when they do, they strike a television-like detector screen. As the electrons stream though, the screen glows, recording the hits.

In experiments like this, electrons passing through the two slits develop an interference pattern on the detector screen, just as the light did in Young's double-slit experiment. The conclusion? A stream of electrons, like light, has a wave-like quality.

Maybe that's not so surprising. Ocean waves are made of particles. They consist of countless water molecules moving collectively in an undulating pattern. Might not the same be true of a beam of photons or electrons? Well, it turns out that nature isn't that obliging to the entreaties of everyday experience.

Let's run the experiment again, but this time let's turn the electron gun down so low that it emits one electron at a time every few seconds. Let's also close up one of the slits. As you would expect, as one electron after another hits the detector screen, a little glowing spot builds up at the impact spot. It also turns out that not all of the electrons make it through the slit. In fact, only about 2% of them do. Continuing the experiment, we open up the other slit and block the first one. Again, as expected, a little glowing spot builds up on the screen in line with the open slit, and about 2% of the fired electrons actually are recorded as hits.

Now let's make things interesting. We open both of the holes and fire one electron at a time. There should be no interference pattern this time. Common sense dictates that some electrons will go through one slit, others will go through the second slit, and the result will be two glowing spots on the detector screen. But as I'm sure you've already guessed, that's not what happens. Over time, instead of two glowing spots, a series of dark and light stripes builds up. We've seen this before. It's an interference pattern.

Now this is really weird. An electron seems to go through one of the slits — we don't know which one — and hits the screen. Successive electrons fly through one at a time, hitting the screen, and the interference pattern that builds up suggests that each one's course is somehow being interfered with in a manner that's characteristic of a wave phenomenon. It seems as if the individual electrons are somehow interfering with each other, even though they're going through one at a time. And as they do, they interfere constructively in some places, building up bright bands, and destructively in others, resulting in dark bands.

Well maybe individual electrons are little wave packets that go through both slits at once. In this case, each electron would interfere with itself, and one by one the interferences would add up, creating the visible pattern on the detection screen. So to test whether this is right, let's make one more modification to our experiment. When we fire an individual electron, let's try to determine which slit it goes through. With a bit of luck, we'll see individual electrons going through both slits at once. We can run this test using a special detector that ''watches'' the slits. Common sense says that watching the slits should make no difference in the pattern that appears on the detector screen. But you probably realized long ago that common sense isn't the best guide when it comes to quantum physics. And in fact, when the slits are "watched," the interference pattern disappears. It's replaced with two bright spots in line with the slits.

So when the cat's away, the electrons will play — they seem to go through two slits at one time and build up an interference pattern indicative of a wave phenomenon. Or else they somehow interfere with each other even though they're going through one at a time. But when those pesky little electrons are watched, they behave themselves, going through one slit or the other, minding their own business, and building up normal little spots.

Well now *that's* a relief.

Okay, not really. Whether an electron behaves like a particle depends on whether it's being watched? How can this be?

For the experimental apparatus to detect an electron as it goes through one slit or the other, or both, it has to interact with the electron in some way. To see why this is so, consider a baseball game. How do you follow the trajectory of a ball thrown by a pitcher? You see it with your own eyes, of course. But what allows you to see it? Light shines on the ball and reflects into your eyes. Since the ball is so big and carries much more momentum that the photons, for all intent and purposes the ball is unaffected by the light. But the same is not true of an electron, because it's very, very, very tiny, and weighs very, very, very little.

So we're able to watch a baseball because photons are bouncing off it and into our eyes. Similarly, to "watch" an electron as it's fired at a screen with two slits, the sensing device must get information from it by bouncing photons off it and measuring the return signal. But since an electron is so small, bouncing photons off it actually changes the experiment. One way to conceptualize this is to imagine that an "unwatched" electron is actually a little wave. When it's "watched," the bouncing photons prod the little electron wave, causing it to "collapse" into a little electron object, which then obligingly races through one slit or the other just like any self-respecting particle would do.

In 1926, Max Born proposed that the little wave representing an electron actually is a *probability* wave. Then the Austrian physicist Erwin Schrödinger took this idea further, developing a wave equation for electrons. The basic idea is that where the probability wave is strongest, the electron (or any particle) is most likely to be found—and this is precisely where classical physics would say it should be found. Places where the probability wave is weakest—theoretically the wave fills the entire universe—represent locations where it is least likely to be found. But just because an electron is not likely to be found in a particular place doesn't mean that it would *never* be found there. Probability means that there's a calculable chance it could be found in an unlikely place.

Schrödinger's wave functions, as they came to be known, proved very accurate at describing the behavior of particles, including what happens when they are shot through slits, as in our hypothetical experiment. And according to this view, when an electron is unwatched, its wave function spreads out. Fire a succession of electrons at double slits in a screen and each spread out wave will interfere with the others and itself, producing the interference on the detector screen. But when you watch as they approach the slits, the electron wave functions collapse, meaning each one's position can be known more precisely, and each electron seems to go through one or the other holes, producing no interference.

In the macroscopic world, probabilistic effects can be ignored. You can say with certainty that the ink making up this sentence is right here on the page, not staining your clothes or splattered on the wall. But you cannot say with certainty where the electrons and other particles that make up the ink are located. All you can do is say that there's a very high probability that they're right *here*, and not on Pluto.

What are the ultimate limits to predicting, even probabilistically, the position and motion of a particle like an electron? In 1927, the German physicist Werner Heisenberg answered that question with what came to be known as the *uncertainty principle*. To understand it, let's return to our experiment.

By bouncing a photon off an electron as it goes through the screen, you are in effect trying to figure out exactly where it is. Is it going through the right slit or the left slit? When a photon used for detection bounces off the electron, it returns to the sensor with data that tell you where the electron was—right slit or left. But the photon's precision in this task is limited—and you just have to accept this as a physical fact—by the distance between two of its adjacent wave crests: the photon's wavelength. The longer the wavelength, the less precise is the positioning information returned by the photon. The shorter the wavelength, the more precise.

So if in the experiment you use long wavelength light to detect the position of electrons, you can't tell very accurately where the electrons were. On the other hand, long wavelength light carries less energy than short wavelength light. (Remember that long wavelength light has a lower frequency; using Planck's equation, $E = h\nu$, you can see that the lower the frequency, or $\nu$, the lower the energy.) So by using long wavelength light, you cause less of a disturbance to an electron as it passes through the screen. And this means it can return more precise information about where the electron was actually *heading* and *how fast*. Conversely, short wavelength light can more precisely pinpoint the electron's *position*. But the higher energy of a short wavelength photon disturbs the electron's movement more. Thus it is less able to pinpoint where the electron was heading and how fast.

Werner Heisenberg demonstrated this principle mathematically. His equation states that the more precisely one wants to pinpoint the position of a particle, the less precisely one is able to know its momentum (where it's heading and how fast). And the more precisely one wants to pinpoint the momentum, the less precisely one is able to know its position.

Heisenberg's uncertainty principle also applies to measurements of a particle's energy. One can never know precisely how much energy a particle has at a precise moment. The more precisely you try to pin down a particle's energy, the longer it takes to complete the measurement. If you try measuring the energy perfectly, it will take forever, which of course means you can never complete it. All of this raises an interesting question: what happens while a measurement is being made? By definition, before the measurement is complete, there's no way of knowing how much energy a particle has. So for all anyone knows, the energy can vary widely.

Let's take this to an even weirder level. Remember that the double-slit experiments showed decisively that, depending on how you look at them, photons and electrons can take on either a particle-like or wave-like quality. So now let's think of the particles as waves. Just as a reminder, the quantum physics revolution showed that photons can be described as probability waves that make up an electromagnetic field. In this sense, the photon as particle-like entity is the irreducible bit—the quantum—of the electromagnetic field. Similarly, electrons can be described as probability waves that make up an electron matter field. And in this sense, the electron is the irreducible bit—the quantum—of the electron-matter field. The vacuum, supposedly devoid of matter and energy, is actually filled with these fields. And because of uncertainty, during an interval short enough to prevent any kind of measurement, such fields can fluctuate wildly. That means a field can go from zero energy to lots of energy out of the clear blue. (Or maybe I should say out of clear uncertainty.) When this happens to an electron-matter field, a pair of particles, an electron and a positron (a positively charged electron), can pop into existence, travel a short distance, come together and annihilate, in much less than a trillionth of a second. When they annihilate, the energy they ''borrowed'' from uncertainty to come into existence is released. And as long as all of this happens in too short an interval for a measurement to be made, there's nothing illegal about it from the point of view of physical law. Energy conservation is not violated, because the exact amount of ''borrowed'' energy is ''paid back'', so the net change in energy is zero. And the particles aren't real in the sense that they don't survive long enough to be detected. This is why such particles are called *virtual* particles.

Fluctuations in an electromagnetic field, in this case produced by an accelerating electron, can also produce virtual particles: *virtual photons*. Again, if the fluctuation happens fast enough, a virtual photon can flit into existence and then disappear before it can be detected. As far as the laws of physics are concerned, the energy books remain in balance, and nothing is amiss.

The fun part is that this is happening all the time, all around us, up in space, in the supposedly empty vacuum, everywhere in the universe, in fact —and without our ever knowing it, at least not directly. Consider just a single electron. Because the universe is not at a temperature of absolute zero (at which all motion would stop), the electron has a temperature, which basically means it jiggles. This vibration creates an electromagnetic field. The electromagnetic field, in turn, gives birth to a cloud of virtual photons that surround the electron. Each virtual photon can, in turn, produce virtual electrons and positrons. And these produce virtual photons, and so on down the line. Now ramp that up to all the electrons in the universe. Each is surrounded by virtual photons flitting in and out of existence; these are surrounded by virtual electron–positron pairs, which are surrounded by virtual photons.

When examined from the point of view of quantum mechanics then, a vacuum is not pure nothingness. It is boiling over with wildly fluctuating quantum fields that give birth to virtual particles. These pop up like sea spray and then vanish before they can be seen.

Anyone could be forgiven for thinking this is ridiculous. After all, by definition virtual particles flit in and out of existence so fast that the events can't be witnessed. And if they can't be witnessed, how in the world can anyone say they actually happen? Well, it turns out that it's relatively easy to detect this hyperactive virtual activity through indirect means. For example, in 1947 physicist Willis Lamb of Columbia University measured two different states of the hydrogen atom previously thought to have exactly the same energy. But Lamb showed that they had slightly different energy. The reason? slightly different contributions of energy from virtual particles. (The phenomenon is known as the *Lamb shift*.)

In another kind of experiment, two metal plates are held parallel to one another with only a very tiny gap between them. If the quantum view of virtual particles is right, quantum wave fluctuations outside the plates—the kind that give rise to virtual particles—should be vigorous. But within the gap, the fluctuations should be more subdued because only a limited number of wavelengths can fit in the tiny space. Because of this, there should be more fluctuations, and therefore more virtual particles, outside than inside. This should tend to push the plates together with a slight force. And when the experiment actually is carried out, that's precisely what happens. (The phenomenon is known as the *Casimir effect*.)

The cloud of virtual photons that exists around every electron in the universe is key to understanding a fundamental aspect of nature: the repulsion of like charges, such as negatively charged electrons. The interaction between two electrons is actually pretty simple. For sake of argument, let's say that one electron is moving toward another, like a cue ball that's been shot at the 8-ball on a pool table. In pool, the cue ball ricochets off the 8-ball, sending the 8-ball flying. Similarly, when one electron gets close enough to the other, the like charges of the two particles repel each other, and they ricochet away. Interestingly enough, the comparison here is deeper than mere analogy. Billiard balls bounce off each other (as opposed to passing right through each other) because electrons on their surface repel each other.

What is the nature of the repulsive force between negatively charged electrons? According to quantum mechanics, the repulsion involves the exchange between electrons of virtual photons.

Let's keep things simple by considering the emission of just a single virtual photon by an electron instead of an entire cloud. As one electron approaches another, a virtual photon borrows energy from uncertainty, pops into existence and skids across the distance separating the two particles. In the process, the emitting electron recoils, just as a gun recoils when it's fired. When the virtual photon reaches the other electron, it's absorbed. In the process, the photon's energy is imparted to the electron, which dutifully zips off in another direction, deflected from its previous path.

That's simple enough, but it turns out that the analogy works only for repulsion of like charges. It does not explain *attraction* of *unlike* charges. Take, for example, the attraction felt between a negatively charged electron and a positively charged proton. In the quantum mechanical view, an exchange of virtual photons still explains this interaction—it keeps electrons dancing around atomic nuclei. But the picture of ricocheting pool balls doesn't work. So to understand the attraction between electrons and protons, think of the photon as a courier that carries a message between the particles: *move toward each other*.

This basic idea of virtual photons acting as messengers that tell charged particles how to interact is the heart of quantum electrodynamics, or QED. Developed by Richard Feynman and others, the theory of the electromagnetic force, or ''interaction,'' as physicists often prefer, explains why this book doesn't slip right through your hands. Atoms are mostly empty space, so by all rights the atoms in the book should pass right through the atoms in your hand. It's the repulsion of like charges in those atoms, stemming from the electromagnetic interaction, that keeps this from happening.

In the mid-1970s, scientists incorporated QED into a more expansive theoretical framework called the standard model of particle physics. The model was designed to describe the elementary particles and the forces that govern their interactions: the electromagnetic interaction, the strong interaction and the weak interaction. (Gravity, the fourth fundamental force of nature, is the odd man out—as we'll see shortly.)

As with electromagnetism, the strong interaction arises from exchanges of virtual particles—in this case, exchanges of *gluons* between quarks within atomic nuclei. Without this exchange, atoms couldn't exist. That's because the strong interaction is what glues quarks together to form protons and neutrons.

The part of the standard model that describes the strong interaction is called *quantum chromodynamics*. With QCD, scientists redrew the old picture of the proton and neutron. Gone was the idea of a hard little lump of matter, replaced by something ethereal, effervescent and insubstantial: a quantum cloud less than a trillionth of an inch in diameter, crackling with technicolor lightning.

The cloud gets its identity as a proton or neutron from different combinations of three *valence quarks*. Like moths fluttering around a light bulb, they race about in an aimless frenzy. A proton is made from two *up* valence quarks and one *down*. Since an up quark has a charge of $+\frac{2}{3}$, and a down quark a charge of $-\frac{1}{3}$, the result is a net positive charge: +1. A neutron is made of two down valence quarks and one up, yielding a net charge of zero.

The valence quarks of a proton make up just 2% of its mass. So while they're important for establishing the identity of the proton as a positively charged entity, most of the mass has to come from another source: more virtual particles, which are teased out of the quantum mechanical vacuum by the valence quarks themselves. These virtual *quark and antiquark pairs* can consist of more up and down quarks as well as other kinds known as strange, charmed, bottom and top.

These terms are completely arbitrary. For example, an up quark has no quality of *upness* to it. And a strange quark isn't really strange. The labels are mnemonics—easy-to-remember names that represent abstract qualities best described by mathematics.

Now we come to the gluons. As described by QCD, they keep the valence quarks from straying too far from each other through something called a *color field*. Here's how it works. Each of the valence quarks can take on one of three *color charges*: red, green or blue. (Again, the labels are arbitrary. Think of these colors as representing different settings for the quarks.) As the quarks continuously exchange gluons with each other, they also swap colors. So when a gluon goes from, say, a blue quark to a green quark, the color settings reverse: the blue quark turns green and the green quark turns blue. (It makes no difference whether the quarks are up or down.) And this swapping of the quarks' color charges, mediated by the gluons, is what keeps them bound together.

There is a strange aspect to this phenomenon. As quarks draw closer together, the strong force binding them together actually weakens. Pull them *apart*, however, and the force *strengthens*. Here's why: the gluons surrounding, say, a momentarily red quark carry a red color charge—and they amplify the red color field around it. Now if you pull on this red quark, trying to separate it from its two partners, more red gluons bubble up out of the vacuum, strengthening the field even further. The quark may exchange a gluon with one of its partners, changing its color, but the color field still reacts the same way. The harder and harder you pull on the quark, the more gluons bubble up and the stronger the color field becomes. And if you try to wrench an individual quark completely free of a proton or neutron, the color field will behave like a rubber band, causing the quark to snap back into place. As a result, it is impossible to isolate a quark from a proton or neutron.

Combine the perspectives of QED with QCD, and an atom seems far more vibrant than the naïve picture of electrons orbiting sedately around a nucleus, like planets around a star. If you could take a fantastic voyage into the heart of an atom, it might look like this: swooping in from afar, the nucleus is obscured by boiling cumuli of electrons, virtual photons, and virtual electrons and positrons. Buffeted by this quantum turbulence, you dive down fast, hoping to find the nucleus. When you do, it becomes clear that looking for a firm landing site on a proton or neutron is pointless. *It's atmosphere all the way down* — wisps of virtual quarks and antiquarks in the nuclear stratosphere, and denser clouds of gluons lower down. Within the gluon cloud, you see red, green and blue lightning bolts flashing continuously between the trio of valence quarks, which change color with each strike.

The standard model describes one more force: the weak interaction. Its role seems like the most esoteric of the fundamental forces. The weak interaction is responsible for radioactive decay in such substances as uranium and plutonium. But appearances are deceiving, because without such decay, there would be no nuclear weapons and no nuclear power plants. In fact, without the weak force, fusion reactions within stars would not occur. And without fusion, there would be no elements in the universe heavier than hydrogen, helium and lithium. That would mean no stars, no planets, no normal chemistry, no life. Like the other forces, the weak interaction is mediated by virtual messenger particles: the neutral $Z^0$ particle and a pair of oppositely charged particles, the $W^+$ and the $W^-$.

Gravity is the fourth fundamental force of nature, and gravitational attraction between masses can be described in quantum mechanical terms too, with the force being transmitted by a theoretical messenger particle known as the *graviton*. But there's a problem with quantizing gravity in the form of this irreducible bit of a gravitational field. Putting aside the fact that the graviton has not yet been observed, this quantum mechanical view of gravity is diametrically opposed to the description given by general relativity, in which gravity emerges from the warping of the *smooth*, *unquantized* fabric of spacetime. Moreover, although the quantum description of gravity is similar to the descriptions of the other forces, it is not part of the standard model of particle physics. This is obviously a problem — one that is driving scientists today in a search for what has been called the Theory of Everything, or TOE, which would explain all the fundamental building blocks and forces of nature in one unified way.

Richard Feynman, who advised his students not to reject the strange things nature was trying to tell them, was one of the pioneers of the standard model of particle physics. Born of Russian and Polish Jews in the Far Rockaway section of New York City, he went on to become one of the most influential — and loved — physicists of the twentieth century. Feynman displayed his brilliance early. As a Ph.D. student at Princeton studying under the direction of John Archibald Wheeler, he developed an approach to quantum mechanics that would later figure prominently in Neil Turok and Stephen Hawking's attempt to devise a theory for the ultimate origin of the universe. It would also prompt Albert Einstein to reiterate one of his most famous statements.

Wheeler dubbed his student's innovation the "sum-over-histories" approach, and to understand it we need to return one more time to the double slit experiments. Recall that when a single electron is fired at the screen, it's not possible to say exactly which path it takes. The best one can say is that some paths are more likely than others. To calculate which one is likeliest when a particle goes from Point A to Point B, Feynman proposed that one should take into account every possible path available in the universe to complete the journey. In fact, in Feynman's technique, the assumption is that the particle takes *all* of these paths simultaneously, and the one we actually observe as being "correct" is the result of a process of interference.

Let's contemplate this for a minute. For the electron to reach the screen, some paths are very direct: straight or nearly straight lines. Others are very indirect: curving, angling or even looping lines. In theory, an electron shot from the gun can zoom around the Sun and back before angling through the slit. Some indirect paths are obviously impossible, because they would require the electron to travel faster than light. Others are simply improbable. And it turns out that when all the possible paths are summed using Feynman's mathematical technique, all of these probabilities interfere with each other. In some cases, the interference reinforces the probability of particular paths; in others, the probabilities tend to cancel. What's left is the most probable path. And experiments have show that this is the same trajectory one would predict using standard, classical physics.

Einstein was not enamored of this probabilistic core of quantum mechanics. In a letter in 1926 to the physicist Max Born, he said that quantum theory "yields much, but it hardly brings us close to the secrets of the Ancient One. In any case, I am convinced that He does not play dice.''[6] More than a decade after Einstein had made that now famous statement, John Wheeler visited the master to tell him of Feynman's sum-over-histories approach. Writing in his memoir, *Geons, Black Holes and Quantum Foam*, Wheeler relates what happened: '''Professor Einstein,' I concluded, 'doesn't this new way of looking at quantum mechanics make you feel that it is completely reasonable to accept the theory?''' In response, Einstein said, ''I still can't believe that the good Lord plays dice.'' (But according to Wheeler, he added this: ''Maybe I have earned the right to make my mistakes.'')[7]

Einstein, who once described himself as ''a deeply religious nonbeliever,'' objected to quantum mechanics because it violated his spiritual belief in the rationality of the universe — it's foundation in simple laws of physics. But as writer Nancy Ellen Abrams and her husband, cosmologist Joel Primack, point out, quantum physics is not magic. ''It now seems that God plays dice,'' they write, ''but the universe is nevertheless rational since the game has rules.''[8]

Feynman is credited with helping to work out those rules. He took the probabilistic outlook of quantum mechanics seriously and, unlike Einstein, even took joy in it. ''I'm rather delighted that we must resort to such peculiar rules and strange reasoning in order to understand Nature, and I enjoy telling people about it,'' Feynman told his students. ''There are no 'wheels and gears' beneath this analysis of Nature; if you want to understand Her, this is what you have to take.''[9]

\*　　\*　　\*

Together, quantum physics and relativity forever altered human thinking about the nature of reality: space could bend, producing gravity; and at its heart, nature was inherently fuzzy — impossible to pin down. By the 1930s, the physics of these two very different insights were ready to be tapped by scientists searching for a theory of cosmic origins. By 1930, the Belgian priest and astronomer, Georges Édouard Lemaître, had proposed his exploding ''superatom'' hypothesis, the first big bang theory of cosmic origins. At about the same time, a young Soviet physicist named George Gamow was beginning to make a name for himself. He was able to hold the seeming contradictions of general relativity and quantum physics together in his head at the same time. More remarkably, he saw that they could be used to lead the way toward a fully realized theory of cosmic origins, which would later become known as the big bang—the subject of the next chapter.

# 3

## HOT BIG BANG

### *Producing the Primordial Soup*

*I am happy that the Big Bang theory passed this test, but it would have been more exciting if the theory had failed and we had to start looking for a new model of the evolution of the universe.*

— John Bahcall[1]

*We are the read-out of the big bang.*

— Neil Turok[2]

The big bang may have happened some 14 billion years ago, and it may have been the biggest explosion the universe has seen in all the years since, but that hasn't stopped John Harris and his colleagues from trying to recreate it in the lab.

Billions and billions of big bangs, actually.

This may seem like hubris, but Harris attributes his interest in recreating the origin of the universe to something other than a desire to play God. There is, of course, the usual fascination with nature. And then there is a passion for tinkering, which Harris traces to a teenage preoccupation with cars.

''Basically, my dad and I used to take engines apart,'' Harris says.[3]

Today, as a Yale University physicist, he's got a much bigger and more interesting piece of equipment to tinker with. After all, producing billions of big bangs in the lab and monitoring how they give birth to baby universes is a bit more complicated than rebuilding the engine of a Triumph TR4 sports car.

Harris does his tinkering today at the Relativistic Heavy Ion Collider. Set within the pine forests of the Brookhaven National Laboratory on Long Island, RHIC is a complex of sophisticated machinery, including a particle accelerator and four particle detectors that monitor the mini big bangs. Chewing through some $480 000 worth of electricity during each week of operation,

RHIC's 1740 superconducting magnets accelerate bunches of gold ions in opposite directions around a two-lane, 2.4-mile underground atomic racetrack. Once the bunches are pumped up to 99.997% of light speed, they are steered toward head-on collisions with each other at six intersections where the lanes cross.

Although each bunch consists of a billion ions, most of the gold nuclei actually miss each other. (It's no simple matter to throw strikes with atomic nuclei.) So 5 million times a second, ions simply whiz by each other. But occasionally—by which I mean about 1000 times a second—there's a collision. "When you throw a billion things at a billion things, every once in awhile you get a crash," notes Flemming Videbeck, a colleague of Harris at Brookhaven.[4]

Gold nuclei are tiny, so even at near light speed, the force of each crash is as puny as that of two mosquitoes running into each other. Even so, the ions pack a tremendous amount of kinetic energy, which is all focused onto an exquisitely tiny region. As a result, the temperature soars in each collision zone to more than a trillion degrees, recreating the extreme conditions of the real big bang. And in a sense, for each of those fleeting instants, and on a very tiny scale, the collision represents the creation of what can be described as a tiny cosmos.

"We are really producing micro-universes," says Dimitri Kharzeev, a theorist working on the project. "And you can explore these universes one by one."[5]

None of these micro-universes ever has much of a future. Each one expires in just 0.000 000 000 000 000 000 000 01 seconds. But they do not die in vain. Just as the real big bang was accompanied by an explosive outpouring of primordial particles, the birth of each micro-universe at RHIC is attended by a spray of thousands of particles. Each one is a tiny messenger bearing crucial information about the mini-big-bang that created it. And it is the job of RHIC's experimentalists, including John Harris, to use four detectors arrayed in strategic locations around the accelerator's racetrack to collect the messages for analysis. The goal? To understand how the particles were created and thereby gain some insight into how our own universe grew from a baby into the cosmos we see today. More specifically, physicists are interested in how the bubbling hot stew of primordial quarks produced in the big bang congealed to form *hadrons*, the elementary particles that feel the strong force, including the protons and neutrons of ordinary matter.

When RHIC became a serious possibility in 1990, Harris hand sketched a particle detector for use at the facility. Today, his drawing has become reality as the 1200-ton, house-sized Solenoidal Tracker at RHIC, or STAR detector. At its heart is a 1500-cubic-foot particle tracking chamber, the largest in the world. The chamber is equipped with huge magnets that suck 3 to 4 megawatts of electrical power from the grid — enough to supply the needs of more than 3000 homes, as well as 76 million picture elements capable of tracking several thousand particles at a time.

When gold ions collide, charged particles spray into the cylindrical tracking chamber. The magnets torque them into curving trajectories, which are precisely monitored by the picture elements. With the resulting torrent of data, Harris and his colleagues can calculate the velocity of each particle — information that helps them determine the kinds of particles produced and the physical conditions attending their birth.

Their abiding hope is that the particles created in collisions will bear the signature of a state of matter not seen in the universe for 14 billion years: *quark–gluon plasma*, or QGP. If the modern view of the big bang is correct, this ultra-hot stew of free-roaming quarks and gluons should have been cooked up in the expanding fireball as the first stage in the construction of the universe we know. All the ordinary matter in the universe — all the protons and neutrons that have gone into all the atoms and molecules that make up every galaxy, star, planet and person — should have congealed from this primordial QGP stew.

How did scientists like Harris and his colleagues get to the point where they could even contemplate recreating the origin of the universe in a machine? The answer to that question is the story of the big bang. And the story begins with George Gamow, one of the scientific giants of the twentieth century.

Gamow was born on March 4, 1904, in Odessa, then a city in the Soviet Union and now part of Ukraine. To encourage a budding interest in science, Gamow's father, a secondary school teacher, gave his son a telescope as a thirteenth birthday present. Gamow's interest blossomed, and in 1923 he enrolled at Leningrad University, where he pursued research in quantum mechanics. While at the university, he sat in on lectures by Alexander Friedmann, who talked about his models of an expanding universe. Gamow, always intellectually restive, became interested in this too. In his very first scientific paper he tried to combine relativity with concepts from quantum mechanics.

Gamow quickly made a name for himself, developing a quantum theory of radioactivity to explain a process in which an alpha particle (two protons and two neutrons) quite mysteriously escapes from inside an atomic nucleus. For an alpha particle, an atomic nucleus is the equivalent of San Quentin: a seemingly escape-proof prison. To break out, the alpha particle would have to jump over an energy wall that seems too high. Yet somehow, it manages. Building on quantum mechanical ideas of uncertainty and probability, Gamow proposed that the alpha particle makes a Harry Houdini-like quantum leap from one side of the prison barrier to the next *without physically passing through it or over it*. The process is called quantum tunneling, and it's quite real. It's used in electronics and in a powerful device called a tunneling electron microscope, which can be used to image individual atoms on the surface of a test material. Back in 1928, Gamow's explanation of alpha decay by quantum tunneling laid the groundwork for theories of how elements are synthesized within stars through thermonuclear processes.

Gamow followed this brilliant success with a suggestion about the nature of atomic nuclei that would form the basis for work on atomic fission. He proposed that the nucleus of an atom, consisting of protons and neutrons, was more like a liquid droplet than a hard billiard ball. The American physicist John Wheeler, and Niels Bohr, a Dane who helped pioneer the modern picture of the atom, extended Gamow's theory, using the liquid droplet idea to provide a physical picture of fission. In this process, a speeding neutron cleaves a nucleus in two. The fragments weigh less than the original nucleus, with the missing mass having been converted into pure energy (thanks to the equivalence of mass and energy). Moreover, fission yields additional neutrons, which can split more nuclei, thereby setting up a chain reaction. If uncontrolled, a chain reaction releases a huge burst of energy—an atomic explosion. Wheeler and Bohr proposed that Gamow's liquid droplet nucleus would absorb the incoming neutron, and the sudden addition of energy would distort the nice spherical shape into a dumb-bell. The two ends would then split at the neck.[6]

By 1932, Gamow was a bright light in the emerging field of nuclear theory. But rigid Marxist–Leninist ideology sent a political storm front bearing down on the culture of free and open cooperation in science that Gamow had thrived in. Local officials of the Communist party began to suspect Gamow's loyalties. They asked his employer, the Radium Institute in Leningrad, for intelligence about the young physicist. The answer? Gamow was apolitical, "relatively undisciplined" and "a typical representative of the literary–artistic bohemia."[7] Hard-line Communists did not view such individuals very kindly, and Gamow knew it. So with his wife, he attempted to paddle an inflatable boat across the Black Sea, from the Crimean peninsula to Turkey—an incredible distance of 170 miles. Driven by a political storm to attempt this Herculean effort at defection, Gamow was then pushed back to shore by a natural storm at sea. Then, in 1933, he got a break: he was allowed to travel to Brussels with his wife to attend a physics conference. The couple remained in the West, and soon immigrated to the United States, where Gamow became a professor at George Washington University in St. Louis. Once there, he continued his pioneering work in nuclear physics.

Gamow and other physicists began zeroing in on the fusion processes responsible for the prodigious outpouring of energy by stars like the Sun. In this case, however, Gamow did not make the key contribution. That came from Hans Bethe and a former student of Gamow, Charles Critchfield. In 1938 they discovered the main energy source of Sun-like stars. Within the super-dense medium of a star, protons are squeezed together so tightly they can overcome their mutual repulsion and fuse. Critchfield and Bethe proposed a scheme called the proton–proton cycle in which protons fuse to form helium nuclei. In the process, the helium nuclei wind up being lighter than the protons that went into their construction. The "missing" mass is emitted in the form of heat energy, gamma rays and particles called neutrinos.

Some scientists working in this field couldn't help but note that the super-hot and dense insides of stars might offer insight into the conditions that existed in the early universe. One of those scientists was the German physicist, Carl Friedrich von Weizsäcker. At the end of the 1930s, he began moving beyond nuclear physics in stars to consider cosmological questions — namely, how nuclear processes could have forged elements in the primordial densities of the early universe. He envisioned "a great primeval aggregation of matter" crushed by gravity to a very small size. And he called this super-hot, super-dense object a "star." The origin of elements within it caused an explosive release of energy, sending fragments flying and explaining the origin of the cosmic expansion. "...we ought at least reckon with the possibility that this motion has its cause in a primeval catastrophe of the sort considered above," he wrote in a paper outlining his theory.[8] It was not unlike Lemaître's big bang origin from a primeval atom. But Weizsäcker went only so far. He did not attempt to link his idea to general relativity — in other words, to the geometrical structure of the universe.

By 1942, the idea of a hot, dense and explosive origin for the universe had begun to take root. At a theoretical physics conference in Washington that year, the consensus, as described by George Gamow in the conference report, was that the mother of all nuclear explosions occurred at the beginning of time, resulting in the origin of the elements and the expansion of the universe.[9] But a conference statement is not a theory. To turn the idea into a detailed and numerical description of cosmic origins, Gamow set out to combine nuclear physics with Friedmann and Lemaître's relativistic equations of an expanding universe.

Gamow's first take, "Expanding Universe and the Origin of Elements," was published in 1946 in the journal *Physical Review*.[10] In his paper, Gamow described the early universe as an expanding hot gas of neutrons. And he proposed that the radioactive decay of neutrons launched a series of nuclear reactions that produced all the nuclei present in our universe today.

Gamow had always been a big picture kind of guy, and so he had not yet figured out the precise details of what would later come to be called *big bang nucleosynthesis*. That work soon fell to Ralph Alpher, Gamow's graduate student. Alpher, the son of Jewish immigrant parents, had been interested in science since he was a young boy. Before his bar mitzvah, he even tangled with his rabbi over the biblical story Genesis, arguing in favor of a scientific explanation of cosmic origins. Later, Alpher enrolled part time at George Washington University, working during the day to earn a living and taking classes in chemistry at night. But friends soon convinced him to switch to physics, telling him that jobs in chemistry were closed to Jews.[11]

Taking up the details of Gamow's 1946 theory, Alpher began not at the instant of creation, which would have presented him with the incalculable singularity, but the universe at about 1 second of age. In the dictionary, Alpher had stumbled across the world "ylem," derived from a Greek word originally meaning wood, timber or material, and in later Greek meaning *matter*. Alpher applied ylem to Gamow's primordial neutron gas, and the moniker stuck. Gamow took it up himself and popularized it in books for the general public.[12]

Working feverishly—literally so, since he had come down with a case of the mumps—Alpher calculated how the ylem was transformed into ordinary atomic nuclei. The first step was the decay of neutrons into protons, electrons, antineutrinos and photons. Next, protons and neutrons fused to form deuterium nuclei, a form of hydrogen; then deuterium nuclei married to make helium nuclei.* Thus was born the first detailed description of how an explosive cosmic beginning could produce the stuff of today's universe—ordinary elements.

Next came a bit of mischief from the irrepressible Gamow. He had noted that "Alpher" and "Gamow" sounded very much like "alpha" and "gamma," two of the first three letters in the Greek alphabet. The missing second letter was "beta," so the Russian physicist decided to put *three* names on the scientific paper describing the work: Ralph Alpher, Hans *Bethe* (pronounced "beta") and George Gamow. But Bethe, one of the greatest physicists of all time, had contributed nothing to the paper. Nevertheless, Gamow simply could not resist the idea of an $\alpha$ (alpha), $\beta$ (beta), $\gamma$ (gamma) paper, so he submitted the manuscript with Bethe's name on it. (In parentheses next to

---

* Later, when things cooled sufficiently, electrons could combine with atomic nuclei to form stable atoms. But we'll deal with that phase of the big bang later in the book. Right now, we're focusing on the early development of the modern big bang theory.

Bethe's name he wrote, enigmatically, "in absentia".)[13] It was published in *Physical Review* in 1948.

The date was April 1.[14]

Alpher's paper was not a bad piece of work for a graduate student. And there was more to come. He teamed up with physicist Robert Herman, and the two proposed that photons produced as neutrons decayed would have survived and eventually streamed free in every direction, today permeating the heavens as an afterglow of creation. Alpher and Herman calculated that this mist of background radiation should have cooled dramatically as the universe expanded over the course of billions of years, and they estimated its current temperature at about 5 Kelvin.[15]

So here was a concrete prediction of the theory of primordial nucleosynthesis — a prediction that could be checked observationally. If such a cosmic background radiation were to be discovered, that would lend support to the idea that the universe began in a singular explosive event, resulting in the synthesis of elements through nuclear processes. But in the late 1940s, the prediction could not be checked, because the technology of the day was not up to the task of detecting radiation at such icy cold temperatures.

As Gamow, Alpher and Herman continued to work on the theory of nucleosynthesis, they confronted a problem. Despite their best efforts, they had been unable to find a credible way to produce anything heavier than a helium-4 nucleus, which consists of two protons and two neutrons. They were faced with an atomic stumbling block. There simply is no nucleus consisting of *five* particles that is stable. So even if nucleosynthesis built a five-particle nucleus, it would quickly decay before it could be used as the raw material for still heavier nuclei. One could imagine making a nuclear jump from helium-4 to nuclei with *six* or more particles, but this was considered improbable.

So the theory of primordial nucleosynthesis was just a start, not the final word. And the truth is, it didn't gain much traction, at least not at first.

"*COSMETOLOGY!*," sneer the critics of cosmology. And it's probably fair to say that back in the late 1940s, when Gamow, Alpher and Herman were working on their theory, the science of facials and cosmetics really was on somewhat firmer ground.

To start with, there weren't many practitioners of the dark art and, among the few who were working in the field, there still wasn't much of a consensus view. So the big bang theory was not exactly planted in fertile ground. Its failure to explain how the heavier elements were created didn't help. And to make matters even worse, shortly after Gamow and his colleagues proposed their theory, a charismatic opposition gelled. It was led by the English physicists Fred Hoyle and Thomas Gold, and an Austrian compatriot named Hermann Bondi. The three scientists proposed an alternative theory for the origin of the universe called the steady state theory. It accepted cosmic expansion but stated that there was no beginning to it. Instead, on the largest scales the universe remains essentially unchanged throughout time. According to the theory, as the universe expands, new matter is continually created out of sheer nothingness, filling the new volume. As a result, the broad characteristics of the universe will look the same to any observer at any location, and they will *always* look exactly the same from now until forever. In other words, the universe is eternal, both into the past and the future.

Fred Hoyle was motivated by an intense dislike of Gamow's theory, a feeling that prompted him to give it what he thought was a derisive label: *the big bang*. Little could he imagine that the term would stick, and not in the negative sense he had intended.

Why did Hoyle dislike the theory so much? The infinite curvature of spacetime at the initial singularity was one reason he gave. Go back in time to the singularity, and the laws of physics break down. This, Hoyle argued, violates a central tenet of general relativity: the laws of physics are the same throughout both space and time. Seen in this light, the steady state theory was an attempt to preserve the beauty and universal applicability of general relativity while accommodating an expanding universe. So it seems that today's movement to find a theory even deeper than the big bang, motivated in part by the ugliness of the singularity, has roots dating back to Fred Hoyle and his colleagues.

The Achilles heel of the big bang—the enigma of the singularity—has not gone away. That evidently killed the idea for Fred Hoyle. But as we shall see, history was to show that the enigma did not make the theory invalid. It simply meant that the big bang was not the final answer.

Throughout the 1950s and into the 1960s, attention turned away from these questions. George Gamow moved to the University of Colorado and withdrew somewhat from the limelight. Fred Hoyle went on to propose how the heavy elements, missing from Gamow's big bang, could be cooked up by fusion reactions within the hearts of stars and then spewed out into space in supernovae explosions.[*] There things sat until the spring of 1964, when two freshly minted Ph.D.s working at Bell Telephone Laboratories in New Jersey ran some experiments with a 20-foot radio antenna. The scientists, Arno Penzias and Robert Wilson, then made one of the most important astronomical discoveries of the twentieth century.

They kept picking up a hiss, some sort of radio background noise, in their horn-shaped antenna. No matter what time of day they checked or which way they pointed their horn, the strength of the signal remained the same. In fact, no matter what they did, including cleaning off pigeon droppings that had accumulated on the device, they couldn't eliminate the noise. A little later, theorists at Princeton concluded that the hiss literally was cosmic. The horn was picking up a mist of microwave photons permeating all of space—the afterglow of the big bang.

With this discovery of what would come to be called the cosmic microwave background radiation, Alpher and Herman's prediction had checked out.[16] Here was compelling evidence that George Gamow and his colleagues had been right, and Hoyle and his colleagues wrong. It didn't settle the argument then and there. That would take more detailed measurements. But with the discovery of the background radiation by Penzias and Wilson, the big bang finally was on its way toward widespread acceptance.

In the decades since, scientists have greatly extended the theory, building a more encompassing cosmology that combines big bang nucleosynthesis with more recent insights. The modern cosmology has several parts, but overall it describes the evolution of the universe in this way. A primordial patch of spacetime

---

[*] Since stars first blinked on after the big bang, they have been very busy transforming the light elements created in the big bang into heavier elements, such as iron, oxygen and carbon. According to Rebecca Bernstein of the Carnegie Observatories, stars have already processed an astonishing 30% of all the baryons created in the big bang.

inflates exponentially, giving rise to a big bang fireball containing the building blocks of matter; nucleosynthesis within the fireball creates the light elements; the fireball leaves a trace of itself in the form of the cosmic microwave background radiation; and matter accumulates gravitationally into galaxies and bigger structures. Many scientists simply pin the big bang label on the entire theoretical superstructure. But as Alan Guth, the discoverer of cosmic inflation, points out, the original big bang theory picks up the story of cosmic origins at about one second *after* the ultimate beginning. Gamow and his colleagues devised the theory as an explanation for how the elements were synthesized in the fireball; they left unaddressed the question of what ignited the flames. Alan Guth's spectacular answer was *inflation*. So he often refers to inflation as happening *before* the big bang.

Let's pick up the story immediately after inflation, when the universe is just $10^{-32}$ seconds old. (I'll save the details of inflation for the next few chapters.) The fireball is still expanding, but no longer at an exponential rate. The universe is inconceivably hot—almost $10^{28}$ K, or nearly 10 billion billion billion degrees. The fireball is seething with energetic particles of radiation and matter. What were these first constituents of the universe and how did they get there? To answer these questions, we need to return to the early days of the quantum revolution.

In 1928, the British physicist Paul Dirac was looking for a way to describe the electron mathematically. His approach was to combine Einstein's special relativity with quantum mechanics. At the time, the electron was one of only two particles known by physicists. (The other was the proton.) When Dirac tried to solve the equation, he saw that there was not just one but *two* possible ways to do it. One solution described a negatively-charged particle, an ordinary electron. The other showed something unexpected: a mirror-image particle of equal mass but *positive* charge—an *antielectron*.

The discovery of what would later be dubbed the positron yielded a remarkable insight about nature. Along with particles of matter she makes particles of *antimatter* as well.

In 1932, Carl Anderson of Caltech showed that Dirac was right. Working with a cloud chamber, a device used at the time to track particles, he found in the atomic debris from cosmic rays a positively-charged particle with the same mass as an electron. Other detections soon followed. (Anderson won a Nobel Prize in physics for his work.) Today, every particle is known to have an antimatter sibling. A proton has its antiproton, a neutron an antineutron, a quark an antiquark, and so on. (The exception to this rule is the photon, which is its own antiparticle.)

When the siblings get together for a family reunion, the laws of physics demand a suicide pact. Mutual annihilation is the inevitable result when a pair of matter and antimatter particles come together.[17]

Now remember: Matter is essentially frozen energy. Einstein's equation, $E = mc^2$, shows the two are equivalent. So total destruction of matter is the same as total destruction of energy, which is illegal. It violates the sacred principle of conservation of energy, which dictates that energy may not be created or destroyed. So when antimatter particles annihilate, the energy they represented does not simply go *POOF!*, out of the universe forever. It is released in a burst of high-energy photons.

Because of the equivalence of energy and matter, nature also allows Einstein's equation to be worked in reverse. Focus enough energy on a spot and you can conjure matter and antimatter particles. *Which* particles you produce depends on *how much energy* you bring to bear. To conjure a matter–antimatter particle pair, you must use an amount of energy equivalent to the particles' masses when they are totally at rest. Physicists measure this so-called *rest mass energy* of particles in terms of electron volts. The rest mass energy of a proton–antiproton pair is 1 billion electron volts. So to bring a proton-antiproton pair into existence you'll need 1 billion electron volts of energy, denoted by physicists as 1 GeV.

That sounds like a great deal of energy. For macroscopic entities such as us, however, it's not. Even a *trillion* electron volts, or 1 TeV, is not that much energy. As an illustration of this, just picture for a moment a small macroscopic entity — an ant, weighing about a tenth of a gram, moving across the top of a typical desk. And let's just say that it takes about 30 seconds for the ant to make the trip. The total energy for the trip actually amounts to a little shy of 1 TeV. Keep in mind that all those many electron volts are being parceled out to an enormous number of particles within the ant. A rough calculation suggests that an ant consists of about $10^{22}$ neutrons and protons. As a result, the 1 TeV of energy is *shared by all of those nucleons*. So as the ant walks across the table, each nucleon gains only a tiny amount of energy.[18]

On the other hand, for microscopic entities such as a single proton, 1 TeV is an enormous amount of energy. If you were to bring 1 TeV to bear on a single proton, you'd raise its temperature to trillions of Kelvins To do something like this—to focus that energy on something so small—you'd need the Tevatron, the highest energy particle accelerator in the world.

Located at the Fermi National Accelerator Laboratory near Chicago, the Tevatron is quite different from RHIC at the Brookhaven laboratory. Instead of gold ions, it accelerates protons and antiprotons in opposite directions at near light speed. Within the Tevatron, these particles are banged together in experiments that are helping physicists probe the basic building blocks of matter.

In 1995, physicists using the Tevatron crashed counter-rotating protons and antiprotons head-on. Each direct hit was accompanied by a 2 TeV flash of energy, from which hundreds of particles congealed. Among the many uninteresting particles produced in this way there was one kind that got everyone's attention. Predicted by theory but never detected before, this particle was the *bottom quark*.

Recall that protons and neutrons are believed to consist of triplets of ordinary up and down quarks. But theory predicts other, more exotic kinds of quarks, including the bottom quark. With its discovery, physicists added another piece of supporting evidence for the theory of quantum chromodynamics. And they also got a surprise. The bottom quark was much heavier than expected, suggesting that some as yet unknown aspect of physics may await discovery.

If physicists can conjure a fundamental building block of matter with magnets and electricity, the big bang certainly would have done the same thing. The energy of the very early universe was so high that particle–antiparticle pairs—electrons, neutrinos, quarks and their antimatter siblings—should have condensed out of it like raindrops in a thundercloud. Of course, the instant they appeared, they came together and annihilated in showers of gamma rays. But no worries. There was plenty of energy to go around. So more pairs were born, more died, and the process repeated over and over, causing the big bang fireball to crackle with the fireworks of creation and destruction.

But like any fireworks burst, the fireball expanded and cooled and the crackling began to subside. Eventually the energy was too low for any additional quark–antiquark pairs to be made. Those left standing at this point came together and annihilated. (The same thing happened with other particles, such as electron/positron and neutrino/anti-neutrino pairs, but at different temperatures.)

What was left? Today, there are something like $10^{78}$ ordinary atoms within the observable universe. (Hydrogen atoms constitute the vast majority.) As for antimatter, there's none of it in the Earth, none in the Milky Way and, as far as astronomers can tell, there are no antimatter galaxies. (Although scientists say the possibility of antimatter superclusters can't be ruled out.) Physicists do create antimatter in their particle accelerators, but these particles quickly annihilate with ordinary matter. From this evidence—and just by looking around at all the intact stuff around us—it seems obvious that there must have been some excess of quarks over antiquarks (and electrons over positrons, etc.) when the universe cooled enough for pair-creation to stop. If this asymmetry had not existed, each and every particle would have annihilated with each and every antiparticle, leaving nothing behind to make ordinary matter.

One possibility is that the excess of matter over antimatter we see today was simply one of the initial conditions of the universe—a built-in feature. But physicists consider that unlikely. The favored alternative, first advanced in 1967 by the Russian physicist and political dissident Andrei Sakharov, is that as the early universe cooled, physical laws showed a very slight preference for matter over antimatter. When the last mass annihilation occurred, some matter particles could not find antimatter particles to annihilate with, so they survived. At early times, these particles were loners because most of the energy in the universe resided in the ultra-hot photons of the fireball's radiation. But as temperatures dropped and the photons lost energy, the matter particles came to dominate. (Today, the cosmic microwave background radiation is only a smidgen above absolute zero. That's certainly not the case for the particles that make up you and me.)

Some hard evidence suggests this scenario is right. In 1964, for example, James Cronin and Val Fitch discovered that a decay-prone particle and its anti-matter sibling (the $K^o$ and anti-$K^o$) did not decay at precisely the same rate, leaving a slight excess of one over the other. Physicists take this as support for the idea that a slight asymmetry is built into the physical laws governing the characteristics and interactions of all particles, an asymmetry favoring matter over antimatter.

Ever see a friend evaporate in a spray of gamma rays? Of course not. From the evidence of our everyday lives, there is no reason to believe that clouds of antimatter lurk around the next corner, ready to zap us without warning. However it may have happened, ordinary matter of the kind you and I are made of appears to have inherited the cosmos. Yet in this case, appearances are deceiving. The best evidence suggests that another kind of matter, mysterious and invisible, yet gravitationally irresistible, dominates the universe: *dark matter*. As we'll see in some detail in chapter 5, several compelling lines of evidence have convinced cosmologists that most of the matter in the universe consists of *this* stuff. The best candidates for the dark matter are invisible exotic particles that have never been seen or created in a particle accelerator, and which interact weakly, if at all, with ordinary matter. But they would feel and exert gravity, giving them a starring role in the origin and evolution of structure in the universe.

Many physicists believe the best dark matter candidates are particles described by a theory known as supersymmetry, which proposes that every particle has a *superpartner* (in addition to an antimatter sibling). So quarks have their squarks, electrons have their selectrons, and so on. And scientists say that some of these superpartners could well comprise the dark matter. Like ordinary quarks, these supersymmetric dark matter particles and anti-dark-matter siblings would have popped into existence from the ultra-hot radiation field in the big bang fireball and then annihilated. As long as the temperature exceeded about $10^{15}$ K, creation and destruction would have continued. But when the fireball cooled sufficiently, no additional supersymmetric particles would have been created. Annihilation then would have destroyed most of the remaining particles, leaving behind some survivors. Since these particles are very stable and do not interact with other kinds of matter, they would have survived intact to this day.

And what about ordinary matter, the stuff made from atoms? The building blocks of atoms—electrons and quarks—were made in the big bang fireball in less than a millionth of a second. Somehow, then, quarks had to come together to form protons and neutrons and thereby build atomic nuclei.

Before a millionth of a second, quarks and gluons can play freely in the radiation fields. This is the epoch of the quark-gluon plasma. But when the universe cools sufficiently, gluons can call the quarks home—into the interiors of protons and neutrons, where they will remain united in triplets forever. Unless, of course, RHIC manages to tease some out to play once again.

From a millionth of a second to 1 second into the life of the universe, protons, neutrons, as well as electrons, neutrinos and photons, roam the expanding arena of spacetime. Conditions are too hot for protons and neutrons to get together in nucleonic matrimony. So these hadrons remain single, and this epoch is known as the hadronic epoch.

At about a second we reach the epoch explored by Gamow and the other pioneers of big bang nucleosynthesis. The temperature is just right for neutrons and protons to fuse in the series of reactions that produce the nuclei of deuterium, helium and lithium. But electrons will actually have to wait about 300 000 years to cool down enough to be drawn into comfortable orbits around the waiting atomic nuclei. During this long interval, the universe is filled with a hot plasma of nuclei, electrons, neutrinos and photons. When the electrons finally join with the nuclei, forming the very first atoms, matter is ready to take the first steps toward gathering itself up into galaxies, stars and planets.

Cosmologists find this picture of matter congealing from the ultra-high energy of the big bang fireball theoretically compelling. With a dose of inflation at the beginning, and a gravitational kick start from dark matter once atoms have formed, the model seems to explain a lot about our universe. But how do cosmologists really know whether any of this is true?

The big bang nucleosynthesis part of the model has checked out spectacularly well. In the decades since Gamow and his colleagues first proposed the theory, scientists have made ever more precise calculations to describe nucleosynthesis, identifying a complex chain of nuclear reactions within the first three minutes of the universe. These calculations have yielded ever more precise predictions of just how much of these elements should be present in the universe today. These predictions have offered the most compelling tests yet of the hot big bang, and the theory has come through with flying colors.

In the argot of particle physics, the predictions concern so-called *baryonic* matter, which is made from the class of particles known as baryons. Protons and neutrons are baryons. If the theory is right, within the first 300 seconds of the universe, big bang nucleosynthesis should have had the following effects.

About a quarter of the baryons in the infant universe should have been incorporated into the nuclei of helium-4 atoms; trace amounts should have gone into the isotopes deuterium, helium-3 and lithium-7; and the rest, some three quarters of the primordial baryons, should have remained as free protons, which later united with electrons to form hydrogen atoms. As far as big bang nucleosynthesis goes, that was it—no other baryonic matter was made by fusion in the primordial fireball. All the elements heavier than lithium were synthesized later, within the nuclear furnaces of stars. So by taking an inventory of the baryons in the universe, by seeing what percentage seem to be incorporated into hydrogen, helium, deuterium and lithium, scientists should be able to put the big bang theory to the test.

Over the years, astronomers have made observations of the relative abundances of the light elements. For example, helium-4 has been studied in nearby dwarf galaxies, whereas deuterium has been examined in distant protogalactic clouds. And it turns out that the predictions of big bang nucleosynthesis have been borne out by observations. But there's a catch: theory and observations match only if there is a certain density of baryons in the universe. To be more precise, there's a match if all the protons and neutrons in today's cosmos add up to between 2.6% and 4.6% of the density of matter needed to prevent the universe from expanding forever. So what is the density of baryons actually out there? By analyzing the cosmic microwave background radiation, scientists have been able to pin this down within a certain range. And this range nicely overlaps that which is needed for the predictions of big bang nucleosynthesis to match observations.[19]

Another recent confirmation of the big bang picture came in 2000, when astronomers led by Raghunathan Srianand of the Inter University Center for Astronomy and Astrophysics in Pune, India, took the temperature of the universe—more accurately, the cosmic microwave background radiation—when it was about 2.5 billion years old. If the radiation really is a relic of the primordial fireball, it should have started out very hot and cooled dramatically in the billions of years since. So here is a prediction of the big bang theory: the cosmic microwave radiation should have been hotter in the past.

To carry out their observations, Srianand and his colleagues used the 8.2 meter Kueyen telescope at the Paranal Observatory in South America. Their target wasn't the radiation itself, which would have been too difficult to detect at this epoch in the distant past. Instead, they observed a far away quasar and measured how its light was altered as it passed through the background radiation. The quasar emitted the light when the universe was about 2.5 billion years old. So observing it today is like traveling back in time to this epoch.

Based on the subtle alterations in the spectrum of the quasar's light, Srianand and his colleagues were able to determine the background radiation's temperature. Their result fell within a range of 6 to 14 K above absolute zero. This is consistent with the big bang theory, which predicts a temperature of 9 K at 2.5 billion years of age.[20]

In a review of the work in the journal *Nature*, Princeton physicist John Bahcall commented that "the Big Bang theory has survived a crucial test." By showing that the universe was hotter when it was younger, Srianand and his colleagues had achieved what Bahcall called "a landmark result."[21]

The big bang has passed other tests as well—so many that astronomers and cosmologists have no doubt that the universe was born in a big bang. "I believe the hot big bang theory will be viewed as one of the great intellectual triumphs of the twentieth century," says Michael Turner, a physicist at the Fermi National Accelerator Laboratory and a professor of physics at the University of Chicago. Based on what he describes as a simple mathematical model of Einstein's equations, first developed by Friedmann and Lemaître and then extended by others, the theory "describes accurately the evolution of the universe from a fraction of a second after the bang until today."[22]

Alan Guth, the discoverer of cosmic inflation, agrees: "Cosmology has come a long way," he says. "While in the 1950s cosmologists debated whether the universe began in a big bang or whether the steady-state model was right, now we have a detailed theory of the big bang that makes a number of demonstrably correct predictions."[23]

Even so, scientists are restive. Cosmologists have only recently begun to put inflation to the test in stringent ways. There are also more precise tests of the standard big bang theory to be done. And scientists have yet to explore the many details of how events unfolded in the early universe. Exactly how, for example, did the transition from quark–gluon plasma to ordinary matter unfold? Scientists are now beginning to probe that transition experimentally with the Relativistic Heavy Ion Collider at Brookhaven.

Many of the defining characteristics of our universe are believed to have emerged at early times in transitions like these—*phase* transitions, to be more precise. A familiar phase transition is the one made by liquid water freezing into ice. When that happens, the symmetry of randomly arranged water molecules is broken: the molecules align themselves in preferential directions. Similarly, the four fundamental forces of nature— the strong, weak and electromagnetic forces, as well as gravity— are believed to have emerged from a single, symmetrical *superforce* in a series of symmetry-breaking phase transitions. Another way of thinking about this is that the forces "crystallized" from the one superforce.

"You can look at a crystal like a diamond and say it's made of carbon, but how do you form a diamond?" says Wit Busza, an MIT physicist on the team of scientists working with the PHOBOS particle detector at RHIC. "Our universe is like a diamond. The forces probably emerged through symmetry breaking, just as the atoms in a crystal align in preferential ways, breaking the symmetry and forming the diamond. But how did this happen? If you want to understand the gem, you have to understand that."[24]

Unfortunately, though, the phase transitions believed responsible for the crystallizing out of the fundamental forces happened at energies far beyond current technologies. On the other hand, theory suggests that the quark–gluon plasma crystallized into ordinary protons and neutrons in a phase transition that should be within reach of RHIC. So the experiments give physicists a way of probing one of the most fundamental aspects of nature.

"Before this transition," Busza says, "the universe consisted of a soup of pointlike particles—quarks, gluons, electrons, positrons, neutrinos, etc. Roughly when the universe was the size of the solar system, suddenly this transition occurred and we got matter of the kind we're used to. But why is the matter the way it is?"

For example, why does the proton have its specific mass? The experiments at RHIC may be useful here because the theory of strongly interacting particles, quantum chromodynamics, doesn't actually give an answer; according to Busza, the equations of QCD cannot be solved to reveal the mass of the proton. Since the defining characteristics of protons and neutrons are believed to have been born in the phase transition from the quark–gluon plasma, recreating that stuff in the lab and watching what happens to it as hadrons form may provide some crucial insights.

But there is an important caveat. Very little is known about how the quark–gluon plasma became protons, neutrons and other hadrons. So it's entirely possible that the shift didn't even happen in a phase transition.

According to Glen Young, a physicist with the Oak Ridge National Laboratory working on the PHENIX detector team, such a shift represents a *discontinuity*, a sharp break in the characteristics of a system. And matter does not have to shift form in that way. For example, if you heat a gas, the atoms will be stripped steadily of their electrons. In this way, as heat is added the fraction of the gas that is ionized becomes bigger and bigger. "That's not a phase transition, a major discontinuity," Young says. "It's a slow, steady, boring process."[25]

So when the quark–gluon plasma turned into hadrons, how did it happen? A sharp break or a slow, boring shift? Theory suggests the former. But it will take experiments like those at RHIC to find an answer.

That answer could be important in cosmology, Glen Young says. If the shift happened in a certain kind of phase transition, latent heat would have been given off, and that could have had a discernible impact on the rate of element formation in the early universe.

For Wit Busza, the ultimate goal of RHIC's experiments is pretty basic. "What's the purpose of science?" he asks. "One is to describe nature. We want to be able to describe how those hadrons were made."

Making a quark–gluon plasma and watching how the hadrons form is no easy task. The strong force carried by the gluons is the strongest force in nature. Gluons keep quarks bottled up so tightly that not one should have roamed free of other quarks since the big bang.*

The *gluons* of QGP are a product of the quantum mechanical vacuum. As you may recall from the previous chapter, this vacuum is anything but empty. Look closely enough at the vacuum surrounding an electron and you would see roiling clouds of virtual photons and other particles popping in and out of existence. Within protons and neutrons, quantum uncertainty also allows gluons to bubble up from seeming nothingness. According to quantum chromodynamics, the theory of the strong force, the farther apart quarks are stretched, the more gluons materialize, and therefore the stronger the binding force becomes.

The effect is not unlike what happens when you stretch a rubber band: the more you pull on it, the more it wants to snap back together. Stretch quarks apart enough and the frenzied exchange of more and more gluons between quarks simply becomes irresistible. When that happens, the quarks snap back together, preserving their union for eternity. In this way, "the vacuum forces quarks to be together," says John Harris of Yale, the spokesperson for the scientific team working with RHIC's STAR detector. "It doesn't allow a single quark to be on its own."

But let's take the analogy of the rubber band a little further to see whether there can be any exception to the no-free-quarks rule. In quantum chromodynamics, the rubber band is a quality of the vacuum, and it just gets stronger and stronger the harder you pull on it. But what would happen if you deposited enough energy

---

* There is one possible exception: within the superdense interiors of neutron stars.

onto the rubber so that for a fleeting instant *you melted it*? That might just allow quarks to go free. Then, when things cooled down again, the band would regain its snap and pull the quarks back together again.

In a nutshell, this is the idea behind the gold ion collisions at RHIC. "We're trying to deposit so much energy on the vacuum that we actually melt it," Harris says.

The energy comes from the collision of gold nuclei within RHIC's four detectors. So let's follow how that happens.

Gold atoms are ionized, meaning stripped of their electrons, and then are guided through a series of machines that help bring them up to speed. The ions are then injected into RHIC's dual racetrack, forming two beams, each about half a millimeter in diameter, going in opposite directions. The beams are not continuous. It's easier to keep the gold ions in bunches — up to 56 at a time. And each bunch consists of billions of nuclei.

Each one of those nuclei consists of 118 neutrons and 79 protons, which means that it is relatively massive compared to many other kinds of atoms. (Compare this to a hydrogen nucleus, which consists of just a single proton.) And those 79 positive charges give RHIC's magnets some traction on the nuclei so they can be accelerated to near light speed.[26] This acceleration to relativistic speed has two important effects. The first is that the nucleus actually *shortens* along the line of flight, flattening into a pancake. (This is a product of Einstein's special relativity.) The second is that the acceleration pumps an enormous amount of energy into each gold nucleus. Because of the equivalence of energy and mass, this pumps up the *mass* as well.

Now the gold nucleus is 100 times heavier, and all that extra mass is packed into just 1% of the original volume. As a result, when two such nuclei collide, the density of mass within the collision zone is 20 000 times greater than that of an ordinary gold nucleus. As the collision unfolds, protons and neutrons run into each other head on — and actually go *through* one another. More collisions occur between other nucleons and the debris created by the first crash. The result? a fiery atomic wreck, with the remnants of each gold ion racing away from each other. The temperature at the crash site between these remnants soars to a thousand billion degrees, about 100 000 times hotter that the temperature of the core of the Sun, creating a mini-fireball — a microuniverse. If all goes well, the heat melts the gluon rubber bands, and quarks roam freely within the fireball.

"There is so much energy in the vacuum that it's melted, but in a confined space," Harris says. And it consists of much more than the fractured remains of the original protons and neutrons. Energy pumped into the colliding nucleons by the superconducting magnets has been converted into new mass — not just heavier gold nuclei but *new particles* – within the fireball of this mini big bang. Some of these race off unaffected by the collision. Captured by a detector, they can be interrogated for any information they may have about the quark–gluon plasma, such as the precise temperature at which it formed.

As particles from the collision expand, they cool. So the energy within the collision zone decreases. When the energy drops below a certain value, the quarks can no longer roam free. They join up, forming hadrons — protons, neutrons and other particles that feel the strong force.

"This is the transition from the quark–gluon plasma to hadrons that occurred in the early universe," Harris says. "It is a transition that we don't really know anything about." By capturing and studying particles emerging from the transition, Harris and his colleagues hope to gather information about how it unfolded.

What I've described is the expected sequence of events in just *one* head-on collision between gold ions. When RHIC is up and running, billions of ions are colliding each second. Creating that many microuniverses may seem like overkill. But if physicists are to trust that what they learn is statistically significant, it's necessary.

"When you study these microuniverses, you will be able to understand the laws of nature that govern their creation and evolution," says Dimitri Kharzeev, a Brookhaven nuclear theorist. "But you have to accumulate *enough* of them. It's not enough to see a single event to understand the properties of the physical laws that govern the creation and evolution of the microuniverses."

RHIC's four main detectors, STAR, PHENIX, PHOBOS and BRAHMS, each use a different approach to monitoring the creation and evolution of a microuniverse. That's because no one is really sure what's going to happen, so they need to cover all the bases.

"One thing science teaches us is that we do not know what nature has in store," says Busza of the PHOBOS team. "Whenever we study matter or nature in new ways, something surprising happens."

And surprises are actually *hoped for*. If nothing new comes out of the experiments, the results would just reinforce what's already known—a valuable thing, but ultimately not as significant as learning something completely new about nature. Since the goal is, in some sense, to be surprised, "we don't know what we want to study," Busza says. "And in a collision, nobody really knows how often something interesting will happen." Interesting results may happen frequently or very rarely. Moreover, the scientists don't know where the interesting things will turn up. So the approach has been to design different kinds of detectors to look at collisions in different ways. The hope is that at least one method will turn out to be the right one for discovering something new and interesting.

"In one sense, all the experiments do the same thing: look at the surface of the fireball as revealed by the particles emerging from it and try to determine what happened inside," says Flemming Videbaek, a physicist at Brookhaven working on the BRAHMS detector. Like the massive magnetic fields that create eruptions on the surface of the Sun, the quarks within the mini-big-bang created by a collision may modify the surface of the fireball. To make sure they capture the interesting stuff, each detector looks at the fireball in a different way.

The STAR detector has been designed to look in detail at the particles streaming away from a collision between gold nuclei. "STAR concentrates on measuring as many charged particles as possible," Harris says. "Most are hadrons." (Although STAR can detect other kinds of particles as well.) If just a relative handful of particles in a collision carry news of something interesting into the detector, Harris and his colleagues will have a chance of finding it.

STAR's heart is a device called the Time Projection Chamber. It tracks and identifies particles emerging from the collision zone. STAR's large magnets produce intense magnetic fields that force the particles onto curving paths. By measuring the degree of curvature, Harris and his colleagues can determine the momentum of each particle. In essence, Harris and his colleagues are using

STAR to determine what's moving where and how fast from the collision zone. From this information, they can take the temperature of the fireball and hope to calculate the temperature at which quarks make the transition to hadrons.

There is, however, one downside to this big coverage of the fireball: STAR cannot monitor every single nucleus-on-nucleus collision that occurs. Says Busza, ''STAR's strength is that they get a good view of everything that happens in a single collision, but they can only analyze a few collisions a second. And that means if interesting things happen in a very small handful of collisions overall, STAR could miss it.''

That's a possibility because of the uncertainty built into the quantum realm, Busza says. At phase transitions, like the transition from quark–gluon plasma to hadrons, physical quantities fluctuate. So what happens in the fireball one time might not happen the next.

In contrast to STAR, PHOBOS will monitor collisions at a very high rate. ''We hope to keep up with most of the reasonably central collisions,'' meaning the ones that happen directly head on, Busza says. The weakness of this approach is that with each collision, the detector will have a less detailed view than STAR's of the particles produced. ''So we'll get complete coverage of the collisions, but a much cruder view.'' By recording more events when the accelerator is operating, however, Busza and his colleagues think they will be in a good position to find the rare one that may be really interesting.

''That's why we look at many events—to see if all are equally boring or to see whether one event is different from another,'' Busza says.

The PHENIX detector is in one way much like STAR: it's incredibly massive, weighing some 4000 tons. ''But we don't try to see the whole collision,'' says Glen Young. ''Instead, we go after a good ID of less common particles.'' PHENIX looks for photons and leptons, particles that do not contain quarks and do not feel the strong force. Among these particles are ordinary photons, virtual photons and muons. Because they are unaffected by the strong force, these particles emerge unchanged from the collision zone, carrying pristine information about what happened there. Like an MRI machine that images organs within the human body, PHENIX looks *inside* the fireball, probing directly the hot, dense matter formed in a collision.

Moreover, "PHENIX looks at a smaller region than STAR and thus produces a finer-grain picture," Young says. With this fine detail, PHENIX can identify, for example, individual gamma rays and determine their energy. STAR can only measure the total energy contributed by gamma rays to the fireball.

Finally, there is BRAHMS (the Broad Range Hadron Magnetic Spectrometer). "With BRAHMS, we can look closely, very accurately, from a few angles," Flemming Videbaek says. "It's like moving the microscope around." The detector measures the momentum, energy and other characteristics of only a small number of particles emerging from a specific set of angles during each collision. And it's the only detector that looks at the fireball along the line of flight of colliding gold ions, which means it's the only one with a very good view of the actual collision region itself.

What has RHIC achieved so far? Collisions between gold nuclei produce jets of particles that spray out from the side. If QGP forms, there should be fewer high-energy particles in those jets—a phenomenon dubbed "jet quenching"—than what physicists would otherwise expect. That's because the plasma is sticky and therefore slows the particles down. And in early 2001, RHIC scientists announced that they had tentative evidence of jet quenching, offering a hint that the quark–gluon plasma had been produced.

"The only thing one can say is that the jet quenching we've seen is suggestive and consistent with what you'd expect from a quark–gluon plasma," Harris says. "It's not incontrovertible proof; the results could also be consistent with other explanations."

RHIC will have to create many more of Dimitri Kharzeev's "microuniverse" before the scientists will have enough information to say anything more definitive. "We need to rule out all other explanations," Harris says. And to do that, more data are needed.

"We know roughly the bounds of what we are going to observe," Kharzeev says. "But we will certainly learn a lot from the experiments. We don't know enough to predict precisely what the experimentalists are going to find."

On a broader level, Kharzeev is hoping RHIC will fill in some major gaps in our understanding of nature. "We don't really understand the strong interaction.'' he says. ''Ninety-nine percent of the mass of the universe is due to it. By observing how it emerged, we'll understand the emergence of this force— the strongest force in nature.''

And based on what they find, scientists like Kharzeev hope to reveal something brand new about how the stuff that would later find its way into galaxies, stars, planets and us first emerged and took on the qualities we're familiar with.

# 4

# INFLATING THE UNIVERSE

## *It Takes GUTs*

*All are nothing but flowers in a flowering universe.*
— Nakagawa Soen-roshi, Zen Buddhist

*All that people see — heaven and Earth — and all that fills it, all
these things are external garments of God.*
— Ascribed to a Lubbavitch rebbe

By the 1970s, the realms of spacetime and the quantum had been explored in detail. With general relativity, Einstein had remapped the nature of reality on cosmic scales, pointing the way toward the big bang. Meanwhile, the pioneers of the quantum revolution had revealed the fuzziness of nature at small scales. And with these achievements as a guide, scientists began blazing a trail backward through spacetime.

In their explorations of the very early universe, they sought answers to some fundamental questions. Immediately after its birth, how did the universe evolve? And how did this evolution lead to the universe we observe today? The explorers recognized that the very early universe was filled with highly energetic elementary particles. So their starting point was the emerging theory of the elementary particles and their interactions—a theory that would come to be called the standard model of particle physics. Even in the late 1970s, when major aspects of the theory still had not been fully tested, scientists recognized that the standard model went a long way toward explaining why the universe is the way it is.

''In the nineteenth century, and even well into the twentieth century, there was no theory for what gave atoms their specific properties—or why there should even be atoms or nuclei at all," says Frank Wilczek, a physicist at the Massachusetts Institute of Technology. "Now we have a theory that tells us about the 'what questions.' In fact, it gives us very compelling answers about them. And it also begins to allow us to understand in a very deep way new kinds of questions about *why*."[1]

The standard model contains two basic classes of elementary matter particles: quarks and leptons. As we've seen, the quarks interact via the strong force and make up things like protons, neutrons and us. The leptons, which do not feel the strong force, include the electron and neutrino. The standard model also describes fundamental forces in terms of force-carrying particles. These "bosons", which include gluons of the strong force and photons of the electromagnetic force, mediate the interactions between the matter particles.

"So over the past 20 years or so we've identified these two basic sets of constituents of matter, the quarks and leptons," says Chris Quigg, a physicist at the Fermi National Accelerator Laboratory near Chicago. "And we've found that the interactions among them are responsible for all the diversity and change that we see in the universe."[2]

In this way, the standard model went a long way toward answering Frank Wilczek's 'what' and 'why' questions. But it did not go *all* the way. It did not answer, for example, a key question about the fundamental forces of nature. Why do they vary considerably in their strengths and ranges of action? The search for an answer to that question would lead scientists toward a radical new theory about the origin of the universe, the theory of cosmic inflation.

To see how that happened, let's take a minute to examine just how much the fundamental forces vary in strength.

The strong interaction, which locks quarks together to form protons and neutrons, is the Arnold Schwarzenegger of the forces, wielding almost 100 times more muscle than the electromagnetic force. But the electromagnetic force, which keeps negatively charged electrons dancing about positively charged protons in an atomic nucleus, is no weakling. It's 1000 times stronger than the appropriately named weak force, which you may recall is responsible for radioactivity.

"We know that electromagnetism and the weak interactions have a common origin," Quigg says. "And yet the manifestations of those two things in everyday life—at least for us in our laboratories—are quite different.'' In addition to their strengths, this includes their ranges of influence. ''Electromagnetism has a range of influence that is infinite,'' he says. ''Start a photon off into the universe and it keeps going. Whereas the range of influence of the weak interaction is very tiny, confined to a range of about $10^{-15}$ centimeters. So what makes those two things so similar and yet so different?''

Before we get to a possible explanation, we need to consider nature's fourth fundamental force: gravity. If you've ever lifted weights, you might reasonably conclude that gravity is at least as brawny as the other forces. But it's not the strength of the gravitational force *per se* that makes weight lifting a challenge. Gravity seems strong when you lift something heavy because each and every atom in the weight is mutually attracted to each and every atom in the Earth. And all that mutual attraction adds up to a lot, requiring a person with the strength of, well, Arnold, to overcome it with ease. On the other hand, when gravity goes one-on-one with any other force in the atomic gym, it is worse than hopeless. For example, the electromagnetic force tries to push two closely held electrons apart with a *million billion billion billion billion times greater strength* than gravity tries to pull them together.[3]

Yet despite gravity's wimpiness, it manages to shape the entire cosmos. It can do this because all the energy and mass in the universe exerts a gravitational influence—a force that operates across vast distances. And even though electromagnetism's reach is similarly far flung, negative and positive charges pretty much balance out in the cosmos. With little net charge left over, electromagnetism cannot dominate the universe as gravity can.

So gravity stands alone—unexplained by the standard model and the true weakling among the fundamental forces.

''One of the profound mysteries of physics is why gravity is so feeble at low energy,'' says MIT's Frank Wilczek. ''Why is it such a feeble force between you and me? I don't feel your gravitational attraction, whereas if I run into you, I definitely feel your presence.''

In the 1970s, scientists realized that something beyond the standard model was needed to answer these questions, some theory that could unite the forces of nature into one theoretical framework. This quest was not really new or even unique to physics.

"A theme common to all of science is the idea of unity – of making not just individual stories to explain different phenomena here and there, but trying to find one whole picture that explains everything at once," says Fermilab's Chris Quigg. "This endows every measurement with a greater meaning than it would have on its own because you can see it resonating with other things."

By the mid-1970s, a possible path to unity opened when scientists realized that the strengths of the non-gravitational forces were not fixed in stone. They vary depending on the temperature. In 1974, a trio of Harvard physicists, Howard Georgi, Helen Quinn and Steven Weinberg, found that as conditions become hotter and hotter, the strengths of the forces begin to converge. One implication was that given a high enough temperature, the strengths of the forces should come to heel equally, and the distinctions between them should disappear. This suggested that the forces were related in some fundamental way. Georgi and Sheldon Glashow followed this lead, devising a new theory, a Grand Unified Theory, or GUT, in which the three non-gravitational forces were seen as manifestations of a more fundamental force.

In the grand unified view of things, all of the particles, fields and forces (minus gravity) are melted together at extremely high temperatures into a primordial soup. Within this roiling cosmic broth, the ingredients lose their distinct identities. Such energies are well beyond the reach of even the world's most powerful particle accelerators – but not the mathematical formulas of theory. "These calculations are quite concrete, and we can use them to extrapolate far beyond the energies at which measurements have been made directly," says Wilczek.

Those calculations show that the required temperatures would easily have been achieved in the early universe. At about $10^{-35}$ after the Ultimate Beginning, or when the universe was a hundreth of a billionth of a trillionth of a trillionth of a second old, all of the particles, fields and non-gravitational forces should have been stewing at an incredible $10^{28}$ K. That would be ten billion billion billion degrees above absolute zero – enough, according to grand unified theories, to cook up the primordial soup.

So now we return to the cosmological question posed earlier, but reworded slightly. What happened to that primordial soup as the early universe expanded and cooled?

According to the GUTs, as the universe cooled from the unification temperature, it went through a series of phase changes. And with each of those changes, one of today's familiar forces "froze out" of the unification force. Although the transitions occurred about 14 billion years ago under unimaginably extreme conditions, they are not very far removed in one respect from our everyday experience. They were very much akin to the transition made by liquid water when it freezes. Such transitions are examples of a fundamental physical process called *symmetry breaking*.

In its liquid form, water consists of molecules arranged in every which way. Unlike a crystal, in which molecules align along specific axes, there is no preferred orientation for water molecules in a liquid. For this reason, the arrangement is considered symmetrical. Examine a drop of liquid water in extreme close-up so you could see the molecules and it would look the same no matter how you rotated the sample. But cool the water below its freezing point and you would break the symmetry. As the water froze, the molecules would begin to line up in a lattice-like structure characterized by preferred directions: the axes of the lattice. Because of this asymmetry, the sample would now look different depending on how you looked at it.

In the very early universe, phase changes in the primordial soup did not involve molecules or atoms, which didn't exist yet. Instead, they involved transitions from a universe dominated by the grand unified force to one dominated by the multiple forces we know today. According to grand unified theories, as the temperature of the universe dropped below $10^{28}$ K, the strong force and a combination of the weak and electromagnetic forces "crystallized" in a phase change out of the symmetrical grand unified force. Later, when the universe was $10^{12}$ seconds old, the temperature had dropped to $10^{15}$ K. With this further cooling, the electromagnetic and weak forces, which until then were unified as an *electroweak* interaction, broke apart and went their separate ways, each crystallizing into the individual forces we know today.[4]

But what actually caused the symmetry to be broken?

The answer, according to the grand unified theories, is something called the Higgs mechanism, which is named after Peter Higgs of the University of Edinburgh. In the extremely hot early universe, when the fundamental matter and force-carrying particles of today's universe were indistinguishable in the symmetrical soup, wildly fluctuating Higgs fields are theorized to have permeated spacetime. Like everything in nature, these fields wanted to run downhill, meaning they wanted to evolve to the state of lowest possible energy. As they did so in the cooling universe, according to the GUTs, the Higgs fields lost their symmetry by randomly arranging themselves in something that can be described metaphorically as a crystal lattice. Once that happened, the perfect symmetry of the early universe was broken, and the properties of the Higgs fields were frozen into place. This had a profound impact on the heretofore indistinguishable fundamental particles, breaking *their* symmetry and thereby endowing them with their masses and unique characteristics.

To understand this metaphor, it helps to have a clear mental image of a crystalline lattice. Picture the outlines of a rectangular box. In a crystal, the corners of the box are the positions of the atoms or molecules, and the lines connecting them are the bonds between them. In the metaphorical "Higgs crystal" that forms after a GUT phase transition, imagine that one axis is longest, another is shortest and the third is of an intermediate length. The lines tracing the outline of the box represent the different "directions" the fields chose randomly as they froze out during phase transitions in the primordial soup. In essence, the lines are the "crystal axes" of the Higgs fields.[5]

Now let's put some particles within the lattice of the Higgs crystal and set them moving. In grand unified theories, how they move is determined by their interactions with the Higgs fields. Some particles are capable of interacting only with fields that trace out the *short* axes of the crystal. Because of this interaction, they are forced to move only in this direction. For sake of argument, let's say that this direction is "against the grain." Because these particular particles can go only the "hard" way, they travel slowly and only for short distances. And as Higgs fields drag on them, like mud impeding the progress of marching soldiers, these particles gain a kind of inertia, which is a property of things with *mass*. So by interacting with the Higgs fields in this way, these particles gain an identity as slow and heavy and capable of moving only short distances. In the GUTs, such entities are none other than the $W$ and $Z$ particles—the carriers of the weak interaction.

There are still some other particles in the box, and some of these can interact only with Higgs fields comprising the longest axis. As a result, they can move only in this direction. This is *not* against the grain, so these particles move unimpeded for long distances through the crystal. Since no inertia is imparted, these particles have no mass. Consequently, they race about exuberantly—at the speed of light. And by interacting with the Higgs fields in this way, they have gained an identity as *photons*, the carriers of the electromagnetic interaction.

Finally, a third set of particles is capable of interacting with Higgs fields that make up the intermediate-length axes of the crystal. This interaction also allows easy travel. Moving unimpeded, these particles are massless and fast, like photons. But they're interacting with different Higgs fields, so they have a unique identity—as *gluons*, which are the carriers of the strong interaction.

With this same mechanism, the Higgs fields also endow other particles with their identifying traits. They cause some particles to walk and talk like electrons, others like neutrinos and still others like quarks.

In quantum theory, a particle can be thought of as an irreducible, compact bit of the field's energy. So what of the Higgs field? Shouldn't it be manifested as a particle, just as the electromagnetic field is manifested as a photon? The answer, of course, is yes. Consequently, one prediction of the GUTs is the existence of Higgs particles associated with the emergence of the forces through the process of symmetry breaking. Zap the vacuum with enough energy and a Higgs particle should tumble out. How much energy? The GUT Higgs particle is predicted to be tremendously massive. Since energy and mass are different forms of the same stuff, that means a Higgs particle should pack a huge amount of energy—on the order of $10^{16}$ *billion* electron volts, or GeV. Expressed in terms of a temperature for the particle, that amount of energy is way beyond comprehension. Consider that just *one* GeV is equivalent to *10 000 000 000 000 Kelvin*, a million times hotter than the center of the Sun.

As they are doing at the Relativistic Heavy Ion Collider on Long Island, scientists routinely conjure fundamental particles, creating them out of pure energy in collisions within a particle accelerator. To produce a Higgs particle involved in grand unified symmetry breaking, a particle accelerator would have to deposit $10^{16}$ GeV of energy onto a volume the size of an elementary particle. That's an enormous concentration of energy—well beyond what existing accelerators can attain. According to Alan Guth, this feat of energy concentration would require an accelerator 70 *light-years* in length.[6] Obviously, such a machine is not a likely prospect any time soon.

On the other hand, the standard model of particle physics describes a lower-energy Higgs particle. This one (and its field) endows certain particles with their masses, and with breaking the symmetry of the electroweak interaction into the electromagnetic and weak forces. The phase transition responsible for this symmetry-breaking event is theorized to have occurred at a lower temperature, and so this Higgs particle—the last remaining particle of the standard model to be detected—would pack much less energy. Theory and experiments suggest that it should come in at something like 100 GeV.

That kind of energy is attainable by an existing particle accelerator, the Tevatron at the Fermi National Accelerator Laboratory, and one now under construction in Switzerland called the Large Hadron Collider. Should experiments at one of these accelerators yield the elusive Higgs particle, a fundamental aspect of nature—the Higgs mechanism for endowing particles and forces with their unique characteristics—would be confirmed. And scientists would feel more confident in the GUTs' explanation of the emergence of forces.

With this as background, we're almost ready to move on to cosmic inflation. But first we have to consider gravity. Theoretically, at temperatures higher than those of the grand unified theories, gravity should have been unified with the three other forces in a single *superforce*. Scientists can describe this process in the language of quantum physics. They can say that the gravitational field has a fundamental, indivisible quantum, or particle, of energy—a *graviton*—that is the carrier of the gravitational force, just as the massless photon is the carrier of electromagnetism.

And the graviton should have gotten its identity in a symmetry-breaking phase change just like the other particles. But here's the rub: scientists have not come up with a successful grand unified theory of how this happened.

Gravity has long been understood in the context of general relativity. To unify gravity with the other forces would require a *quantum theory of gravity*. And scientists have tried to formulate such a theory. Just as the classical theory of electromagnetism contains waves — light waves — so does classical general relativity: these are *gravity waves*. And just as scientists have quantized light waves, coming up with a description of the photon, they have also quantized gravity waves. The result is the graviton. So it might seem that devising a quantum theory of gravity should not be too difficult. But unfortunately, it has been.

According to Andrew Hamilton, a cosmologist at the University of Colorado at Boulder, the problem arises because gravitational waves carry energy. Since energy produces gravity, gravitational waves react back on themselves. This back reaction also contains energy, which increases the back reaction. This increases the energy, which increases the back reaction, and so on. "The disturbing result is that the self-gravity of gravitational waves diverges — it becomes infinite," Hamilton says. "All traditional attempts to quantize general relativity run aground on this reef."[7]

Because of this problem, scientists have not been able to devise a successful description of how gravity emerged in a symmetry-breaking phase change. So while scientists theorize generally about gravitons emerging in a symmetry-breaking phase change, neither general relativity nor a grand unified theory explains in detail how this could have happened.

For the explorers of the very early universe, however, all is not lost. Even at the very tiny scales characteristic of the early universe, they can still use general relativity to explore the nature and evolution of spacetime. In *Geons, Black Holes and Quantum Foam*, John Archibald Wheeler illustrates this by taking readers on a fantastic conceptual journey through smaller and smaller dimensions of space and time. When the spatial dimensions are the size of a proton and the time dimension is less than that required for light to traverse this tiny distance, "we see the expected frenzied dance of particles, the quantum fluctuations that give such exuberant vigor to the world of the very small," he says. "But of such effects on space and time we see nothing. Spacetime remains glassy smooth."[8]

For probing the nature of the universe when it was this small, therefore, scientists can still map spacetime with general relativity. And they can still chart the behavior of fields, particles and their interactions using quantum mechanics. The problem, of course, is that when they are dealing with spacetime, scientists have to use one theory, and when they are dealing with particles and fields, they must use another.

You might be wondering what happens to spacetime when we follow the evolution of the universe back even farther, to much tinier dimensions of space and time—to a point where gravity and the other fundamental forces should have *all* been united in a single superforce. Such a dimension of space is called the *Planck length*. It measures just $1.6 \times 10^{-35}$ meters, and when the universe was this size, it was just $10^{-43}$ seconds old. For help in understanding just how small this is, Wheeler invokes a mind-boggling image: Planck-length spheres lined up to stretch across a single proton. How many spheres would it take? The same number of protons that would be required to span the state of New Jersey!

Down at the Planck scale, violent quantum fluctuations should have churned spacetime into a burbling, jumbled mess. "Spacetime foam," is how Wheeler famously described this quantum stuff. If you could travel in it, there would be no up or down, no in or out, no back or forth, and no past, present or future. Here, spacetime should be "not merely 'bumpy,' not merely erratic in its curvature; it should fractionate into ever-changing, multiply connected geometries," Wheeler says.[9] Here is where gravity finally meets the quantum—and general relativity fails. And at the earliest moments of the universe, when it still had not emerged from its Planck-scale cocoon, all interactions in nature, including gravity, should have been One. The trouble is, there's no unified theory that can describe this bizarre world or explain how gravity went its separate way. Such a theory, the so-called Theory of Everything, or TOE, is often held up as the Holy Grail of physics.

Many scientists believe a class of ideas known as *superstring theory* may lead them toward the TOE. The theory challenges the notion that the basic matter particles (the quarks and leptons) and the force carriers (the photons, gluons and $W$ and $Z$ particles) cannot be shattered into smaller components.

''One of the basic questions before us is whether quarks and leptons are truly structureless,'' notes Fermilab's Chris Quigg. ''Will we find when we examine them in greater detail that they have gears and wheels running around inside or some sort of internal structure that might enable us to understand why they have the characteristics they do?''

Instead of gears and wheels, superstring theory invokes Planckian bits of vibrating, one-dimensional *string*. According to the theory, each of the fundamental building blocks consists of a bit of string that vibrates in a characteristic way that determines the identities of each of the matter particles. And each carrier of a fundamental force is made up of its own bit of uniquely vibrating string. This includes the graviton, the carrier of the gravitational force. So string theory naturally encompasses gravity.

For string theory to work, a context of higher dimensional spacetime is required—up to 10 dimensions, in fact. All but the familiar three of space and one of time are said to be curled up so compactly that we are not aware of them.

The catch is that the mathematics comprising a full, precise and consistent superstring theory have so far proved stubbornly resistant. In fact, there are five *different* kinds of string theory. Moreover, the existence of superstrings still has not been tested experimentally. On the other hand, the five flavors of string theory, plus a theory known as *supergravity*, now appear to be manifestations of a single, overarching framework known as *M-theory*. The theory describes, in addition to one-dimensional bits of vibrating string, two-dimensional membranes—the *branes* that collide in Neil Turok and Paul Steinhardt's theory of the origin of the universe—as well as even higher dimensional objects.

There are a number of reasons to believe that scientists may be on to something here, perhaps even the Theory of Everything. For one, M-theory seems to bring gravity's strength into line with the other forces at the very high temperatures of the early universe. For another, the different string theories and supergravity have overlapping relationships, called *dualities*, which form the realm of M-theory. ''Not to take this web of dualities as a sign we are on the right track would be a bit like believing that God put fossils into the rocks in order to mislead Darwin about the evolution of life,'' Stephen Hawking says.[10]

Although the search for a Theory of Everything goes on, and grand unified theories still are incomplete, the concept of symmetry-breaking phase transitions in the very early universe is accepted by scientists as a good explanation for how the fundamental forces emerged. But as any ice fisherman can tell you, phase-transitions don't necessarily happen everywhere simultaneously. With the arrival of winter, a lake freezes unevenly, with some parts icing over first and others later. So with the emergence of grand unified theories in the 1970s, scientists began to wonder what consequences patchy phase transitions might have had in the early universe.

In 1976, physicist Tom Kibble of Imperial College, a mentor to the young Neil Turok, proposed that patchy phase transitions would have had a strange effect. At the boundaries between the patches, spacetime would have been scarred by imperfections dubbed *topological defects*. One such defect is a strange object known as a magnetic monopole. Unlike a normal magnet, which has north and south magnetic poles, a monopole is a magnetic particle with just a single pole. According to Kibble's description, such particles are essentially one-dimensional droplets of the hotter, older phase that made it into the cooler, less symmetrical phase. His description of their production, now known as the *Kibble mechanism*, did not explicitly incorporate grand unified theories. But other scientists, including a young particle physicist named Alan Guth, became interested in the connection between the GUTs and magnetic monopoles.

Working with Henry Tye of Cornell University in May of 1979, Guth began to tackle a question that ultimately led him to the theory of cosmic inflation. Under the assumptions of the grand unified theories, Guth wondered how many magnetic monopoles should have been produced by phase transitions in the early universe. The question had important implications. According to the GUTs, monopoles should be extremely massive, weighing in at $10^{16}$ times the mass of a proton. Consequently, if large numbers of monopoles were produced in the early universe, their collective gravitational effect could have been disastrous, possibly causing the universe to re-collapse shortly after its birth.

As Guth and Tye got to work, they learned that John Preskill, a young graduate student then at Harvard, was trying to make a similar calculation. And in short order, he had beaten them to the punch. Based on the GUTs, Preskill estimated that the number of monopoles produced in the early universe should equal the number of protons and neutrons. This would have been gravitationally catastrophic. To avoid a re-collapse, and to account for the universe we see today, there would have to be some 100 trillion times fewer monopoles around.[11]

Obviously, something was wrong. And working with Tye, Guth began looking for ways to avoid what he came to call the "monopole menace."

Monopoles are produced because of communications problems in the early universe. To complete the phase transitions in an orderly and uniform manner, Higgs fields throughout the universe would need to influence each other—to communicate. To picture this, consider dignitaries milling about in a White House ballroom. They are waiting for the entrance of the president, but no one knows which door he will enter through. So people are facing every which way, randomly. This is a symmetrical state analogous to the condition of the Higgs fields before a phase transition.

Suddenly, an honor guard at the south door to the ballroom snaps to attention. As he does so, people nearby pass the message to neighbors: the president will be arriving through the south door. Those people pass the message to their neighbors, and in this manner, the message quickly sweeps from one end of the room to the other. Before long, everyone is facing the south door, awaiting the entrance of the president. The symmetry has been broken.

Now there is a speed limit on how fast the message can get from one side of the room to another: the speed of light. Remember that nothing can travel faster than this. Obviously, this is not an issue for people milling about in a ballroom. But let's imagine that the people waiting in the ballroom represent the Higgs fields in the observable universe, the portion of the cosmic whole that we can actually see.* Today, the observable universe is about 28 billion light-years across, and the "Higgsians" within it are all facing in the right "directions." More precisely, they long ago settled into the values that determined the masses of particles and the strengths of the fundamental forces. But early on, at the time of GUT phase transitions, the observable universe was scrunched down into a much tinier volume measuring about 1 meter across. (The universe as a whole may have been much bigger.) And the Higgsians were arranged randomly, just like the people waiting in the White House ballroom.

You might suspect that a message sent by a Higgsian on one side of the room—one side of what would become the observable universe—could easily travel to another Higgsian on the other side. After all, the message could travel at up to the speed of light, and it only had to traverse a few feet. But the speed of the message was only part of the equation. The other part was this: how much *time* would the Higgsians have had to communicate with each other between the Ultimate Beginning and the phase transition? Since the universe at this point had just come into existence, the answer was *not very long!* And curiously enough, calculations show that there would have been insufficient time for a message to make the trip. Even traveling at light speed, a message could have traveled only $10^{-27}$ meters. So Higgsians on opposites side of the room would not have been able to influence each other. And as a result, they could not have completed a phase transition at the same time.

What does this have to do with monopoles? Because the Higgs fields at opposite ends of the nascent observable universe could not have completed the phase transition at the same time, it happened in a patchy way. And that's what should have given rise to monopoles. But Guth and Tye realized that it didn't necessarily have to be so. If the phase transition took *long enough*—say, if something *delayed* its completion—then Higgs fields everywhere would have had enough time to "communicate," allowing them to align and thereby wipe out most of the

---

* The universe may actually be much bigger than the part we can observe. We can only see light from distant ends of the cosmos that has had enough time to reach us. Light from beyond about 14 billion light-years away in any direction has not yet had enough time to arrive at our telescopes. In this sense, it is beyond what cosmologists call our "horizon."

patchiness. In turn, this would have resulted in vastly fewer magnetic monopoles.

But how could this have happened? Guth realized that a phenomenon of phase transitions called *supercooling* could have done the trick. It is not at all uncommon. Liquid water, for example, can undergo supercooling before freezing. When this happens, the water can rapidly cool to more than 20°C (36°F) below the freezing point before the phase transition actually happens and ice forms. So during supercooling, the transition is delayed. If Higgs fields could supercool like this, their freeze-out would have been delayed too, giving them more time to communicate and align themselves once the transition was completed. And this was a possible way to avoid the monopole menace.

Realizing that they were on to something, Guth and Tye decided to write up their findings in a scientific paper. As they hurriedly worked through the details of the theory, Tye pointed out that they would have to check an assumption in their calculations: that the supercooling would have no impact on the expansion rate of the universe. This assumption turned out to be spectacularly wrong.

In an ordinary GUT phase change, the Higgs fields roll downhill, if you will, from a relatively high energy level to a lower one. As the universe cools, this is only natural. But Guth realized that for supercooling to happen, the Higgs fields would have to get stuck temporarily, retaining a high energy level and resisting the phase change for a while despite the cooling. Such a state is known in physics as a *false vacuum*. Why *false*? In a true vacuum, the Higgs fields would have an energy level close to zero. But during supercooling, the energy remains temporarily high, so the vacuum is "false."

Guth then came to an astonishing realization. If such a false vacuum state were to have formed, a staggeringly large burst of repulsive energy would have been produced, causing the early universe to inflate. Why would this have happened? It has to do with pressure. If you've ever opened a soda can that's been sitting in the hot sun for a while, you know that the contents have a positive pressure. A patch of false vacuum would be under pressure too. But strangely enough, the physics of false vacuums dictate that it would be a *negative pressure*. That may seem like *suction*, and suction sure doesn't seem like repulsion. But the false vacuum would not have behaved in the way that everyday logic suggests.

To understand why, consider the soda can again. Because it's under positive pressure, some soda bubbles out when you pop the top. This happens because of a pressure *difference* between the contents of the can and the outside world: more pressure inside than outside. But the vacuum of the early universe had a *uniform* pressure everywhere, so there were no pressure differences from place to place. Hence, there was no suction.

On the other hand, energy, like mass, produces gravity. Since pressure comes from energy, it is associated with a gravitational force too. In the case of, say, air at room temperature, the gravity from the air pressure is much lower than that produced by the air molecules themselves. Even so, according to general relativity, the pressure in a room of air, or a can of soda for that matter, is associated with a gravitational force. And since the pressure is positive, this force is the normal, everyday, attractive gravity we're all used too. On the other hand, the *negative pressure* of a false vacuum should yield the opposite: an exotic, beginning-of-the-universe, repulsive, *antigravity*. And since the density of energy in the early universe would have been very great, the negative pressure in the false vacuum—and thus the repulsion—would have been extremely strong.[12]

Guth realized that if the early universe supercooled during a phase transition, this antigravity would have inflated it exponentially—meaning that its size would have doubled at regular intervals. According to Guth, calculations based on the underlying grand unified theories suggested that at least 100 doublings in size should have occurred in the incomprehensibly brief interval of about $10^{-35}$ of a second.[13] To fully grasp the implications of this, consider that a paltry ten doublings of a growing quantity increases its size by about a factor of 1000. And 100 doublings *yields an increase in size by a factor of $10^{30}$*, or a thousand, billion, billion, billion times.

The implications of Guth's discovery for the observable universe were breathtaking. In just a trillionth of a trillionth of a trillionth of a second, it could have expanded by a greater percentage than it did in the following 14 billion years,[14] growing

102

almost instantly from something smaller than a proton to something a bit larger than a basketball. And what about the *entire* universe, not just the observable portion? According to Guth, the universe overall at the onset of inflation was probably much larger than the portion that became observable to us. And since it would have inflated at the same rate as the observable portion, it should have remained much larger—at least $10^{23}$, or a hundred million million billion, times larger, Guth says.[15] If that's the case, there's obviously an awful lot out there that is beyond our horizon.

There is one other thing you should know about the inflation: it would have proceeded at *faster than light speed*. You might object that this violates the cosmic speed limit. But the limit applies only to *stuff*—matter and energy. During inflation, the *fabric of spacetime* itself would have expanded, everywhere at once, not the stuff within it. And as it did, it would have simply carried the contents of the universe along for the ride.[16]

If all that happened was stretching of the spacetime fabric, we wouldn't be here talking about it. Somehow, radiation and matter, the stuff that would later form itself into the universe we know today, had to come from somewhere. And amazingly enough, Guth realized that inflation could account for this as well.

As the universe supercooled during inflation, its energy was sucked up into the stalled Higgs fields. But when the fields finally rolled downhill, they released all that pent up energy. (The effect is analogous to the latent heat released when water freezes.) With this release of energy, every corner of spacetime ignited in the biggest conflagration the universe has ever seen—the fireball of ultra-hot particles and radiation of the standard big bang. In other words, inflation was the thing that put the BANG in the big bang. "Inflation supplies the beginning to which the standard big bang theory is the continuation," Guth says.[17]

If all this seems to suggest that energy was created from nothing, it did to Guth too. He estimates that the total amount of mass in the tiny patch of spacetime that inflated into our observable universe was only *one ounce*. (When I say mass, I mean energy and matter.) From this puny seed, all the matter and energy in the universe is supposed to have grown. So to this day, Guth is fond of calling inflation the "ultimate free lunch."

Amazingly enough, this does not violate the laws of physics. "It may sound as if I wasn't there in my physics class when they talked about the conservation of energy," Guth jokes. "But I was." During inflation, he says, "the total energy of the system is *conserved*." That's because the enormous positive energy that builds up with ferocious speed during inflation creates a precisely balancing amount of negative gravitational energy (the ordinary attractive kind). "And so the net probably is zero," Guth says.[18]

Guth had begun working on the implications of a super-cooled phase transition in the early universe late on the night of December 6, 1979. By 1 a.m. the next morning, he had discovered cosmic inflation. At that point he began to wonder whether there was any reason to think that this unbelievable event had ever happened. After all, just because supercooling *could* have happened, yielding inflation, doesn't mean that it *did* happen.

Right away, Guth saw one factor in inflation's favor: it would have easily eliminated the monopole menace. By stretching the universe to prodigious size, inflation would have swept most monopoles out of the observable universe. If supercooling's delay of the phase change didn't completely do the trick, there could still be many of them out there. But with inflation they'd be spread so thin that there may be no more than a single one left in the part of the universe we can see.

As Guth thought about the theoretical implications of infla-tion, he realized that it could also solve other serious problems plaguing the standard big bang theory. The first is known as the flatness problem. You may recall that Alexander Friedmann, the Russian meteorologist who solved Einstein's equations of general relativity, discovered that there were three possible geometries for the universe: closed, open or flat. Which of these geometries characterizes our own universe depends on the density of matter and energy, because that determines the overall gravitational field. For a flat universe, which is perfectly Euclidean—not curved overall—the average density of matter and energy would have to equal a specific value known as the *critical density*. It amounts to a hundredth of a billionth of a billionth of a billionth of a gram of stuff per cubic centimeter, which works out to an average of five hydrogen atoms in every cubic meter of the cosmos.[19]

104

In cosmological equations, the ratio of the *actual density* of matter and energy in the universe to this *critical density* is denoted by $\Omega$, the Greek letter *omega*. In a flat universe, $\Omega$ equals 1 because the actual density equals the critical density. In a closed universe, $\Omega$ is greater than 1, meaning the density of matter and energy is more than the critical amount—more than enough to curve spacetime in on itself, just like the surface of a sphere. In an open universe, $\Omega$ is less than 1, meaning the density is less than the critical amount, and spacetime is therefore curved away from itself; as I mentioned earlier, you can visualize this curvature as something like the surface of a saddle.

When Guth was pondering inflation for the first time in December of 1979, $\Omega$ was thought to lie somewhere between 0.1 and 2. Although this is a large amount of uncertainty, he had learned from a lecture by Robert Dicke of Princeton that these numbers carried great meaning. For $\Omega$ to have a value within that range, the density of matter and energy in the universe one second after the Ultimate Beginning would have equaled precisely 1, meaning that the universe would have been preternaturally flat. To explain why this is so, Guth invokes the image of a pencil placed on its tip. If it is balanced absolutely perfectly, and there are no disturbances (like shaking or a whiff of air), it will remained perched on its tip for eternity. But if it is placed on its tip ever so slightly out of balance, it will very quickly fall over. Guth says the expanding universe would have been something like this pencil. Any deviation from perfect flatness close to its birth would have been magnified greatly over the 14 billion year history of cosmic expansion.

According to Guth, Dicke showed this numerically in his lecture. If the density of matter and energy at one second was less than 0.999 999 999 999 99 times the critical value, the expansion would have magnified that deviation by a huge amount in a short period of time, before any galaxies could have formed. Similarly, if $\Omega$ was more than 1.000 000 000 000 01 times the critical value, the universe would have collapsed in a gravitational crunch very quickly.[20]

So there you have the flatness problem. To produce anything within the observed range of uncertainty in the density of matter and energy in our universe today, the cosmos at just one second should have been astoundingly flat. Even a very tiny deviation would have meant no galaxies, no stars and no planets—and, therefore, no us. Obviously there *are* galaxies, stars, planets, and us. So it almost seems as if the Hand of Physics tuned the geometry of the early universe with exquisite precision to allow for our emergence. This makes scientists deeply suspicious. Given all the possibilities, it seems extremely unlikely that the universe should have been so precisely fine tuned at its birth. (Well, God could have tuned it that way, but invoking the supernatural is not a part of science.)

One possible way around this dilemma is known as the anthropic principle. As expressed in the early 1970s by Brandon Carter, the principle holds that what we can expect to observe is restricted by the conditions necessary for the development of creatures that are capable of making the observations. So from an anthropic point of view, one could say that the universe had to start out in a state that would lead to galaxies, stars, planets and us simply because we are undeniably here and observing these things. (One might also say, "We're here. Get over it.")[21]

But early in the morning of December 7, 1979, Guth came up with a purely physical explanation for why the early universe should have been so profoundly flat. No matter what the shape of the infant universe was, the exponential expansion of inflation would have driven the part of the universe that we can see today toward perfect flatness. In other words, fine tuning was not necessary. The universe could have started out in any old shape and, with inflation, it would inevitably have wound up looking like it does today.

To help understand why, imagine you are an ant crawling on the surface of a deflated balloon. You can easily detect that the surface is curved very strongly in on itself. Now imagine that the balloon is being inflated. As it gets bigger and bigger, the part of the surface you can see—your observable universe— begins to look for all intent and purposes totally flat, just as the surface of the Earth seems to be flat to us. The intense curvature has simply been inflated away.

Back to the real universe now. The portion of the very young cosmos that would eventually go on to become our observable universe could have started out with a closed or open curvature. Then, spacetime inflates. The expansion rate continually accelerates, as if driven by rocket boosters. In the process, any curvature is inflated away. When the false vacuum decays, the cosmic boosters shut off. Spacetime continues to expand, but it no longer accelerates. It coasts, as described by the original big bang theory. And the resulting geometry, if this simple inflationary model is correct, should be very nearly flat, if not perfectly so.

The morning after his discovery, Guth says he wrote the following in his notebook: "SPECTACULAR REALIZATION: This kind of supercooling can explain why the universe today is so incredibly flat—and therefore resolve the fine-tuning paradox pointed out by Bob Dicke." And he drew a double box around this to emphasize the point.[22]

# 5

## THE GALACTIC COBWEB

### *and the Multiverse*

*Have we, then, been right to call it one Heaven, or would it have been true rather to speak of many and indeed of an infinite number?*
— Plato[1]

*...it is perfectly sensible, and not pretentious, to say that the universe was discovered in this century.*
— Alan Dressler[2]

When a fleck of spacetime the size of a proton inflates almost instantly into something big enough to play catch with, interesting things happen.

As Guth saw, the process drives the universe toward preternatural flatness. But that's not all. In the months that followed Guth's discovery, scientists realized that inflation could go a long way toward explaining why the universe looks the way it does.

So just how does the universe look?

Prior to the 1930s, scientists held to the belief that, at the very largest scales, the universe should look pretty boring. It should be *isotropic*, meaning it should look the same in every direction. And it should be *homogeneous*, meaning matter should be distributed smoothly with no apparent structure.

But even as early as the 1930s, there were some hints that nature did not conform to this banal description. At that time, for example, Fritz Zwicky, a California Institute of Technology astrophysicist, found hundreds of galaxies that he insisted were huddled together in big groups. And in the 1950s, a graduate student named George Abell, working with the Palomar Sky Survey, found many more apparent galaxy clusters. But most astronomers simply wrote off these results as trivial deviations from the real picture, which they felt was as unbroken as the view from a boat becalmed at sea.

Zwicky, as well as a University of Texas scientist named Gerard de Vaucouleurs, argued passionately for a very different picture of space, one graced by galactic archipelagos—clusters and even superclusters scattered in every direction. While Zwicky and de Vaucouleurs favored different kinds of clustering, and they debated these details heatedly, both believed the universe was a much more interesting place than the boring expanse envisioned by most other astronomers.[3]

With improvements to technology, and therefore the view, more evidence of a rich cosmic structure emerged. In the 1970s, Princeton's Jim Peebles found evidence for clustering in a map of the positions of a million galaxies.[4] It was the first galaxy map of most of the sky, and it looks something like a Jackson Pollock painting, with chains of galaxies forming luminous splatters against a dark background. Even then, however, astronomers remained stubbornly skeptical. In fact, Peebles himself doubted that what he was seeing was real. The reason was that it was very difficult, despite improvements in technology, to determine the distances to galaxies, which meant that galaxy maps were two-dimensional. While these seemed to show galaxies drawn together in massive agglomerations, the prevailing wisdom was that this was a chimera created by the inadequate two-dimensional view. Map the universe *deeply* as well as widely, most scientists believed, and it would look boring once again.[5]

In science, cherished ideas die hard. So the debate continued even when the first three-dimensional views supported the case for cosmic structure. Then, in 1985, the debate was put to rest by a detailed, three-dimensional view of a slice of the cosmos by Harvard astronomers.

In that year, Margaret Geller of the Harvard-Smithsonian Center for Astrophysics was working with a colleague, John Huchra, and a graduate student, Valerie de Lapparent, to map the distribution of 1000 galaxies. The idea was to choose a strip of sky and look deeply into that region to plot the positions and distances of galaxies found there. Imagine drawing a line on the night sky arcing about a third of the way across. Such a line defined the boundaries of the scientists' telescopic search for galaxies. For each one they found, Geller, Huchra and Lapparent determined its position and *distance*. In this way, they probed a fan-shaped slice of the universe extending from Earth out to about 500 million light-years. Geller and Huchra were convinced that when this survey was complete, it would turn up absolutely nothing of interest—just a smooth distribution of galaxies refuting the claims of galactic clustering.

Lapparent had been busy writing the computer program that would plot the results, and in the fall of 1985 it was ready. As the map finally emerged from the printer, something seemed dreadfully wrong. The galaxies, plotted as little dark dots on a white background, seemed to be congregating in huge, gently curving structures enclosing big empty regions.

And then, Geller saw the Stickman: the outlines of a bow-legged person luminiously inscribed by hundreds of galaxies gathered together on the dark canvas of spacetime. It looks for all the world like an enigmatic figure from a pictograph drawn by ancient peoples on canyon walls in the southwestern United States.

After checking and rechecking their methods and data, Geller, Huchra and Lapparent could find nothing wrong. As astronomer Alan Dressler describes it in his book, *Voyage to the Great Attractor*, "They were perhaps the first humans to see the pattern so clearly, so undeniably. There would be no going back to the smooth universe. What they had set out to restore, they had destroyed forever."[6]

In her scientific paper about her first galaxy survey, Geller described the overall galactic structure as looking something like soap suds in a wash basin. And among astrophysicists, the suds analogy has held. To my eye, though, the filigree of glowing filaments and sheets looks something like a three-dimensional Pollock—a tangled cosmic *cobweb* shot through gaping voids.

Whether it is described as sudsy or cobwebby, the three-dimensional structure of galaxies revealed by sky surveys pertains only to a very tiny minority of all the matter in the universe. Most of the matter is *invisible*, consisting of *dark matter*, believed to interact with other matter by gravity only. So how has this dark, stand-offish stuff arranged itself? Does it trace out the same pattern of filaments and sheets as the luminous matter? Or does it fill up the seemingly empty holes in cosmic cobweb? Cosmologists would prefer the former—a marriage of visible and dark matter—because it would conform more comfortably with the standard theoretical account of how the universe got to be the way it is. As we'll see in a little bit, a slightly lumpy distribution of dark matter in the very early universe is believed to have kick started the formation of galaxies. But if dark and visible matter don't stick together, this would be harder to explain.

If the holes in the cobweb were actually filled with dark matter, the galaxies, clusters and superclusters tracing out the filaments and sheets would be something like the homes dotting the foothills of the Rocky Mountains not far from where I live. At night, I can't see the mountains even though they are huge. They are just too dark. Even so, I know they're there because I can see the lights flickering from the windows of the many homes on the mountain slopes, and these lights trace out something approaching the profile of the mountains. Similarly, it is entirely possible that galaxies are arrayed on giant "mountains" of invisible dark matter filling the voids in the cosmic cobweb.

Answering the question requires massive surveys of the sky— the logical continuation of the work done begun by Geller, Huchra and others—so that scientists can get a big enough sample of galaxies to feel confident that they're seeing a real pattern. One such effort, known as the 2dF Redshift Survey, has been designed to determine the positions and relative distances of 250 000 galaxies out to distances of hundreds of millions of light-years.[7] Using data on about 130 000 galaxies, Licia Verde, an astronomer at Rutgers and Princeton Universities, and Alan Heavens of the University of Edinburgh, completed a statistical analysis of the shapes of galaxy clusters. These shapes should be affected by the gravitational pull of dark matter. The analysis, announced in late 2001, showed that on large scales, the dark matter must be distributed in the same way as the galaxies.[8]

So it looks like the holes in the cosmic cobweb actually are empty, making it easier to explain how the matter gathered itself together gravitationally to trace out that pattern. But what about other facets of the structure? Does the cobweb have any particular qualities that might help constrain the possible explanations for its origin?

Observations suggest the answer is yes. The structure of the cobweb seems to be *self-similar*, meaning small parts look the same as large ones. This is a *fractal* structure, and it's evident in many aspects of nature.[9] Consider, for example, a watershed. The patterns traced out by creeks running into small rivers look the same as the patterns traced by small rivers running into bigger ones. But what happens when the universe is considered from broader and broader perspectives? Overall, does the fractal structure hold at larger and larger scales — do the fractal structures simply get bigger and bigger? Or is there a fundamental upper limit to the size of structures in the universe, past which the distribution of galaxies begins to look smooth, as has long been assumed?

Cosmologists have dubbed this upper limit the *end of greatness* and, in the spring of 2000, researchers with the 2dF Survey announced that they had found it in a large-scale map of galactic structure.[10] Other kinds of surveys have found similar results. For example, astronomers have mapped so-called radio galaxies, which emit copious radiation in the radio end of the spectrum and which are relatively deep in space. And they've examined a fine mist of X-ray photons, discovered in 1962, called the X-ray background. The sources of the X-ray glow, which permeates the entire sky, are unknown. But astronomers believe they also are far away. As with the galaxies surveyed by 2dF scientists, at scales larger than about 300 million light-years across, the distribution of radio galaxies and X-ray background sources looks smooth.

This view is also corroborated by studies of the cosmic microwave background radiation. As the afterglow of the big bang fireball, the radiation pervades the cosmos. Every cubic millimeter of the universe contains, on average, about 1 background radiation photon. Collectively, these photons comprise a kind of snapshot of what the universe looked like early in its history. "When we look at the cosmic microwave background, we are really looking at a picture of the universe when it was that old," says Scott Dodelson, a cosmologist at the Fermi National Accelerator Laboratory.[11] And if there is any similarity between this snapshot and what the universe looks like today, cosmologists could be more confident that their description of cosmic structure — and their theories for how the universe got to be that way — are correct.

112

The first clear and unambiguous picture of the cosmic microwave background across the entire sky came in 1990 from the Cosmic Background Explorer satellite. COBE took the temperature of the background radiation and found it to be 2.735 degrees above absolute zero. Later refinements using data from the same instrument on COBE pegged the temperature at 2.726 degrees. In 1992, another instrument on the satellite revealed that the background radiation was not precisely uniform. COBE revealed hot and cold spots in the radiation—very tiny deviations from point to point on the sky, amounting to just one part in 100 000.

COBE's map of the microwave background told a story broadly similar to the one that has emerged from surveys of galaxies, radio sources and the X-ray background. Like the grown up universe, the microwave background is *homogenous* on large scales but *inhomogeneous* on smaller scales.

More recently, Ofer Lahav and Sarah Bridle, both of Cambridge University's Institute of Astronomy, used the 2dF Redshift Survey to compare today's cosmic structure to the microwave background radiation. The analysis showed that the distribution of galaxies is remarkably coincident with the pattern of hot and cold spots in the background radiation.[12]

The conclusion? There is a clear connection between the baby features captured in the microwave snapshot of the early universe and the more mature face of today's cosmos. And in both cases, the overall pattern of homogeneity on large scales and rich structure on smaller scales prevails.

"What the universe looks like now is what it looked like then," Dodelson says.

This leads to the next question. Can inflation produce a cosmic structure that conforms to these observational constraints?

In the standard big bang model *before* the advent of inflation, cosmologists simply assumed that the universe began in a homogeneous state, and some other process later sculpted matter into the structure we see today. But why should it have been so? The universe could very well have started out in a highly *irregular* state. And if that were the case, how did things get smoothed out overall?

This is no trivial issue because of a phenomenon first raised in Chapter 4: the horizon problem. In the fanciful thought experiment involving communication between Higgsians in the White House ballroom, Higgs fields on opposite sides of what would become our observable universe were too distant from each other to coordinate their activities. In other words, they were beyond each other's horizon. This gave rise to the monopole menace, which was driven off by Alan Guth with the theory of inflation. Well it turns out that in the standard big bang's description of the very early universe, there is another horizon issue.

When physicists try to push theory to describe events that happened about $10^{-43}$ seconds after the big bang singularity, they reach a boundary between what can and cannot be known, at least with the current laws of physics. Across that boundary is the Planck era, a time when the universe was smaller than a Planck length, or $1.6 \times 10^{-35}$ meters across. As you may recall from the previous chapter, in this Planckian realm quantum fluctuations roil spacetime into a foaming, jiggling, twisting mess. Quantum phantasms — spectral forms made of bits of highly contorted spacetime — dance for an instant and then vanish, only to be replaced by new eruptions. In this hopelessly distorted environment, direction has no meaning, time no direction. And so there simply is no way to define just what state the universe is in.

"You want to define what your universe is," says Andrei Linde of Stanford University. "But the quantum fluctuations of spacetime are so strong they defy your desire to define something definite. Your clock starts running backward and forward, and your ruler starts shaking you."[13] In short, the universe cannot be said to be homogeneous or inhomogeneous because these concepts have no meaning in a place where there is no up or down, no before or after. As a result scientists have no way of predicting what state the universe was in right after the Planck era.

If they could find some wisp of hard, physical evidence from this early era, maybe they could tell. But no such evidence has yet been found. So the early universe might be as smooth as a baby's face or as grizzled as a grandfather's. Who knows?

But the laws of physics and the evidence available in today's universe do allow theorists to come to some very plausible conclusions about other things. For example, scientists have estimated the density of matter within the tiny fleck of spacetime that grew into our observable universe. At $10^{-43}$ seconds, the density was $10^{95}$ times that of ordinary water.[14] To achieve this incredible compaction of matter today, you'd have to stuff more than a trillion suns into the volume of a proton.

With everything packed so close together, you might guess that matter on opposite sides of the nascent observable universe would have had no problem communicating with each other. But calculations show that this actually was not the case, because in addition to being very tiny, the universe was very *young*. And during its brief existence, the matter on opposite sides simply would not have had enough time to "communicate." If the matter started out smooth, this is not an issue. But if it started out lumpy, it could not homogenize — come to a uniform temperature and distribution.

So a good theory must explain the homogeneity astronomers see on the largest scales in the universe. And inflation does just that.

To see how, let's journey back to the Planckian threshold. In the standard big bang theory, the nascent observable universe is a specific size that is determined based on the rate of cosmic expansion and the time that has elapsed since the bang. As we've seen, that size is too big for matter on opposite sides to communicate. With inflation, however, this part of the universe can be orders of magnitude tinier. That's because the burst of exponential growth allows the spacetime of the observable universe to "catch up" in size. Then, when inflation ends, expansion proceeds at the big bang's more sedate pace.

And here is the essential point. Before inflation, the observable universe can easily be small enough for antipodal matter to communicate and smooth out their differences. So with inflation, the horizon problem goes away, just like the monopole menace, and the universe can easily become homogeneous at large scales.

But what about stars, galaxies, clusters and superclusters? Something must start the growth of the cosmic cobweb. It turns out that inflation, coupled with dark matter, can take care of that as well.

Long before Guth's discovery of inflation, scientists exploring the origin of cosmic structure had concluded that quantum fluctuations in the very early universe could have done the

115

trick by producing a slightly uneven distribution of matter from place to place. Since matter tells spacetime how to move, the fabric would wrinkle under the influence of these *density perturbations*, as cosmologists refer to them. These wrinkles would represent slight variations in the strength of gravity, with a wrinkle's depression representing a slight enhancement over the average value, and the high ground representing attenuated gravity. And just as objects like comets and asteroids roll into the large spacetime depression created by the Sun's mass, so too would additional matter particles roll "downhill" into the wrinkles' depressions, forming little clumps that would further enhance the gravity and thereby sow the seeds of galactic structure.

But just any old pattern of quantum fluctuations would not necessarily produce the particular pattern of structure seen in today's universe. So in 1968, Edward Harrison, now of the University of Massachusetts, expanded on the quantum fluctuation idea by determining the kind of wrinkle pattern, or *spectrum*, that would survive and go on to seed galaxy formation. Subsequently, the Soviet physicist Yakov Borisovich Zeldovich independently came up with the same spectrum. The wrinkles, both scientists concluded, should have occurred on many different scales. Moreover, they should be *scale-invariant*, meaning they should look the same regardless of their size. Harrison and Zeldovich reasoned that if a different spectrum of wrinkles had emerged, matter would have clumped together too quickly or too slowly, resulting in either black holes instead of galaxies or no long-lasting structure at all. To get a universe like ours, they said, you need a scale-invariant spectrum of fluctuations and resulting wrinkles.

There was just one little problem. What quantum mechanical process could produce the required pattern of wrinkles? Scientists didn't have a good answer, so for many years they simply assumed a scale-invariant pattern as one of the initial conditions for the origin of structure.

After Guth's discovery in 1980, Stephen Hawking began exploring how inflation could have generated the required quantum fluctuations. In 1982, he proposed that quantum uncertainty would cause the inflation-driving fields to behave ever so differently from place to place. As a result, throughout the baby universe slightly different amounts of radiation and matter would be produced, wrinkling the spacetime fabric and thereby seeding the formation of cosmic structure.

Hawking's proposal was important because it gave hope that inflation theory could become a new paradigm in cosmology — a framework of ideas that could explain how the universe got to be the way it is. But there were serious problems with inflation, not the least of which was that Guth's original version of the theory produced a universe that looked nothing like ours. (We'll get to that problem in just a bit.) And there were other problems as well.

In the spring of 1982, Hawking brought cosmologists together in a conference called the Nuffield Workshop on the Very Early Universe. During the extraordinary three-week gathering, Alan Guth, Andrei Linde, Paul Steinhardt and many other scientists wrestled with inflation's problems and converged on a consensus view for how it could have led to the formation of galaxies. "It was probably the most exciting scientific meeting I will ever go to," says Michael Turner of the Fermi National Accelerator Laboratory. And in the years that followed, these ideas have been refined and worked out in greater detail, creating the new cosmological paradigm. Says Turner, "These ideas have been driving the field for the last 20 years."[15]

So let's turn now to the details of the paradigm.

When exponential expansion of the universe begins, quantum fluctuations in the inflation-driving fields create density perturbations and the first spacetime wrinkles. These are instantly inflated to cosmological scales, ironing them permanently into the warp and woof of the spacetime fabric. As inflation continues, yet more wrinkles are imprinted in the fabric, then others, and then still others. (Yes, inflation happens in a flash, but quantum fluctuations are even faster.) Since wrinkles form at different times during inflation, some have more time to grow, others less. The result? Wrinkles of many different sizes. And since all quantum fluctuations are alike, all scales get the *very same* imprint. In other words, inflation produces the scale-invariant pattern originally proposed by Harrison and Zeldovitch — the very pattern required if galaxies are to grow from the primordial wrinkles.

This was clear to the scientists at the Nuffield Workshop in 1982. Inflation, they saw, could do much more than banish magnetic monopoles, solve the flatness mystery and eliminate the horizon problem. It could also explain the initial conditions needed for the growth of cosmic structure.

*Quantum twitches at the level of elementary particles grow into galaxies.* "This connection between inner space and outer space, it's written on the sky," Turner says. "The distribution of galaxies was the result of quantum noise."

Or as Wayne Hu of the University of Chicago puts it, with evident amazement, "Any small quantum fluctuation in the early universe will grow over the course of 15 billion years into all the structure we see today. That's a really dramatic statement."[16]

A galaxy hosts a kind of interstellar ecosystem, in which the death of stars supplies the raw materials for the growth of new generations of stars and planets. "That's the function of galaxies," notes Alan Dressler, an astronomer at the Carnegie Observatories in Pasadena, California. "They are the reservoirs of the heavy elements that can be made into planets."[17] So a description of how the universe pulled off the nifty trick of transforming quantum twitches into galaxies is an important part of the new cosmological paradigm. As we've seen, inflation is an indispensable part of the modern picture. But it's not the only one. Inflation can establish the initial conditions for galaxies to grow. But then the growing has to happen, and in the modern picture, this is possible only because of an invisible influence.

"The important stuff early on was the dark matter," explains Dressler, who has played no small part in the investigation of this mysterious stuff. The gravity of the dark matter, he says, was what most likely herded ordinary matter together in clumps in the beginning stages of galaxy formation.

Cosmologists have a name for this modern theory of galaxy origins: the *cold dark matter*, or CDM, model. It had its beginnings in 1984 in a ground-breaking paper by George Blumenthal, Sandra Faber and Joel Primack, all of the University of California, Santa Cruz, and Martin Rees of Cambridge. For a scientific paper, it began with uncharacteristic simplicity and clarity:

*"Why are there galaxies, and why do they have the sizes and shapes we observe? Why are galaxies clustered hierarchically in clusters and superclusters, separated by enormous voids in which bright*

*galaxies are almost entirely absent? And what is the nature of the invisible mass, or dark matter, that we detect gravitationally round-about galaxies and clusters but cannot see directly in any wavelength of electromagnetic radiation? Of the great mysteries of modern cosmology, these three may now be among the ripest for solution."*[18]

Their answers were built on decades of investigation and debate about the existence, effects and nature of the dark matter.

Its possible existence was first proposed way back in 1936 by none other than Fritz Zwicky, the Caltech iconoclast who also championed galactic clustering. He was measuring the distance to an apparent group of galaxies dubbed the Coma cluster by taking their spectra and analyzing the redshifts he found. With this technique he found that galaxies within the cluster were moving at considerably different speeds. This meant that while the cluster overall was riding off into the distance with the general expansion of the cosmos, galaxies within it had their own peculiar velocities. This seemed easy to explain: they were being pushed and pulled by each other's gravity.

Based on his estimate of the galaxies' distances from each other, and how much mass each one contained (which would determine its gravitational strength), Zwicky calculated what those speeds should be: a few hundred kilometers per second. But his measurements of the velocities revealed something else entirely: velocities of *1000 kilometers per second*. These guys were moving much faster than their masses alone had suggested. One possibility was that the cluster itself was flying apart in the aftermath of an explosive birth. But Zwicky proposed something much wackier: galaxies in the Coma cluster were moving faster than their masses indicated because they actually contained *ten times more mass* than what was visible.[19]

In the years that followed, astronomers found the same puzzling phenomenon whenever they examined apparent galaxy groups. And so Zwicky's idea gradually seemed less and less wacky. But here was a perfect example of an extraordinary claim: most of the matter in the universe *is invisible*. So it would take overwhelming evidence to convince astronomers that it was actually true.

The evidence began coming in during the 1960s, as astronomers began measuring the rates at which stars within a galaxy rotate around the center. Given the visible structure of a galaxy, it should behave something like a solar system. In our solar system, most of the mass is concentrated in the center—within the body of the Sun, of course—and it drops off toward the edges. This means the solar system's gravitational well is deepest in the center, shallowest at the edges. And this is why planets closer to the center orbit more quickly than those toward the edges. Astronomers call this a Keplerian velocity curve. In a typical spiral galaxy like the Milky Way, the central bulge is a stellar metropolis—the place where the greatest numbers of stars reside. As one moves farther out in the surrounding disk, the population of stars drops off. So it's not unreasonable to assume that most of the mass of a galaxy is concentrated in the middle. And if that were true, rotation speeds should drop off toward the edges in Keplerian fashion, just as they do in the solar system.

But in the early 1960s, an astronomer named Vera Rubin and her graduate students began finding hints that nature wasn't obeying this nice, tidy picture. Rubin, who would go on to play a crucial role in shoring up the case for dark matter, had become fascinated by astronomy as a girl of 12. But later, when she was applying to graduate schools, she learned the sad truth that women weren't exactly welcome in the physical sciences. Princeton in its wisdom told her not to bother to apply because women simply were not admitted to doctoral programs. But Rubin was accepted by Georgetown, and none other than George Gamow took her under his wing as her advisor.

While most astronomers were focusing on the centers of galaxies, Rubin decided to explore the outer reaches. And when she began teaching and doing research at Georgetown, she and her students took on the difficult task of measuring the rotation rate of stars around the center of our own Milky Way. "We saw no Keplerian fall in the velocity," she recalls. "But people objected, saying the data were no good."[20]

Later in the decade, Rubin moved to the Department of Terrestrial Magnetism at the Carnegie Institution. She also turned her attention to the Andromeda galaxy. Using the four-meter Mayall telescope at the Kitt Peak National Observatory, she sampled the light coming from hot clouds of ionized hydrogen gas in different parts of the galaxy, starting in the center and working her way outward to the visible edge. By analyzing the Doppler shifting of the light—how much it was shifted either to the blue or the red end of the spectrum—she was hoping to tell how fast a particular spot on a galaxy was moving either toward the Earth or away from it. And this, Rubin hoped, would reveal the rotation velocities, from galactic center to edge. Once again, she found that they did not fall off.

She was not the only astronomer doing work like this. A colleague at the University of Virginia, Mort Roberts, was measuring radiation emitted by cold, non-ionized hydrogen gas in spiral galaxies. This radiation is in the radio end of the spectrum—invisible to our eyes. But radio telescopes had shown that it extends well beyond the visible edge of galaxies, as defined by starlight.

Sandy Faber, one of the pioneers of the cold dark matter model, was a graduate student working with Vera Rubin at the time. And she remembers the day in 1971 when Roberts came over to share his findings.

"I was a graduate student just hanging around," Faber says. "They had an old-fashioned room down in the basement with an ancient laboratory and a big mahogany table. One day, Vera got a call from Mort. He said, 'Vera, I gotta come up and see you tomorrow.' The next day he comes over and lays out a huge picture of M31, the Andromeda Galaxy. He had been making measurements with one of the Greenbank radio telescopes of the rotation velocity of hydrogen gas. So he started putting numbers down on the photograph, out into empty space beyond the visible part of the galaxy. And Vera's jaw is dropping. She says, 'No Mort. Really?'"

"The rotation velocities were actually staying constant, from the center outward and well *beyond* the visible portion of the galaxy," Faber recalls. "So I'm sitting there watching this, and I can see that these guys are very deeply impressed by this. I'm saying, 'What's the big deal?' And they looked at me as if I were a dunce and said, 'Well, can't you see? There's no matter out there,' meaning there was no starlight. Well of course there *was* matter out there—*dark* matter."[21]

Rubin continued her own measurements of Andromeda, and in 1970 she published a paper detailing her results and suggesting that the rotation velocities did not drop off because dark matter invisibly lurks all around the galaxy, keeping the rotation speeds up through its gravitational influence. But with her instrument, she could only detect light coming from the luminous part of a galaxy. On the other hand, Roberts had been looking at the radio emissions from cold hydrogen extending as far as 100 000 light-years out from the visible edge of galaxies. As he had shared with Rubin, his measurements showed a flat rotation curve too. (Albert Bosma, at the University of Leiden in the Netherlands, had made similar measurements.)

"That was really the beginning," Faber says. "Morton Roberts' measurements of M31 really began to crack it open."

The radio measurements, combined with Vera Rubin's observations of glowing clouds of hot hydrogen, told a consistent story. "It all sort of came together, because both the optical and the radio observations looked so much the same," Rubin recalls.

Even so, some astronomers remained skeptical. Even after Rubin published data on 10 galaxies, each showing flat rotation curves, some astronomers refused to accept it. "I could name some eminent astronomers — but I won't — who just didn't believe it," Rubin says.

It would take several more years for the existence of dark matter to become widely accepted. Help came from two theorists, Jeremiah Ostriker and James Peebles, who showed that a rotating galactic disk without dark matter would be unstable. As Faber points out, "suddenly, there was both a *theoretical* reason to believe that there was an extra matter component of galaxies, as well as an *observational* reason."

Finally, she and Jay Gallagher were asked to review all that was known about dark matter for a scientific paper. "I'm an observer — and historically, kind of a modest observer in the sense that I grew up in a tradition where the goal was to get good numbers," Faber says. "So I approached this paper with the idea that I simply would *not* be convinced about dark matter. 'Over my dead body,' was my attitude. 'Kicking and screaming.' So we went at it as total skeptics. But in the end, we just couldn't escape it." Without dark matter, the universe just didn't add up.

According to Joel Primack, Faber's partner in the cold dark matter model, this review put the whole issue to rest. "Before their review," he says, "the conventional wisdom was that dark matter probably didn't exist and that there was simply this funny evidence that would go away." Some people even thought that the theory of gravity could be wrong. "But after the review, the conventional wisdom was that dark matter *does* exist, and we've got to think about what the implications might be," Primack says.[22]

He and Faber, along with George Blumental, their colleague at U.C. Santa Cruz, and Martin Rees from Cambridge, got to work on that very question. And it became clear that dark matter could do for the universe what the universe, working only with ordinary matter, could not do it for itself: make galaxies.

Ordinary matter consists of *baryons*, run of the mill protons and neutrons that make up the nuclei of all atoms. (Electrons are part of ordinary matter too; but these leptons are so light they contribute just $1/2000$ of the mass of ordinary matter.) In the 1970s, scientists realized that the very high temperatures of the early universe would have prevented baryons from mustering enough gravitational oomph to make a universe that looks anything like our own. Here's why.

For about 300 000 years after quantum wrinkles are imprinted on spacetime, the universe is so hot that baryons cannot settle down with electrons to form neutral stable atoms. So the matter exists as a plasma: a hot gas of electrons, protons and helium nuclei, as well as photons, all bouncing about on their own. Within the plasma, the photons are trapped, hemmed in by the other particles. (These photons ultimately become the cosmic microwave background — a subject we'll explore in greater detail in the next chapter.) The photons carom about with the baryons, pushing hard on their neighbors and thereby creating a pressure that counteracts gravity — a pressure that keeps the baryons from clumping to start galaxy growth.

When the universe cools to about 3000 K, a little lower than the temperature of the surface of the Sun, protons and electrons settle down enough to marry, thereby forming stable nuclei and, crucially, setting the photons free. This, in turn, releases the pressure and allows the baryonic matter to start clumping at the bottom of spacetime wrinkles.

The upshot is that, in a universe with only baryonic matter, it takes 300 000 years for the baryons to begin to gather themselves up into the seeds of galaxies. And therein lies the problem.

Calculations show that with this delay, the seeds take a stagger-ingly long time to sprout into mature galaxies—more than the 14 billion years of cosmic history, in fact, which means there should be nothing like the structure we see today.

"If the universe only has baryons, it's not at all clear why there should be any galaxies," Primack says.

But he and his colleagues realized that with the addition of cold dark matter the cosmos could make galaxies in time. That's because dark matter exerts and responds to gravity but in other ways has nothing to do with the ordinary stuff of the uni-verse. You may be able to imagine the result. Because dark matter particles feel gravity, they can collect inside spacetime wrinkles from the outset. And since they do not feel the photon pressure, they are free to *continue* accumulating, forming the seeds of galaxies early. Then, when the photons fly free, the baryons are free to join the dark matter in the spacetime wrinkles—and the seeds sprout.

But what might the nature of the dark matter be? And exactly how would the early stages of galaxy evolution unfold?

As to the first question, the scientists described in their paper two broad possibilities: *hot* dark matter and *cold* dark matter.[*]

Hot dark matter in the early universe would consist of particles moving at relativistic speeds, meaning close to the speed of the light. A prime candidate considered by the scientists was *neutrinos*, nearly massless particles that interact weakly with ordinary matter. Neutrinos should have been produced in abun-dance during the big bang. But getting fast-moving neutrinos to clump and thereby form the seeds of galaxies would have been difficult at first. For a time, when temperatures were very high, the particles would have streamed away freely, smoothing out any accumulations left by the original quantum fluctuations. Eventually, though, temperatures would drop enough for neutri-nos to clump in spacetime wrinkles. This very slight delay would have had a huge impact on the kinds of structures that could form first in a hot-dark-matter universe, and how the subsequent course of galaxy evolution would have unfolded.

---

[*] Actually, they described an in-between possibility as well: warm dark matter. But the overall picture does not change by omitting it from my description, so I have done so for the sake of simplicity.

There actually were two models for that evolution back when Primack, Faber and their colleagues were working on their 1984 paper. One, championed by Zeldovich and other Soviet scientists, was known as the top-down model. According to this theory, *superclusters* form first and then fragment to form smaller structures, which then fragment, and so on down the line until galaxies are born. In the other model, known as bottom-up, *galaxies* form first and then *merge* hierarchically to make bigger and bigger structures, eventually forming clusters and superclusters.

In their 1984 paper, the scientists argued that the top-down model would prevail in a universe dominated by hot dark matter. They pointed out that when neutrinos slow down enough to begin accumulating, a certain number would find themselves within the horizon—meaning they were close enough to "communicate" with each other gravitationally. The mass of these slower neutrinos within the horizon was found to be roughly $10^{15}$ times that of the Sun, which is just about the mass of a supercluster. The conclusion? In a universe dominated by hot dark matter in the form of neutrinos, superclusters would form first.

But the scientists found a number of reasons to doubt the top-down model, not the least of which was evidence suggesting that galaxies, not superclusters, form first. So they turned their attention to cold dark matter. The matter would have been cold in the sense that the particles moved very slowly. Such sluggishness would have prevented them from streaming away as freely as neutrinos, so they would not have smoothed out the initial clumps of particles left by quantum fluctuations. And that would have allowed the small clumps to grow early and merge into bigger clumps, starting the growth of galaxies and bigger structures from the bottom up. As we'll see in a moment, the scientists realized that this model went a long way toward explaining the origin of large-scale structure in the universe.

But first, how can one possibly have a cold particle in the impossibly hot early universe? And what might these elusive dark matter particles *be*?

One possibility is that the particles are *massive* and therefore lumbering compared to light and speedy neutrinos. In the early 1980s, Primack, working with Heinz Pagels, had come up with just such a particle. The two physicists had been working on aspects of a theory called *supersymmetry*, which proposes that every elementary particle has an alter ego known as a *superpartner*. Most of the partners would live a fleeting existence. But Primack and Pagels found that the lightest superpartners in the theory would be stable. Now don't be misled. Even though these particles would be lighter than their supersymmetric brethren, they would still be massive compared to the neutrino — massive enough to stay cold and slow in the early universe and thereby fit the expected *modus operandi* of a cold dark matter particle.

One such supersymmetric particle considered in the 1984 cold dark matter paper was the *photino*, the superpartner of the photon. Since then, researchers have investigated other supersymmetry candidates for cold dark matter, known collectively as *Weakly Interacting Massive Particles*, or *WIMPs*. These exotica all have fanciful names. In addition to photinos, the WIMPs include zinos, Higgsinos, neutralinos, sneutrinos and axinos. Based on the underlying supersymmetry theory, physicists have calculated that if WIMPs actually existed, roughly 100 000 would pass through every square centimeter of the Earth each *second*. So as you read these words, *millions* of WIMPs could be zipping through your body. Do you *feel* it? Of course not. The *WIMP* moniker is appropriate: these spectral particles would have almost no effect on ordinary matter. (Except of course when enough of them got together in the early universe to trigger the growth of galaxies.)

Another dark matter candidate is a theoretical force particle known as an *axion* that is very light yet still very slow.[23] And then there are the *Massive Astrophysical Compact Halo Objects*, or *MACHOs*. As their name suggests, these are not elementary particles. And they're not really all that exotic because they're made of ordinary baryonic matter, just like you and me. MACHOs include such well-known things as black holes, neutron stars, white dwarf stars, and planet-sized collections of rock and ice. And they are said to flock undetected within galaxy haloes, making up some of the dark matter responsible for the odd rotational behavior of the luminous stuff. But regardless of the influence MACHOs may wield with the galaxies in today's universe, they would not have had any role in the *origin* of galaxies. So I'll leave the discussion of MACHOs at that.[24]

Having identified various candidates for cold dark matter, the scientists then explored how a universe dominated by this stuff would have evolved. The description they offered in their 1984 paper became the foundation of the cold dark matter model. And on this foundation, scientists have since built the modern theoretical picture of galactic origins and evolution.

Sitting in Primack's cramped office on the forested U.C. Santa Cruz campus, he and Sandy Faber described it to me in a discussion that lasted for more than half a day. Although they first worked on the model many years ago, the thrill they felt in having created a beautiful theory seemed to flower again.

The model's explanation of galactic origins begins with quantum fluctuations induced by inflation.[25] Since these fluctuations are random, in some places there is a little more cold dark matter at the outset; in others a little less. The universe is expanding, "but in those places with a little bit *more* than average density, the expansion slows just a bit," Primack says. "And in those places with a little *less* matter, the expansion is just a little faster." That's because a higher density creates a stronger gravitational field, which pulls on the spacetime fabric even as the overall cosmic expansion tries to spread it wider. And a lower density has the opposite effect.[26]

As some areas go slow and others fast, "you get rapid growth in the *contrast* between the high and low density regions," according to Primack. When the contrast becomes great enough, the highest density regions actually stop expanding with the universe and begin to collapse in on themselves gravitationally. And here is where the initial pattern of spacetime wrinkles really begins to have a dramatic effect.

To understand this effect, I want to take you on a quick imaginary journey into the Grand Canyon, which is, in a sense, a gigantic wrinkle in the crust of the Earth. (Yes, there is *less* density of matter within it—just air as opposed to rock. But the analogy still works.) You are standing on the South Rim looking across a gaping void. But as you look closer, you can see that the structure of the canyon is complex. The walls do not simply plunge straight down to the Colorado River. Toward the bottom, their verticality is interrupted by a step. And carved into this step is the *inner canyon* —a steep-walled notch at the bottom, where the Colorado River flows. In essence, it is a canyon within a canyon.

The spacetime wrinkles laid down by inflation were something like this. Initially, the highest densities of dark matter were found in the inner canyons of the early universe: the very narrowest of wrinkles. These densest accumulations were embedded in broader wrinkles with lower densities, which were embedded in even broader wrinkles.

Overall, on all scales the wrinkles look the same. So here's what happens. The first accumulations of dark matter to begin collapsing are the very densest—the ones in the narrowest of wrinkles. With the collapse, these wrinkles stop expanding with the universe. Next comes slightly broader wrinkles. At a certain point they stop expanding too and collapse gravitationally. As you might imagine, there might be two or even more smaller wrinkles embedded in the broader ones. So when the broader ones begin collapsing, the smaller ones merge. As time goes on, this process repeats with bigger and bigger wrinkles. And this hierarchical collapse causes smaller blobs to merge into bigger blobs, and the bigger blobs to merge into even bigger ones.

Now what about the photons and baryons? All this time, they've been busy bashing into each other. When the photons finally fly free after 300 000 years, substantial blobs of dark matter are already in place, creating deep gravitational pockets for the baryons to fall into.

"So now, galaxies are actually beginning to form because the dark matter has collapsed and the baryons are finally collapsing with it," Primack says. Calculations suggest this happens when bright and dark matter blobs have the mass equivalent to about 1 million Suns. At this point, the universe is something like 50 to 100 times smaller than it is today.

The universe is still expanding and cooling, and the blobs keep merging and growing. "But if you were swimming through this universe at an age of 1 million to some tens of millions of years, you wouldn't see a thing," Primack says. That's because during these so-called *dark ages*, the radiation is at $300^\circ$ K— about as warm as a person—putting it in the invisible infra-red region of the electromagnetic spectrum.

At this point, the blobs consist of a mixture of baryons and dark matter particles. Because the dark matter particles are weakly-interacting at best, they pass right by the baryons without jostling them a whit. Meanwhile, the baryons, which have gathered into gas clouds, *do* interact — they collide violently, Faber says, creating shocks. These shocks yield heat, and lots of it. (One way to think of this is that some of the energy of infall is converted into heat.) The estimated temperature is a few hundred thousand to a million degrees. And with such blazing heat, the gas radiates X-rays. What was once cold and dark is now searingly hot and bright.

And the dark ages are over.

At this point, a blob of bright and dark matter is ready to form itself into something that you might call a galaxy. Picture a big halo of dark matter that is deeply warping the fabric of spacetime. Mixed in are hot, luminous clouds of ordinary matter — gas consisting mostly of hydrogen. (The heavy elements have not yet been made because the first stars haven't formed yet.) As the gas clouds radiate energy, they lose it to the outside world. This energy has been the matter's only effective weapon against gravity because, by moving fast, the baryons have resisted the insistent pull from the gravitational well. But as their energy radiates away, the clouds of ordinary baryons become less able to resist gravity. "And so they have no choice but to move into the gravitational well created by the dark matter," Faber said. "It's *this* action which causes the baryons to move toward the center."

The result? Ordinary matter surrounded by a halo of dark matter.

"You could say that the definition of a galaxy is an object in which the baryons have settled to the middle, leaving the dark matter behind," Faber said. "That's what *I* call a galaxy."

Hydrogen gas within the clouds of ordinary matter may begin to collapse at this point, Primack adds, causing fusion fires to ignite and the very first generations of stars to light up. "It's possible," he says. But whether this really happened so early on is anyone's guess.

Regardless of whether stars begin to shine quite so early, these galaxies — or some scientists might prefer to call them proto-galaxies — are quite small. In a continuing process of hierarchical merging and clustering, "they probably get swallowed up in subsequent galaxies," Faber says. And in this way, from the bottom up, clusters and superclusters of galaxies are born.

"The cold dark matter theory put everything together," says Primack. "It gave us a candidate for the dark matter, the generic conditions that the candidate has to work with. It allowed us to work out the spectrum of density fluctuations. And it allowed us to predict from the spectrum what the galaxies would look like, more or less, and also how they would group together into groups and clusters and superclusters."

Yes, but does the model produce a pattern of galaxies, clusters and superclusters that looks anything like the one we see in the universe today? The moment of truth for the cold dark matter model came back in 1984 when Primack, Faber and their colleagues compared the model's pattern to actual galaxy maps made by John Huchra and his colleagues at the Center for Astrophysics. And to everyone's relief and even astonishment, Faber recalls, "There was more or less staggering agreement."

"We were just blown away," Primack says. "For two weeks I think our heads were swimming. It was the most exciting sustained scientific discovery of my career."

But the cold dark matter picture, having been painted in the broad brush strokes of cosmology, leaves out many of the crucial details of how individual blobs merge to produce proto-galaxies, and how proto-galaxies merge to produce more mature structures. How do formless blobs get transformed into something as complex and beautiful as the Milky Way, with many billions of stars crowded into a dense, massive hub at the center, and many others strung like pearls on the graceful spiral arms that orbit majestically around the center?

"Saying we understand the broad brush details of galaxy formation is like saying we know the bigger process of the birth of a baby," says Alan Dressler of the Carnegie Observatories. "What is the bigger process? Male sleeps with female, and you get a baby. That's broad brush. But if you'd like to study the embryo, it doesn't teach you much. I think that's the problem we face now."

In short, we know how the baby is conceived but we don't know the details of growth and maturation. "Everybody believes it's hierarchical," Dressler says. "But *when*? Did it all sort of hierarchically assemble itself before what we call a galaxy existed? Or has it been going on in bits and pieces and dribs and drabs all the way along?"

Arjun Dey, an astronomer with the National Optical Astronomy Observatory agrees that scientists are really just getting started on the details of growth and maturation. "We'd like to know whether our picture is correct overall," he says. But more importantly, "we'd like to know how galaxies acquire the chemistry that they have, how star formation progresses in these galaxies, how they wind up with the shapes and sizes that they have today.... We'd like to know when all of this occurs, so that we end up with the galaxies we have today. And we'd like to know *in what order* do these things occur, whether they all occur simultaneously or whether there is some sequential aspect to it, and whether one of these events triggers one of the others. We don't really know very much about this picture other than from some theoretical indications and a very few observations."[27]

The theoretical indications he refers to include computer modeling of galaxy development in the early universe. These simulations show where galaxies develop and how dark matter is clumped. "With improvements in technology it has become possible to learn much more about the evolution of galaxies," Primack says. But even with powerful supercomputers, these simulations can go only so far. "As soon as you are dealing with gas and stars and supernovae, things become *very* complicated. You must follow the evolution, say, of heavy elements of thousands of galaxies simultaneously. And the trouble is that we don't really understand how much in the way of heavy elements come out of supernovae."

Alan Dressler, a consummate observer, emphasizes that even as computer technology improves further, simulations will remain difficult. "It will be a lot like simulating weather," he says. "The physics will be very easy, but the behavior will be *very* complex. Everybody knows what goes into the weather. The physics is understood. There are no supernatural events. Yet we have the world's most powerful supercomputers and we were still surprised when it rained last Tuesday."

In Dressler's view, theorists have done an excellent job describing how the universe goes from primordial fluctuations to amorphous clumps. "But not from there to a galaxy," he says.

And so observers have been trying to tease out those details. But the challenges are great here too, because those amorphous clumps merged into galaxies during the first billion years or so of the universe. Actually *seeing* how that happened therefore requires that astronomers gather and analyze photons emitted by these baby galaxies more than 10 billion years ago.

Charles Steidel, an astronomer at Caltech who pioneered a technique for doing just that, has had spectacular success studying galaxies whose light began its journey to us when the universe was just a toddler. This means the expansion of the universe has stretched the light considerably, shifting it toward the red end of the spectrum. Astronomers describe the degree of redshifting numerically with a number denoted in equations as "$z$." If $z = 0$, it means the light has not been redshifted. In other words, it has not yet been appreciably stretched by cosmic expansion, so its source must be very, very close by. And therefore we see it as it looks today. When astronomers like Steidel look at a galaxy with a $z$, say, of 1, they are seeing substantially redshifted light, and they see the galaxy as it looked when the universe was about 35% of its current age.[28] With the techniques Steidel pioneered, he has used the Hubble Space Telescope and the Keck telescopes in Hawaii to find and study galaxies with a *redshift of 3*, "meaning they're incredibly dim and incredibly old," he says. The photons Steidel collects from such galaxies were emitted when the universe was four times smaller than it is today, 64 times more dense, and 80% younger.[29]

From these and other kinds of observations, astronomers like Steidel have begun to learn some interesting details. At a redshift of 3, galaxies don't look anything like the Milky Way. "It's not until later, at a redshift of 1, that galaxies start looking modern as opposed to looking like small blobs," Steidel says. "But they appear to be explosive—there are violent winds. They also have a chaotic appearance, and they're smaller than galaxies today."

According to Michael Bolte of U.C. Santa Cruz, observations like these have lent further support to the idea that galaxies evolve through hierarchical merging. "This model is favored now because people behold the universe, and it looks like lots of small things form early on. It seems to be a hierarchical system. And we do see lots of mergers of galaxies."[30]

But there are limits to what astronomers can do with current technology. Even with the Hubble Space Telescope, it's not possible to peer far enough back in time to witness the first blobs merging, the first stars lighting up, and the first supernovae popping off. "That's what the Next Generation Space Telescope is supposed to help with—to reach back not just to the time of these billion-year-old galaxies that we look at today," Alan Dressler says. NGST, NASA's planned successor to Hubble, is being designed to look back to "within the first billion, maybe half a billion years, after the Big Bang." The hope is to see "a first light kind of episode—the first stars, the first supernovae, the first star clusters, and then the first galaxies in the sense that they're smaller units that will grow as pieces fall in."

Given the enormous cosmic expansion that has happened since that time, all the light from such objects has been shifted into the infra-red. Here on the warm Earth, such heat radiation would be impossible to pick out. So after its planned launch some time near the end of the decade, the telescope will be stationed at a point beyond the Moon, where the temperature is very cold. And astronomers hope that from this icy perch the detector will open a new window on the very young universe, when the rich structure of the cosmos first began to take shape.

NGST and other ambitious efforts to explore the early cosmic frontier are being designed to fill in the details of the big picture, in which inflation lays down the primordial wrinkles and then cold dark matter collects in them to kick start galaxy growth. We've seen that computer simulations and observations are starting to lend support to the idea that this process produces cosmic structure through hierarchical growth. And in the next chapter, we'll look in some detail at how inflation itself has fared in observational tests. But before we move on to that, we need to explore one more facet of inflation: how it experienced a near-death experience and emerged making a prediction that may be more profound than anything we've covered so far.

Inflation's close call is known as the *graceful exit problem*. And it involved the ending of inflation—how the false vacuum propelling exponential growth decays into the normal vacuum we live with today.

In Alan Guth's original conception of inflation, the Higgs fields release their pent up energy in the form of hot particles into expanding bubbles of normal vacuum. As these bubbles collide and merge, the volume of normal vacuum grows. Ideally, this would happen smoothly and uniformly, creating a universe filled with a normal vacuum and the big bang's fireball of matter and radiation. Unfortunately, though, Guth's original inflation theory didn't work quite that way. When Guth and others worked through the detailed calculations describing the end of inflation, they found that the random formation of normal vacuum bubbles actually destroys the overall homogeneity of matter. Since our universe looks pretty darn homogenous at the very largest scales, this was a fatal flaw in the theory.

Guth did not try to hide or explain away this flaw in the theory. "One reason Guth deserves so much credit is that he essentially says in his paper, 'Inflation is a beautiful idea, but I can't make it work,'" says Michael Turner. "And he really laid out why it didn't work."

But inflation seemed so beautiful that Guth and other scientists would not abandon the idea. They began to search for what they called a graceful exit from inflation. It was invented independently by Andrei Linde and particle physicist Paul Steinhardt working with a graduate student, Andreas Albrecht.

It's interesting to note that in the 1970s, Linde and a colleague Gannady Chibisov had actually come up with an inflation theory, years before Guth. But like the Russian meteorologist Alexander Friedmann, who believed the expanding universe he had discovered in Einstein's equations had no physical reality, Linde and Chibisov failed to recognize the significance of what they had discovered. According to Guth, they even called their results "garbage."[31] They turned out to be very wrong, of course, and Guth got the credit for discovering inflation. In 1981, after Linde had recognized his mistake, he proposed his solution to the graceful exit problem. He called it *new inflation*. And working on their own but aware of Linde's work, Steinhardt and Albrecht came up with the same solution.

According to new inflation, the Higgs fields don't get *stuck* at a high energy level. Instead, they simply roll downhill *very, very slowly* to the lowest energy value.

To picture the different alternatives, physicists draw a diagram that depicts the evolution of the Higgs fields. This diagram just happens to look very much like a hat. Picture a Mexican sombrero. It's got a wide, uplifted brim, and a peak in the middle (where your head goes) surrounded by a circular well. The sombrero representing the evolution of the Higgs fields in Guth's original inflation theory is a little unusual in that it also has a dent at the top of the peak. In this diagram, the height of the hat's peak represents the energy level of the Higgs fields. Picture them as being represented collectively by a marble. Because the universe was extremely hot in its infancy, this marble is bouncing violently up and down. As the universe cools and the marble starts to settle down, just by chance it gets stuck in the little dent. And that's what triggers inflation. Eventually, by quantum tunneling, the ball burrows out of the dent and rolls down the slope of the hat's peak, winding up in the well and reaching the lowest possible energy. When this happens, inflation ends.

The problem with this version of inflation was that it ended too abruptly, producing all those little bubbles, instead of a universe that looks like ours. In new inflation, the Mexican sombrero is replaced with something more like a sailor's hat. The peak is low and very broad, like a plateau, and there is no dent in the center. The marble representing the Higgs fields starts at the top, with high energy. Since the top of the peak is like a plateau, the marble doesn't get stuck; it simply rolls downhill slowly, making little progress at first. Inflationary growth ensues because the energy level of the marble remains high as the universe cools. As the ball rolls slowly away from the peak, a bubble of normal vacuum forms. And because it takes longer for the marble to reach the well representing the lowest energy value, this bubble has a relatively long time to grow. As a result, it grows very large—large enough, it turns out, to *encompass all of our observable universe*. When the ball reaches the well, it rolls back and forth, up the inside of the brim, back down into the well, up the peak of the hat, and back again. As the Higgs fields oscillate like this, they convert their energy into a hot, homogenous and isotropic blast of photons and matter particles. Other bubbles form and then intersect with our own, producing a much bigger universe overall than in the old inflationary model. But our own bubble has grown to immense size by this time too, so the intersection with others happens extremely far away, well beyond our horizon. When we look out at the universe today, all we see are the remnants of what took place in *our* bubble: a homogenous and isotropic distribution of matter and energy overall.

"I think it was this idea of a slow roll that really turned it all around," Turner says.

But new inflation didn't solve all problems. And it actually raised some new ones. One of those problems was that the creation of a false vacuum with Higgs fields produces quantum fluctuations that are just too large to account for the structure of the universe observed today. The solution was to make the plateau of the sailor's hat even broader and flatter. This worked, but it introduced another problem: Higgs fields just won't behave in the way described by this energy diagram. So it became clear that to get the quantum fluctuations right, Higgs fields would have to be discarded altogether. Physicists modified inflation theory, introducing a new kind of field, called an *inflaton* field, which could behave correctly. (Another name for these kinds of fields, used by Andrei Linde and others, is *scalar field*.)

If all of this tinkering and fine tuning to make the theory come out right makes you suspicious, you should be. It bothers physicists too. Linde tackled this problem with a solution he calls "chaotic" inflation. This theory holds that the primordial scalar fields that drove inflation started off with many *different* values at different spots in the primordial speck of spacetime. This eliminated the need to enter "by hand" the initial conditions for starting inflation correctly: the relatively low energy and the long, slow decline on the plateau of the energy hat. With many fields of many different values to start with, it should come as no surprise that one was just right to drive inflation in a way that produced a universe like ours. Or so goes Linde's argument.

According to chaotic inflation, in some regions of the infant universe, the fields weren't up to the task of driving much growth. These areas remained very tiny. But in other regions, such as our neck of the cosmos, the chaos just happened to result in fields with the right properties to trigger prodigious inflation. These areas may have been very rare, but since they inflated to huge size, they wound up dominating the universe overall.

Today, chaotic inflation is married to a concept called *eternal* inflation. Here's how it works. Inflation ends when the false vacuum decays into a normal vacuum. This decay occurs exponentially, meaning that after about $10^{-30}$ seconds, half of the region of false vacuum decays into a normal vacuum. But that leaves half of the universe *still* in a false vacuum state. This part consequently continues to inflate exponentially. Theory shows that the rate of this inflation is greater than the rate of false vacuum decay. And simple logic dictates that if some areas are inflating faster than other areas are decaying, then inflation ultimately beats decay. There always is some part of the universe that is continuing to inflate.

"For many years we thought inflation would end after a certain amount of time," Guth says. But then he and his colleagues realized this wasn't so. "In almost all versions of the theory, the exponential expansion is much faster than the exponential decay process. So even while this peculiar inflation-driving material is decaying, the total volume is in fact *increasing* with time rather decreasing. So once started, it never ends. Pieces of the inflating region decay and become pocket universes, and an infinite number of these are produced, on and on forever, in an eternal process."[32]

If this is right, we're living in a region in which the false vacuum decayed, giving rise to a standard big bang. But other areas of the universe must still be inflating. And if Linde's chaotic inflation theory is right, there would be other areas in which inflation never really took off in the first place because the primordial fields didn't have the right values. The result is a growing, multi-branching, fractal entity—a *multiverse*—of which our universe is just one tiny part. If the theory is right, the ultimate origin of the multiverse, if there was one, occurred so far back in the inky mists of cosmic time that it may be pointless to ask how it happened.

The multiverse, if it exists, may help explain a deep puzzle: nature seems eerily fine-tuned to allow for our existence. There are many examples of this. Consider, for example, the weak force. Had it been just a tad stronger, all the hydrogen in the big bang fireball would have been synthesized into helium. In that case there never would have been water or long-lived stars.

No water, no stars, take your pick. Either way, it means no us.

On the other hand, if the weak force had been just a smidgen weaker, neutrons formed in the early universe would not have decayed into protons, and we'd be back to the same sad outcome.

The other forces seem to be fine-tuned for our convenience as well. Add just 1% to the strength of electromagnetism and protons would repel one another so intensely that helium could not exist. That would have short-circuited nucleosynthesis — and, once again, us. And if gravity had been just $10^{33}$ times weaker than electromagnetism, as opposed to $10^{39}$ times weaker, stars would be slimmed down by a factor of a billion, and they'd burn a million times faster.[33] Not a happy prospect for life trying to gain a toe-hold in the universe.

Why should these forces, as well as many other aspects of nature I haven't mentioned, all be so finely tuned to allow for our existence? Why didn't some of them turn out to be just slightly different? Is it really possible that in a game of chance with impossibly bad odds, we simply hit the jackpot? Understandably, scientists feel uneasy about all of this.

One can simply explain the coincidences away by invoking the anthropic principle, as many scientists do. As you may recall, the principle states that however improbable it may be, the universe is the way it is because if it weren't, we wouldn't be here asking the question. That certainly makes sense, but it begs the question: how did it get this way?

The multiverse may provide an answer. In creating this ensemble of countless universes, God plays dice. (Apologies to Einstein.) Every roll creates a new universe with a different set of physical laws and constants. The chances that the die will come up snake eyes, creating a universe nurturing to life, are incredibly slim. But God has an infinite amount of time, and with enough rolls, She'll eventually throw a life-friendly universe.

But is the multiverse really science? Because each of the multitudes of other universes likely would have developed laws of physics wholly different than our own, we may never be able to detect whether they really exist. So the theory of the multiverse, derived and supported by logic, attempts to describe not what is *apparent* but what philosophers call an *ultimate reality*. In this way, it resides in a fuzzy realm between physics and metaphysics — between that which is knowable through experiment, observations and other evidence of our senses and that which is forever beyond our ken.

Guth and Linde's vision of a multiplicity of universes beyond reckoning, and a cosmic history beyond comprehension, may also seem like the final humiliating episode in humanity's demotion, a downward spiral that began with the Copernican notion that the Sun, not the Earth, was at the center of the universe. If the proponents of the multiverse are right, we appear to be nothing more than the progeny of a random fluctuation — and one among many — in a primordial field. Purely by chance, mind you, this quantum flicker had just the right qualities to produce a rich, resilient, life-sustaining branch from a profoundly ancient tree. From our perch on the branch, we cannot actually see the tree. We can only infer that it exists from one aspect of the theory of inflation.

If Richard Feynman were alive today, he might respond to these unsettling ideas by reiterating his famous advice about the improbable nature of quantum mechanics: if you want to understand nature, this is the kind of thing you must be ready to accept. And many scientists have, in fact, already accepted the multiverse as a real possibility. "The multiverse is the most important thing about inflation," Michael Turner says. "I have little doubt that 500 years from now our equations will look comical — but not the idea of the multiverse."

And if the multiverse seems like further confirmation of our insignificance in the cosmic scheme of things, consider what we inconsequential humans have accomplished. In just 100 years or so, we have come to understand that quarks, *E. coli* and galactic superclusters share a common cosmic heritage with us. We've learned that everything in the universe, from the smallest component of matter to the very largest structure in the universe, traces its cosmic lineage back to a singular event at the beginning of space and time. And we puny humans have even managed to push our understanding of cosmic evolution up to just the tiniest fraction of a second of that event.

"Einstein said that the most amazing thing about the universe is that it's understandable," says Alan Dressler. "That *is* the most amazing thing about it — that creatures on a little planet have the intellectual capacity to go back and see how they got there. I think that's the most amazing revelation of the last hundred years."

# 6

# PARADIGM SHIFT

## *Inflation Checks Out*

*I have to admit that when physicists go as far as they can go, there is an irreducible mystery that science will not eliminate.*

— Steven Weinberg[1]

*There is more to the universe than the mere vibrations of its particles, just as there is more to music than merely the vibrations of instruments.*

— John Polkinghorne[2]

Inflation was supposed to have stopped a long, long time ago—at least in our corner of the multiverse. But if Saul Perlmutter is right, inflation's back.

In 1998, Perlmutter and his colleagues on the Supernova Cosmology Project announced that cosmic expansion is *accelerating*, fueled by some mysterious, all-pervasive "dark energy" that dominates everything, including gravity. As thin as a marathon runner and fueled by an almost manic enthusiasm—call it bright energy—Perlmutter seems determined to keep pace with the runaway universe. "There is so much to learn," he says. "We dabble here and dabble there, and eventually with enough dabbling we hope to pull together a nice big picture that we really like."[3]

He may call what he does "dabbling," but in reality the work is frantic and grueling. From his office at the Lawrence Berkeley National Laboratory overlooking San Francisco Bay, he directs astronomers around the globe in all-night observing marathons with some of the world's largest telescopes. The researchers target distant galaxies in hopes of capturing stars in the act of exploding. These supernovae flare like fireworks—suddenly and brightly—and then fade quickly. To find them across billions of light-years of space, the astronomers photograph thousands of galaxies in a single night, hunting among the billions of stars in each one for any that weren't there just a few weeks earlier. When a supernova is found, Perlmutter and his colleagues can then sift through its redshifted light for nuggets of information about the expansion of the universe.

Gravity should be clawing at spacetime, slowing its expansion. And that's exactly what Perlmutter and his colleagues expected to find. Their goal was to determine how much the expansion was slowing. So when they discovered that space-time-stretching, dark energy seems to be overwhelming gravity—a conclusion arrived at independently by researchers of the High-z Supernova Team—the news went against all expectations. It meant, for one thing, that inflation was back. And it was also a vindication of the cosmological constant, a term represented by the Greek letter *lambda*, or $\lambda$, that Einstein had inserted into his equations to stabilize his model universe against collapse.

Hailed by the journal *Science* in 1998 as the "breakthrough of the year," the discovery of cosmic acceleration may well go down in history as one of the most important findings of modern cosmology. But in the years since, the dark energy presumed to be causing the acceleration has proved to be both a blessing and curse.

A curse because if it really exists—and the evidence gets stronger year by year that it does—then just what is this stuff?

"I don't have a clue," admits Michael Turner, a cosmologist at the Fermi National Accelerator Laboratory.[4] And even though many of his colleagues have speculated about it, none of them really knows either, he says.

The most plausible possibility is that the dark energy is the background energy of the vacuum. But if it is this so-called vacuum energy, then the elegant equations of theorists demand that it be absurdly higher than it seems to be.

On the other hand, the dark energy has proved to be something of a blessing because it has saved the theory of inflation, widely regarded in the 1980s as a cornerstone of a new cosmological paradigm, from an ignominious fate. In its outlines, the paradigm can be described in a simple equation: cosmic inflation + big bang nucleosynthesis + cold dark matter = our universe. These three big ideas seemed to explain the cosmos in a nice, tidy package. The inflation part of the equation was a compelling addition because of its explanatory power. As we've seen, it seemed to solve some puzzling mysteries: why none of the predicted monopoles have turned up; how the universe could have started out extremely flat; and why the universe is homogeneous overall yet also graced by a filigree of galaxies, clusters and superclusters.

But by the late 1980s and early 1990s, inflation was in trouble. For one thing, the theory was pretty much untested, so all it really had going for it was its remarkable explanatory power. As scientists like to say, extraordinary claims require extraordinary evidence. And some of the evidence was not helping inflation's case. In the late 1980s, astronomers were finding that the density of matter in the universe was less than that required to make spacetime geometrically flat. Recall that the simplest and therefore most appealing models of inflation predicted a perfectly flat universe. So it began to seem that either inflation was wrong or someone would have to find a way, however clunky and unappealing, for exponential growth to produce an *open* universe.

In the mid-1990s, some scientists, including Neil Turok, managed to perfect models of open inflation. These theories were complicated and required ugly fine tuning of the equations. But the alternative — the death of inflation — looked even worse to cosmologists like Michael Turner and Alan Guth. So in 1996, even Guth was ready to accept an open universe from inflation.

But there was one other possibility. Astronomers and theorists knew they really hadn't completed a full accounting of the contents of the universe. Many scientists wondered whether lambda, the vacuum energy predicted by particle theory, might actually exist. And in 1990, Turner championed the idea that it could rescue the old-fashioned notion of a flat universe from inflation.[5] The reason? The energy represented by lambda, although repulsive, would be equivalent to matter (according to $E = mc^2$) just like every other form of energy. And if there were enough of it, this repulsive stuff could flatten the universe.

But at the time, observational astronomy was not yet up to the task of revealing a precise value for lambda, if it existed at all. According to Andrew Hamilton, an expert in relativity at the University of Colorado, the limited evidence available suggested that lambda must be very small. "For me," he says,

142

"that observational evidence, plus a natural distaste for anything so ad hoc as a cosmological constant, combined to put lambda in the category of pigs with wings."[6]

So in the early 1990s, the promise of inflation was being undermined because one of its most important predictions — that the universe should be flat — wasn't checking out. And as for other predictions of inflation theory, testing had barely begun.

It would have been nice if it were possible simply to use a powerful telescope to find some direct evidence of the inflationary epoch — some photons, say, produced at the instant inflation ended. Telescopes are, in one sense, time machines. A Hubble Space Telescope image of a galaxy 10 billion light-years away shows us what the galaxy looked like 10 billion years ago. That's because it has taken that long for the photons to reach us. So to test inflation, why not build a telescope with enough light-gathering ability to peer all the way back across 14 billion light-years to the inflationary era itself? By taking a telescopic time machine back to this instant, astronomers could take a picture of what things looked like and compare it to what's predicted by inflation theory.

Unfortunately, this just wasn't possible. As you may recall, the universe was completely opaque for about 300 000 years following inflation — photons were lost in a kind of fog, unable to fly free to carry information to awaiting telescopes today. The best astronomers could do was look back to 300 000 years after the big bang to see the imprint which inflation may have left on what is known as the *surface of last scattering* at this time. When astronomers actually do this, when they use instruments to collect the first photons ever to fly free of the primordial fog, they are looking at the cosmic microwave background radiation. It is the earliest evidence available for testing inflation.

The photons of the background radiation were freed at a moment cosmologists refer to as *recombination*. Since then, the electromagnetic waves have been streaming outward in every direction. And as the universe has expanded, the waves have been *stretched* and thereby redshifted. The background radiation is so old, in fact, that its light has been redshifted by more than a *thousandfold*.

Despite the enormous stretching of the background radiation, the photons still are accurate reporters bearing stories

143

about the early universe. Actually, to keep my metaphors straight here, I should say they are photojournalists, because the radiation is a snapshot of what the universe looked like in its infancy. The snapshot has been reddened, but its features still are clear.

How do scientists really know for sure that the microwave background is primordial, the actual afterglow of the big bang fireball?

Following the detection of the cosmic microwave background by Penzias and Wilson in 1964, astronomers began measuring the radiation to test whether it was, in fact, a relic of a hot, explosive beginning or simply the product of something closer to home in space and time. In 1965, Jim Peebles of Princeton University figured out that if the universe really had begun with a bang, the radiation reaching us today from that event should have a blackbody spectrum.[*] To confirm such a spectrum, astronomers have to take measurements of the radiation at different wavelengths. But Penzias and Wilson measured the hiss they detected in their horn at just one wavelength. So they could not tell whether the radiation had a blackbody spectrum.

In the 1970s and 1980s, researchers began measuring the microwave background at different wavelengths. They faced a serious obstacle, because only a small fraction of the background photons reaching Earth actually make it through the atmosphere to the ground.[7] To overcome this, researchers used instruments on high altitude aircraft, rockets and balloons. (One of the latter was a behemoth the size of a football field.) This approach had its own limitations, however, because the instruments could stay aloft for hours at a time at best—not enough time to make highly accurate measurements. Even so, the results taken together seemed to show that the background radiation was at least close to blackbody, offering a tantalizing, if inconclusive, hint that it was a primordial relic of the big bang.

In 1974, John Mather, a 28-year-old scientist who had worked on one of the balloon experiments, spearheaded a proposal to NASA for a *Cosmological Background Radiation Satellite*, which could take very detailed measurements of the cosmic microwave

---

[*] As described earlier, radiation with an intensity and spectrum that is determined solely by the temperature, with no contribution from the composition of the stuff that emitted the radiation, is blackbody. It is produced only when a system has come to thermal equilibrium.

background using instruments lofted above the obscuring atmosphere. After many years of planning, design and redesign, NASA funded the project fully in 1982. By then, the satellite's name had been changed to the *Cosmic Microwave Background Explorer*, or *COBE*. It was designed to hitch a ride to orbit on a space shuttle. But when the Challenger exploded shortly after blast-off in January of 1986, COBE was put on ice.

Redesigned to fly on a rocket, COBE finally got its chance just before the end of the 1980s. On November 18, 1989, a Delta I rocket carrying the satellite blasted off from Vandenberg Air Force Base near Lompoc, California. Injected into a polar orbit 560 miles above the Earth, the spacecraft began mapping the microwave background across the entire sky. In January of 1990, after just two months of mapping (and using just 9 minutes of data), John Mather stood before a packed session of the American Astronomical Society's annual meeting in Arlington, Virginia, to show his colleagues what COBE's FIRAS instrument had seen. (FIRAS stands for Far Infra-red Absolute Spectrophotometer.) Toward the end of his talk, he used a transparency projector to unveil the most important data: a curve showing the spectrum of the radiation. With a range of uncertainty of just 1%, the curve was very nearly precisely that of a blackbody, exactly what would be expected if the universe had a hot, explosive beginning. Here was powerful corroboration of the big bang theory. The scientists in the audience rose to their feet and applauded Mather and his colleagues.

These first results from COBE also raised hopes that more detailed measurements from the satellite would show how matter in the big bang fireball began drawing itself together to form the structures of today's universe. For proponents of inflation in particular, this effort was particularly important. With the more detailed measurements to come, they hoped to test the proposal that quantum fluctuations during inflation planted the seeds of cosmic structures by wrinkling spacetime and thereby laying in subtle variations in gravity. Inflation theory predicted that those primordial wrinkles should be manifested as variations in the *temperature* of the microwave background from place to place across the sky. And if exponential growth of spacetime really happened, there should be a particular pattern to those variations—a kind of fingerprint of inflation. If COBE were to find the fingerprint, the theory would pass a second crucial test.

On April 23, 1992, COBE scientists once again stood up in front of a crowded gathering of their peers, this time at an American Physical Society meeting in Washington. Their announcement? COBE's Differential Microwave Radiometer, or DMR, had indeed detected variations in the temperature of the microwave background.

Overall, the FIRAS instrument on COBE had determined that the temperature of the radiation was a uniform 2.735 degrees above absolute zero. (With additional data, the temperature was revised in 1993 to 2.726 K.) But the DMR instrument found very slight variations in this temperature amounting to just one part in a hundred thousand.[8] These *hot and cold spots* in the otherwise homogeneous background radiation were distributed randomly across the microwave sky. So far, so good. But did the pattern of these spots match inflation's fingerprint?

As you may recall, inflation is predicted to iron a particular pattern of wrinkles into the spacetime fabric—a *scale-invariant* pattern. Since this pattern is a critical part of inflation's fingerprint, let's briefly review it.

Inflation creates a scale-invariant pattern because it irons a series of wrinkles into the spacetime fabric over a period of time and stretches them all to cosmic scales. Early generations of wrinkles—the first ones to form—are the very largest, because they had most time to stretch; later generations of wrinkles are smaller, because they had the least time. The result? Wrinkles of many different sizes comprising a scale-invariant pattern, meaning all the wrinkles, big, medium and small and everything in between, should look the same. "That's because all quantum fluctuations are alike," explains the University of Chicago's Wayne Hu. "Inflation stretches them all to cosmological scales— and in the same way, so that all scales get the *very same* imprint."[9]

So here is one aspect of the fingerprint. If inflation really happened the way theory suggests, there should be wrinkles on many scales, and they should be scale-invariant. But since the universe was opaque immediately after inflation ended, how can we actually *see* these wrinkles today?

Just as we can't see the warping of spacetime created by the Sun, we can't actually see inflation's wrinkles *per se*. But we know the Sun has warped spacetime because of its gravitational effect on us, and the same general idea pertains to the wrinkles. We can detect the *effect* they had on matter and radiation, an effect that should have survived to the time of recombination.

Here's how. When the very first wrinkles form, some photons and matter particles by chance find themselves on high spots; others find themselves in low spots. To fly free to reach our instruments, the photons that wind up in the low spots, in the gravitational troughs of the wrinkles, have to *climb out*. In the process, they lose energy and their wavelengths get longer — meaning they become *redshifted*. As a result, photons from these over-dense regions appear on the COBE map as *cold* spots. Conversely, photons from under-dense regions appear as hot spots.

The hot and cold spots discerned by COBE are rather large, each one stretching across more than $7°$ on the sky. This is 14 times the diameter of the Moon as we see it from here on Earth.

Why are they so big? They come from wrinkles in spacetime that formed at the very outset of inflation. As a result, there was ample time for inflation's explosive exponential growth to stretch them from subatomic scales to structures that are actually larger than the very largest galactic superclusters. Matter on either side of each one of these big wrinkles is so widely separated, in fact, that even after 14 billion years, there still has not been enough time for "communication" via gravity. In other words, matter on either side of such a wrinkle has not yet "felt" each other and therefore has not started falling towards each other. As a result, the hot and cold spots revealed by COBE represent wrinkles that have not yet participated in the formation of cosmic structures — galaxies, clusters and superclusters. So they are an unaltered imprint of the earliest stages of inflation alone.

COBE's eyes were not very sharp. They could discern only hot and cold spots bigger than $7°$ across. But theory predicts that there are smaller spots as well. They should exist because they would have formed later during inflation and therefore would have been stretched less than inflation's first wrinkles. According to Wayne Hu, matter on either side of a wrinkle smaller than $2°$ across should have been close enough

147

immediately after inflation to communicate via gravity. There-fore, in the early stages of cosmic evolution the matter would have begun the process of infall. The hot and cold spots resulting from these wrinkles therefore would represent the seeds of galaxies and bigger cosmic structures seen in the universe today. (For a reason that will soon become clear, the hot spots in this case represent over-dense regions, and the cold spots under-dense regions.)

By carefully analyzing these smaller temperature variations in the microwave background radiation, scientists can glean a wealth of information about how the seeds actually sprouted into galaxies—information that can help them check whether inflation offers an accurate description of the origin of cosmic structure.

During the 300 000 years before recombination, the seeds cannot sprout into galaxies. The baryons—the ordinary matter—cannot accumulate along with the dark matter because of the pressure created by the peripatetic motions of the photons within the troughs. Until the radiation is liberated at recombina-tion, this photon pressure tends to push the matter apart—to *rarefy* it—even as gravity tries to pull it together in the little space-time wrinkles. So within the troughs, the plasma of photons and ordinary matter is pushed together and pulled apart, pushed together and pulled apart, over and over again, in a tug of war that lasts for 300 000 years.[*] When the photons fly free, the pres-sure disappears, and the baryons can finally fall into the dark matter's gravitational pockets, allowing the growth of cosmic structure to begin in earnest.

Before that moment, the early universe literally rings out with a primordial sound that reverberates within the hot plasma. The oscillations of compression and rarefaction within the plasma are very much like the sound waves that fill the body of a guitar when a string is plucked. A vibrating guitar string causes air mole-cules within the instrument to compress and rarefy, compress and rarefy, and these oscillations are sound waves. Transmitted to the

---

[*] Since the dark matter does not interact with photons, it doesn't feel the pressure and therefore does not oscillate like the ordinary matter. It just sits there waiting for the time when the baryons can finally join it at the bottom of the wrinkles.

air around the guitar, the sound waves travel to our eardrums, causing us to sense the sound of the plucked string. Although the hot plasma of the early universe was quite different to ordinary air, the acoustics should have been broadly similar, Wayne Hu says.

The sound from a guitar or any other musical instrument is characterized by a fundamental tone and a series of overtones. With a guitar, the fundamental tone is determined by the length of string that's allowed to vibrate when it's plucked. The overtones come from different vibrational modes of the string. And it is the overtones that help make a guitar sound distinctive, unlike, say, a banjo. Similarly, the sound that reverberates through the plasma should have a fundamental tone and a series of overtones, Hu says. And this has an important impact on the cosmic microwave background radiation we see today. The *spectrum* of these sound waves should supply some fine detail to the fingerprint of inflation—*acoustical* detail that can be used to test the theory.

How can cosmologists ''listen'' to this sound to study that acoustical detail? And what might it tell them? ''We don't actually listen to the sound,'' Hu explains. ''We *see* it.'' That's because the temperature of the photons in the smaller wrinkles rises and falls with the oscillations. And when the photons finally fly free at recombination, those that just happened to be compressed at that instant comprise the smaller hot spots in the cosmic microwave sky (the ones smaller than 2° in size); those that were rarefied comprise the smaller cold spots.

''In the cosmic microwave background, we actually see compressions and rarefactions from a *sound wave*,'' Hu says. ''It's just like a normal sound wave in air. But we're seeing it because the gas temperature depends on its density.''

Today, the spectrum of hot and cold spots smaller than 2° in size on the sky should carry the specific imprint of this acoustical process. And what is that spectrum? According to the details of inflationary theory, there should be a dominant kind of spot representing the fundamental tone, and smaller spots representing the overtones. The fundamental spots should be wider on the sky than the overtone spots, and they should represent the highest peak of deviation from the average temperature of the background radiation. The overtone spots should be smaller in size, and their variation from the average temperature should be less pronounced.

Inflation produces this particular pattern because it irons wrinkles of many different widths into the spacetime fabric. Basically, the narrower the wrinkle, the more oscillations that can take place within it before the photons fly free and release the pressure. In wrinkles of exactly the right width, there is only enough time — no more and no less — for the plasma to compress just *once*. In narrower troughs, there is enough time for *one compression and one rarefaction*. In still narrower troughs, there's time for a compression, a rarefaction and then a *re*compression. And so on. Troughs with just one compression *produce the fundamental tone*, which shows up in the background radiation as a spot with the largest variation in temperature. And the narrower troughs, with their more rapid oscillations, produce the overtones — spots with somewhat lower temperature variation peaks.

If you're thinking that what I've described is a departure from a perfectly scale-invariant pattern, you are exactly right. The fundamental and overtone spots represent *peaks* in variation from the average temperature — little anomalous crescendos produced during the plasmatic symphony of the early cosmos. And according to Hu, this exception to the rule of perfect scale-invariance is a key prediction of inflation theory. If inflation really happened, observers should detect those peaks. If they don't, inflation would fail a crucial test.

Of course, an even more stringent test would be possible if the theory were to predict the actual *size* of the fundamental and overtone spots. And, in fact, the theory does make one such prediction. If the universe is geometrically *flat*, as predicted by the simplest models of inflation, the fundamental spots should measure $1°$ across on the sky. On the other hand, if the universe is positively or negatively curved, meaning closed or open, the fundamental spot should be bigger or smaller than $1°$. That's because cosmic microwave background photons have followed the shape of spacetime in their journey to us, and that shape affects how the spots appear to us. If the photons took curving paths, the sizes of the hot and cold spots would, in a sense, be *lensed*, making them appear either enlarged or diminished. The effect is roughly analogous to what happens to light when it passes through, say, a binocular lens. When you look at a distant object, the lens bends the light in such a way as to enlarge it. If you turn the binoculars around and view the object through the wrong end, the light bends in such a way as to make the object appear smaller.

So here, finally, is inflation's unique fingerprint in all its telling detail. It should consist of hot and cold spots at many scales. All spots bigger than about 2° across should be perfectly scale-invariant. Below 2° in size, the spots should show a slight departure from perfect scale-invariance. There should be a peak in variation from the average temperature—a kind of hot and cold spot that represents the fundamental tone; there should also be lesser peaks representing the overtones. And if the simplest and most attractive model of inflation is correct, the flatness of the universe should yield fundamental spots that are about 1° wide.

Before COBE project researchers stood up to present their findings at the American Physical Society meeting on April 23, 1992, scientists had little convincing hard evidence connecting the big bang to the formation of stars, galaxies and larger cosmic structures. All they had were theories, and cosmic inflation was but one of several. But thanks to the new data from COBE, that was about to change.

The Differential Microwave Radiometer on COBE had measured the background radiation over the entire sky to an accuracy of one part per million. At a press conference for reporters, DMR project leader George Smoot would compare this to measuring the height of Mount Everest to within an inch.[10] To accomplish this feat, the DMR took more than 200 million measurements. COBE scientists had been kept busy for a year processing that avalanche of data and checking for errors. When they were done, they produced a map of the background radiation that has since become an icon of cosmology, and possibly one of the most famous scientific images of all time. Elliptical in shape (like a map of the globe portrayed in a Mollweide projection), the map depicts what we would see on the dome of the sky if our eyes could see the primordial photons streaming from the big bang. Hot and cold spots of many different sizes are portrayed in psychedelic colors: orange, pink, blue and turquoise.

The spots in COBE's map are big: the largest stretch across 10 billion light-years, or 59 billion trillion miles of space, much bigger than a galaxy or even a supercluster. So here is the unaltered imprint of inflation's quantum twitches.

Before their peers at the American Physical Society meeting, the COBE scientists presented the project's findings in all their rich, technical detail. But the data everyone was waiting for, including Alan Guth and his fellow pioneer of inflation theory, Paul Steinhardt, was actually a simple graph. Plotted on the vertical axis of the graph was a measure of the degree to which hot and cold spots departed from the average temperature. And plotted on the horizontal axis was a measure related to how big these hot and cold spots appeared to be on the dome of the sky. As simple as the graph was in concept, it was monumentally important because it would reveal whether the pattern of hot and cold spots matched inflation's fingerprint.

Guth says both he and Steinhardt were overwhelmed by a "sense of awe" when they saw the results: a nearly perfect match between inflation's prediction and what COBE observed.[11] The pattern of hot and cold spots were scale-invariant.

"What we have found is evidence for the birth of the universe and its evolution," DMR project leader George Smoot told the swarm of reporters gathered at the press conference afterward.[12] The next day, the New York Times carried a story about the findings on its front page. "Scientists Report Profound Insight on How Time Began" read the headline. "They have found the Holy Grail of cosmology," said Michael Turner in the accompanying story.[13] In the Washington Post's page one story, Stephen Maran of the Goddard Space Flight Center described COBE's map of the microwave background in biblical terms: "They've found the first evidence of shape arising in the universe. It's like Genesis."[14] And in the days following the announcement by the COBE team, Stephen Hawking went even further. Quoted by a Reuter's news service reporter, the famous Cambridge physicist said, "It is the discovery of the century, if not of all time."[15]

Given some of the explosive findings that would emerge in the years to come, including the discovery of cosmic acceleration by Saul Perlmutter and his colleagues, sentiments like Hawking's may seem a little over blown. But in the context of what was known at the time, the enthusiasm is understandable. Before COBE, the cosmic microwave background radiation was little more than static—a faint whisper bearing indecipherable stories of our cosmic origins. With COBE's precise measurements, the whisper took on a tangible form: an actual picture of the primordial fireball from which everything in the universe emerged.

On a scientific level, COBE connected the dots between the theory of the hot big bang and the formation of structure in the universe. "It was a very big milestone," Hu says. "It really told us that our model for how things form in the universe was correct: Gravity takes small fluctuations and turns them into galaxies. Before then, there were many other models for the origin of structure, and all were part of cosmological thinking back then. COBE eliminated most of them." And by finding a scale-invariant pattern to the hot and cold spots, COBE offered the first observational support for the idea that it was inflation that caused the initial fluctuations.

Finally, COBE ushered in a new era in cosmology, in which increasingly precise measurements would put theories to ever more stringent tests. In the aftermath of the release of COBE's findings, some scientists chose to emphasize the pressing need for such tests. Because COBE could only measure spots larger than $7°$ across, it left the finer details of the cosmic microwave sky shrouded in mystery. So even though inflation had passed its first test, many more were in the offing.

COBE's data comprised a "single datum point," said Princeton's James Peebles in a follow-up story in the *New York Times*. It revealed structures too big to correspond to the first sprouting of galaxies, clusters and larger structures, he pointed out. So more definitive support for inflation would have to await more detailed observations. "The inflationary theory is not contradicted by COBE's findings, but neither is it proved by them," Peebles concluded.[16]

Even as inflation was surviving its first trials, astronomers were showing that the density of matter was less than the critical value needed to flatten the universe. You may recall that cosmologists measure this critical value in terms of a ratio denoted in equations by the Greek letter omega, $\Omega$. The ratio compares the measured to the critical density required for a flat universe. If $\Omega = 1$, it means the measured value equals the critical value, and the universe is therefore flat. Well, much to the dismay of inflation theorists, $\Omega$ was turning out to be somewhere in the vicinity of 0.3, or 30% of the critical value.

"Inflationary theory seemed to be very successful in explaining things," Saul Perlmutter says. "But it seemed to insist that the total amount of mass and energy in the universe had to come out to the critical amount. When astronomers went out and did the measurements, the numbers were coming in low."

At that point, according to Perlmutter, theorists proposed that the astronomers hadn't done a good enough job of accounting for the stuff of the universe. But the astronomers took better and better measurements, and the results kept showing that the density of matter and energy in the universe was low. "They were *always* coming in low," Perlmutter says.

At this point, some theorists began thinking outside the box. One of those theorists was Neil Turok, who went to work on the problem of low matter density with two collaborators: Martin Bucher, a colleague at Cambridge, and Alfred Goldhaber of the State University of New York at Stoney Brook. With much finagling, they contrived a way for inflation to produce an open universe—one with a dearth of matter, and therefore gravity. In their theory, it takes a *double dose* of exponential growth to do the trick. The mechanism works like this. First, as in traditional inflation, scalar fields temporarily get stuck, causing a patch of spacetime to nucleate a bubble of false vacuum. This bubble inflates, as in the old picture. But as the fields begin to decay, they get stuck for a *second* time. When this happens, a *second* bubble of false vacuum forms within the first bubble and grows exponentially. This is the second bout of inflation. And bizarrely enough, Turok and his colleagues were able to show that the space inside this second inflating bubble, which corresponds to *our universe*, was geometrically open.

It worked, but it wasn't very elegant. Turok himself calls the double dose theory "terribly baroque." Others came up with different open models. But all were just too elaborate and *special* for the tastes of physicists, who have a strong preference for simple, *generic* solutions.

"The open inflation models all seemed to require rather special assumptions about the nature of the inflaton field," Guth says. "Of course, people have argued about just *how* special these assumptions are."[17] But there was no denying that the fields needed a suspicious amount of tinkering to get them to work.

So here is where inflation stood until 1998. A seductively complete solution to a plethora of fundamental problems, and one that had yet to be proved wrong. But reality was stubbornly refusing to conform to theory. The real universe just didn't look like the one that should have been produced by the most attractive inflationary model. It was too open.

But theorists kept at it. "Some started saying that maybe we have to consider the possibility of something else," Perlmutter says. "Something like *the cosmological constant*."*

One of the theorists pressing this view was Fermilab's Michael Turner. In fact, he and a few colleagues began the crusade in 1984, in a paper they wrote "to solve the omega problem," Turner says. Even back in those early days of inflation theory, the universe wasn't looking flat. So Turner and his colleagues proposed that the missing density of matter and energy could come from two possible sources: relativistic particles (meaning ones racing about at the speed of light), or the cosmological constant, Einstein's repulsive energy represented in equations as the Greek letter lambda.

By 1990, relativistic particles had been ruled out. Turner then made what he describes as a Sherlock Holmes argument: "When you rule out what's impossible, whatever is left must be true. So what was left that could flatten the universe and not cause any problems? The answer is lambda." With this in mind, Turner devised a recipe for a universe that would fit most closely the predictions of inflation theory. In a second paper, he called it his "best-fit universe." It should consist, he said, of 5% ordinary matter and energy, 25% cold dark matter, and 70% repulsive, cosmological constant energy. "But lambda was considered so stinky that nobody wanted to consider it," Turner says.

---

* Cosmologists have so many different names for the stuff that's propelling accelerated expansion that it can be confusing. So here are some things to keep in mind. When cosmologists use the term *dark energy*, they are referring to the mysterious stuff powering cosmic acceleration. The term *cosmological constant* refers to the term they enter into their equations to represent the *effect* of the dark energy. The Greek letter *lambda* is the actual *symbol* they use in those equations. And *vacuum energy* is one hypothesis for what the dark energy is. Confusion arises because cosmologists sometimes use all these names interchangeably.

He tried making the argument again in a 1995 paper titled "The Cosmological Constant is Back."[18] (Two other scientists, Paul Steinhardt and Jerimiah Ostriker published a paper at about the same time making a similar argument.) And he tried it once again in 1996 at a meeting at Princeton University. As Turner describes it, the meeting turned out to be a "beauty show" for different ideas. "My argument basically won the show. But there was one blemish: Saul Perlmutter."

At that time, Perlmutter's Supernova Cosmology Project had found and analyzed eight supernovae—too few to say anything definitive, Turner says. But Perlmutter presented preliminary findings anyway, findings that actually seemed to *rule out* cosmic acceleration.

Perlmutter and his colleagues, as well as Brian Schmidt and his colleagues in the High-$z$ Supernova Team, were using telescopes to take pictures of large swaths of the sky. Comparing these with images taken earlier, the researchers were hunting for stars that had exploded as supernovae in the intervening time. In particular, they were looking for a kind of supernova that can be used as a "standard candle" for determining its distance from Earth. Known as a type Ia supernova, it begins as a white dwarf—a kind of star that packs the mass of our Sun into a volume no larger than the Earth. With this impressive density, a star like this is ravenously hungry. If it has a smaller stellar spouse, as many such stars do, it reaches out across space and wrestles matter away from its companion, getting fatter and fatter in the process. When a white dwarf becomes sufficiently gorged, it explodes. And here's a key point: all white dwarfs destined to become type Ia supernovae have about the same mass before they explode. With the same mass, they pop off with *exactly the same brightness*. Astrophysicists are able to calculate this brightness, and this is what makes them so valuable to scientists like Perlmutter and Schmidt.

Because a type Ia supernova has a standard brightness, whenever one is discovered it's possible to use a simple formula to find out how far away it is.* Using this technique, the two

---

*Recall that the light from a star falls off in intensity by a standard amount as it moves farther and farther from its source. So once you know the apparent brightness of a type Ia supernova—how bright it looks to us here on Earth—you can compare it to its known actual brightness and use a simple equation to determine the distance.

teams built up catalogues of type Ia supernovae at many different distances. And with sensitive instruments, the astronomers then took careful spectral measurements of the exploded stars. This revealed how much their light had been redshifted, or stretched, by the expansion of the universe.

"The farther away they are, the farther back in history you're looking," Perlmutter says of the supernovae. "So the basic technique is that we find a bunch of different supernovae at different times back in history, and then, using the redshift, we look to see how much the universe has expanded at *each of those different times*. So in this way we get to see how much the universe has stretched 1 billion, 2 billion, 4 billion, 6 billion years ago."

In 1998, with enough supernovae under their belts to say something with confidence, the two groups knew that the expansion of the universe was accelerating. "We found that the universe has been doing more of its stretching more recently, and less of it in the past," Perlmutter says. "Our conclusion was that the expansion of the universe is speeding up. It was a very, very straightforward measurement."[19]

For proponents of inflation, these results helped rescue the theory from the ugly fine-tuning needed to produce an open universe. The discovery of cosmic acceleration showed that Turner and other lambda advocates had been right. And based on the degree of acceleration observed by the supernovae teams, the researchers calculated that the dark energy fills in what's missing to make the universe flat.

"If the mass density of the universe is indeed around a third, the dark energy turns out to be just enough to make the universe flat," Perlmutter says. And that means the simplest versions of inflation will work just fine.

In the *New York Times*, John Noble Wilford wrote that the discovery of cosmic acceleration appears to "make a prophet of Dr. Turner and others who were beginning to share his views."[20] Of course Turner wasn't really a prophet. Neither did he did pick the cosmological constant out of the ether. As we have seen, it was a Holmesian deduction. And even though Einstein discarded the idea, it had never really died because quantum mechanics predicted that it should exist.

The quantum mechanical vacuum, as you may recall, is far from empty. Virtual particles pop in and out of existence all the time, so it is really a roiling sea of energetic activity — literally energetic, because all those virtual particles *add energy* to the vacuum. And it is a *repulsive* energy, just like Einstein's original cosmological constant, and just like the energy believed to have fueled the first bout of cosmic inflation. "It would tend to make space want to inflate," Perlmutter says.

But there's a major problem with this picture. "The problem is that based on quantum mechanics, there should be *way too much* repulsive energy," Perlmutter says. In fact, simple calculations predict an *infinite* amount of vacuum energy. To rescue quantum mechanics from this absurd result, theorists in the 1960s attempted to make certain corrections. But even considering these corrections, the vacuum energy from virtual particles should be 120 orders of magnitude larger than the energy contained in all of the matter of the universe.

The theory of supersymmetry could offer a way out of this dilemma. It proposes that there is a partner to every particle, and that these marriages produce positive and negative amounts of energy that cancel perfectly. In this view, there is no vacuum energy left over to produce a net cosmological constant. But with the discovery of cosmic acceleration, another possible conclusion is that the energy does not cancel perfectly, Perlmutter says. If a slight bit of positive vacuum energy were left over, it could produce the observed acceleration.

In other words, maybe the super*symmetry* is *broken*. "There are a lot of broken symmetries in the universe," Perlmutter notes. Indeed, we've already learned that without symmetry breaking, the four fundamental forces would never have broken out of the one superforce. So a slightly broken symmetry producing enough vacuum energy to account for the cosmic acceleration is a real possibility, he says.

If only things were that simple. It turns out that the supersymmetry would have to be broken to an exceedingly precise degree to produce the amount of repulsive energy needed to re-inflate the universe at the observed rate. So here we go again. Fine tuning just won't go away. "No other symmetry that we know of is broken in such a fine-tuned way," Perlmutter says.

The problem arises because the density of dark energy in the universe should have evolved in a completely different way than that of ordinary energy and matter. "Dark energy is a quality of space," Perlmutter points out. "So the more space you have as the universe expands, the more dark energy you've got." On the other hand, the volume of space increases at the same rate. So the *density* of the dark energy *stays the same* with cosmic expansion. Not so the other stuff of the universe. No more of this matter and energy is being created, so as the universe expands its density should go *down* with the increased volume. Given that the universe has been expanding for 14 billion years, give or take, the dark energy has had a long time to evolve to a density that's drastically different from that of the other contents of the universe. But it's not drastically different. The densities of the dark energy and the other contents are very similar—about 70% versus 30%.

Why should this be so? "It's not obvious," Perlmutter says. "There is no trivial answer. But if you could come up with some *other* kind of energy that tends to *track* the mass density, that would seem more comfortable." One possible answer championed by some theorists has been dubbed *quintessence*: a dark energy that *changes* over time—a cosmological constant that isn't constant.

For Michael Turner, the mystery of the dark energy is vexing. "Monday, Tuesday and Wednesday of every week I say, 'HELLO, we know the vacuum energy has to be there.' All of the other things people discuss, like quintessence, don't even have that virtue." On the other hand, quantum mechanical calculations give that "runaway answer—infinity," he says. So the other days of the week he doubts that vacuum energy can be right.

But he does not despair. Far from it. "In terms of a problem for theorists to solve, this one is just about the right size," Turner says. "We don't have to throw away our entire framework. General relativity can accommodate an accelerating universe. But we do have a very, very tall problem. Any taller and we *would* have to throw away general relativity."

Cosmic acceleration was so surprising, so *radical*, and ultimately so mysterious, that some scientists had a hard time accepting it at first. The general consensus was that even though both groups had gone about their work with great care, the results might be unreliable because they were based on a relative handful of measurements of very distant supernovae. Even with additional supernovae observations, that issue would not go away completely. But in March of 2002, scientists using a different way to check for dark energy and the cosmic acceleration it may be causing lent confidence to the supernovae findings. The research, conducted by 27 scientists at 14 institutions, is based on a simple notion. By comparing the seeds of structure in the cosmic microwave background radiation with the structure of the universe today, it should be possible to determine how the universe got from point A to point B. In other words, from wrinkles to galaxies. And if dark energy has caused the expansion of the universe to accelerate, that should be evident from the comparison.

Led by George Efstathiou of Cambridge University, the scientists analyzed variations in the pattern of clustering within a sample of 250 000 galaxies mapped by the Anglo-Australian Observatory in Australia. (The galaxy mapping is part of the Two Degree Field Galaxy Redshift Survey, or 2dFGRS.) The researchers then compared the clustering to the spectrum of hot and cold spots in the cosmic microwave background. From this they determined that dark matter plus ordinary matter and energy comprise just 30% of what's needed for the universe to have gotten from A to B—from the wrinkles to the clusters. The rest, they concluded, is made up of dark energy.[21]

"It seems that Einstein did not make a blunder after all," Efstathiou says. "Dark energy appears to exist and to dominate over more conventional types of matter."

So it's looking more and more that the dark energy balances the books, yielding the flat universe produced by the simplest, most favored models of inflation. Based on this information alone, the theory is not ruled out. But neither is it yet ruled in. It's always possible that some process other than inflation made the universe flat. For scientists to feel more confident in inflation, they need more specific tests—which is just what the finer details of the cosmic microwave background radiation can provide. And in recent years, those details have begun to come into view.

The first to be observed was the overtone spot—the biggest temperature peak predicted by inflation theory. It was detected in 2000 by researchers working on a project known as BOOMERANG: Balloon Observations of Millimetric Extragalactic Radiation and Geophysics. The experimenters deployed a 1.2 meter primary telescope mirror for gathering cosmic microwave background photons, and an array of detectors chilled to just 0.28 degrees above absolute zero. The instrument, weighing some 3100 pounds, was lofted to an altitude of 120 000 feet over Antarctica by a balloon. Unlike COBE, which scanned the entire sky, BOOMERANG focused on a small portion, taking detailed measurements. Circling the icy continent for ten and a half days, it measured the microwave background across 3% of the sky, enabling researchers to produce a map 40 times more detailed than COBE's.

The precision paid off. In BOOMERANG's map, the fundamental spots are unmistakable—and they measure exactly 1° across. This finding buoyed researchers, because it provided yet more evidence that the universe is geometrically flat. But the relief came with a twinge of anxiety because something crucial was missing: the smaller temperature peaks of overtone spots. This is where the most telling details of inflation's fingerprint should be found.

''BOOMERANG certainly offered the first high-precision data, meaning lots of statistics,'' Hu says. ''It marked the first time that we really nailed things—the first peak. But it really got us worried about the secondary peaks. These have always been the most important thing because the most ironclad proof of our understanding is in those peaks.''

In January of 2001, a telescope high in the Chilean Andes called the Cosmic Background Interferometer, or CBI, picked up a tantalizing hint of the second peak. This was encouraging, but scientists weren't absolutely confident of the results.[22] So they had to hold their breath for several more months. Then, researchers on the BOOMERANG team, along with scientists participating in two other research programs, announced at another meeting of the American Physical Society that their instruments had ''heard'' the elusive overtones. Based on 14 times the data used in their first analysis, the BOOMERANG team found the predicted second peak and evidence for a third one. The fundamental spot and two overtone spots were also detected by another balloon-borne experiment, the Millimeter

Anisotropy EXperiment Imaging Array, or MAXIMA. A third successful detection of the overtones was made by researchers using a ground-based Antarctic telescope called the Degree Angular Scale Interferometer, or DASI, which measures the microwave background radiation in a completely different way than the balloon-borne experiments.

With these findings, scientists could breathe again. Mike Turner was in a celebratory mood when the results were announced: "This is really a great party," he was quoted as saying in a news report in the journal *Science*.[23]

Turner and other scientists were happy for many reasons. First, the findings were a vindication of the acoustic oscillation model of the early universe. "They are incontrovertible proof that we understand what's going on—that the early universe underwent these oscillations," says Wayne Hu.

The findings also supported inflation's description of what got the oscillations going. "Here was a prediction of inflation, and it has been verified. I was very relieved," Hu says.

But there was even more. The new observations of the cosmic microwave background helped researchers resolve a conflict between two different approaches for tallying the amount of matter in the universe. One accounting method, based on measurements of the relative amounts of different kinds of atoms, found that ordinary, non-dark matter comprises some 4% of all the matter and energy in the universe. But an accounting based on earlier analysis of the cosmic microwave background had pegged the number at 6%.

The background radiation could be used to complete such an audit because dark matter subtly influences the outcome of the tug of war between gravity and photon pressure within the troughs of spacetime wrinkles. This, in turn, has an effect on the relative sizes of the fundamental and overtone spots in the microwave background. BOOMERANG, MAXIMA and DASI measured those spots with unprecedented precision, enabling the research teams to correct the accounting of ordinary matter based on the microwave background. The new tally? Non-dark matter is 4% of the cosmic bottom line.

With the discrepancy resolved, it looks like the universe is made of 4% ordinary matter/energy, 26% dark matter and the rest dark energy or quintessence. Or let's put it another way, with a sharper edge to it. The matter we can actually *see* — the ordinary baryons in galaxies, stars, planets and us — comprises just a small fraction of the cosmic total. And the energy we can actually *feel* — including the light that makes the universe sparkle and the gravity that keeps us grounded — also makes up just a small fraction of what's actually out there. In other words, the universe consists mostly of dark matter and dark energy, stuff we can't see, stuff that has no impact on our everyday lives.

Scientists have learned an enormous amount about the universe in the past few decades. But it seems literally true that the more they learn, the more that becomes mysterious.

"We've learned in the past few years that one of our ideas of what would be elegant and simple may be right — that the curvature of space is flat," Perlmutter says. "This is the ordinary Euclidean space we all grew up knowing and loving. But another one may have been wrong. We did not imagine that there might be a few different components to the energy budget of the universe, and that one of them would be this bizarre cosmological constant, or dark energy."

So even though inflation has withstood some serious testing, the theorizing and the observing will continue. "We're through round one," Turner says. "But we're not out of the woods yet. We have not proven inflation."

As you read this book, the first results may well be in from a satellite that is measuring the microwave background with the greatest precision yet. The Microwave Anisotropy Probe, or MAP, was rocketed into orbit on June 30, 2001. According to Wayne Hu, it should shore up the observations of the secondary peaks, putting inflation to an even more stringent test. Some years later, the European Space Agency's Planck mission should take even more detailed measurements.

Researchers are also searching for a different kind of signal in the microwave background. Photons should have become polarized when they scattered off electrons in the primordial plasma, just as they do when they reflect off an ordinary surface like a lake. Scientists have already detected hints of this polarization, and in coming years they will be examining the finer details of what went on, Hu says. One possibility is that gravity waves, giant undulations in the fabric of spacetime predicted to have been launched by inflation, left a distinct pattern in the polarization of photons. ''This will really test the physics of inflation and help us choose between it and other models,'' Hu says.

Turner stresses that there is still much to learn: ''We know much, but we understand little.'' Theorists have done a good job describing the mechanism of inflation. ''We've defined it well,'' he says. ''On the other hand, the theory is based on speculative ideas.'' Theorists invoke an inflaton field as the driving force behind inflation. ''But we don't know what the inflaton is,'' Turner says. Until physicists manage to make an inflaton particle in the lab—a challenge he describes as ''orders of magnitude beyond extreme''—some doubt, however small, will remain about inflation.

Even so, Turner is confident that the new paradigm of cosmology—inflation + big bang nucleosynthesis + cold dark matter = our universe—will continue to survive testing. ''We've identified the basic features of the universe,'' he says. ''Acceleration ain't going away. Dark matter ain't going away. Dark energy ain't going away. And the consensus view is that inflation is getting at much of the truth.''

These findings have been the product of what Turner describes as ''the most exciting period in cosmology ever. And if we manage to raise our level of understanding, no one will be able to deny that this is truly the golden age. Cosmology is showing enormous promise. It is a child prodigy. So 20, 50 years from now, will we be able to say that not only do we know a lot but we understand a lot?''

Turner leaves the question hanging, but it's clear that his answer is an unequivocal 'yes.'

# 7

# BEFORE INFLATION

## *The Pea, the Brane, & the Search for the Ultimate Beginning*

*What was there before the Primordial Soup? The Primordial Soup comes from nothing. What is nothing? Nothing is something. Nothing has energy. Nothing can change.*[1]

— Rocky Kolb

*It looks to me that probably the universe had a beginning, but I wouldn't bet everything on it.*[2]

— Alan Guth

Was there an Ultimate Beginning to everything?

If you believe, as some scientists do, that the multiverse exists and is eternal both into the future and into the past, you can dispense with this question and move on.

But if you accept the idea that even the multiverse, or just our universe, if that's all there is, had to have an Ultimate Beginning, then you are stuck with a paradox that has bothered philosophers for millennia. Something can't come from nothing, yet defining an ultimate origin for the universe means describing how *everything* could have come from nothing.

By nothing, I don't mean some pre-existing plenum waiting for a universe to fill it up. I mean absolute non-existence. In this context, even the phrase *from nothing* is a little misleading, because it suggests a journey or a transformation from one state to another. If the universe had an ultimate origin, there would have been no prior state, no pre-existing vacuum, no previous time or place before.

We are used to the idea that something had to cause the universe to come into being, that the energy of the cosmos had to come from somewhere, and that spacetime had to emerge from some prior state or system. But once you invoke a cause, a source of energy, and a pre-existing state or system, you're saying that the universe didn't come from nothing. It came from something else. And just how did *that* begin?

For seekers of an Ultimate Beginning, quantum mechanics offers some help. Within the quantum realm, probabilities and statistics replace classical causation. For example, something doesn't really *cause* virtual photons to pop out of the quantum mechanical vacuum around an electron. Because of the uncertainty principle, there is simply a calculable probability that they will. Similarly, probabilities, not classical causation, likely held sway in the infant universe as well — because it was smaller than an atom and therefore existed in the quantum realm.

"In quantum theory it is a principle that anything can happen if it is not absolutely forbidden," says Stephen Hawking.[3]

So might the universe have started as a quantum fluctuation? In a seminar in the late 1960s, a Columbia University physicist named Edward Tryon proposed just that to his colleagues. They laughed, thinking it a joke. But in 1973 he formally proposed a theory in which the universe began from nothingness as a vacuum fluctuation. "Our universe," he said, "is just one of those things which happen from time to time."[4]

But even if quantum mechanics offers a way around the causation issue, how can all the energy in the universe come from nothing? Doesn't that violate the law of energy conservation? In his paper in the journal *Nature*, Tryon pointed out that the huge *positive* energy of the universe would be perfectly balanced by all the *negative* energy represented by gravity. According to this way of thinking, we live *in a zero-energy universe*, so there would have been nothing to prevent it from coming from zero-energy nothingness. In other words, our universe may well be a case of nothing coming from nothing.

Even so, one thing did have to exist beforehand: the laws of physics. After all, those laws appear to govern the universe. And even in Neil Turok and Stephen Hawking's universe-from-nothing proposal (which we'll investigate shortly), the theory does not generate *those*. They need to be pre-existing for the theory to work. In other words, no matter how you cut it, there must be some prior "state," even if it is just a set of physical laws.

"We don't really understand where the laws of physics came from," admits Alan Guth.[5] And ultimately, science may prove incapable of dealing with this problem.

When the concept of the multiverse emerged from inflation theory, some scientists became hopeful that the question of an Ultimate Beginning could at least be avoided. As we saw in Chapter 5, inflation theory suggests that once exponential expansion of spacetime gets going, it's eternal. Always, somewhere, a pocket of spacetime continues to inflate. At the very least, this idea sweeps the question of the Ultimate Beginning away into the distant corridors of cosmic history. According to eternal inflation, *our* part of the multiversal whole—our observable universe—had a beginning. It was born in a burst of exponential growth triggered by inflaton fields already in place in the multiverse. And whether the multiverse itself had a beginning could just be ignored.

In fact, some scientists put faith in the idea that the multiverse itself had no beginning—that it was eternal both into the future and into the past. But Alan Guth is not so sure. "It's called eternal inflation, but we really know that it's semi-eternal, that is, it's eternal into the future," he says. "Pushing it back into the past, it's a whole different question to ask. Does one expect to find the beginning someplace, or might it go backwards in the past forever? There really is no crystal clear answer to that question at this point. But the indications seem to be that even with so-called eternal inflation, you cannot push it infinitely far into the past. Somewhere, there has to be a beginning."[6]

So here are the two possibilities cosmologists wrestle with: describing the evolution of our own pocket of an eternally existing universe, or trying to do what seems to be very difficult, if not impossible: devising a universe out of nothing.

Neil Turok has tried it both ways.

Turok began his career as a mathematical physicist working as a graduate student under the direction of Tom Kibble of Imperial College in London. At the time, Kibble was one of a number of physicists exploring the possibility that spacetime could be marred by defects generated during phase transitions in the early universe. If they had ever existed, these little tears and punctures could have strongly influenced the course of cosmic evolution.

We've already encountered one such hypothetical "topological defect": the monopole. But there were others, including one dubbed a "cosmic string"—a one-dimensional line-like object that, despite its diminutive girth, would be extremely massive. (Do not confuse cosmic strings with the superstrings theorized to comprise elementary particles; these are different objects entirely.) According to the theory of topological defects, the early universe would have been tangled up in a dense web of these cosmic strings.

Kibble was interested in how those strings would evolve. Theory suggested that when two strings crossed, or when one string crossed over itself, they would chop themselves up, forming loops and smaller bits of string. Kibble showed Turok some scientific papers about this hypothetical phenomenon. And according to Turok, he said, "Look, there's this puzzle. If you have a network of strings, like spaghetti, they might chop themselves up into loops." But would the loops chop themselves up into more tiny fragments and disappear? Or would they be long-lived? If they were long-lived, if they managed to survive through the big bang, "they could be the seeds for galaxy formation," Turok says.

The strings would have accomplished this by enhancing gravity. This excess of gravity around the strings would have drawn matter in, enhancing the gravity even further and thereby setting off a self-reinforcing process that ultimately would have caused galaxies and galactic clusters to coalesce.

"I worked extremely hard on this problem over the weekend," Turok recalls. "I was excited at being given a physically interesting problem that had implications for the real world."

By the end of the weekend, Turok had come up with a solution. Strings would survive. Kibble checked it—and just didn't buy it. "So I checked his checking and found that my solution was right," Turok says. "In the end, we wrote a paper about it together. And that was the start of my career."

Turok, along with other scientists, continued working on cosmic strings. And over the years, a fuller picture of how they could have seeded the formation of galaxies has been sketched out. This picture is a possible alternative to inflation's mechanism for creating structure.

According to these theories, as the universe evolved, the strings and loops would have become very sparse as they chopped themselves up. According to Turok, mathematical modeling suggests that there should be just 10 long strings stretched across enormous distances in the cosmos today, and possibly 1000 smaller loops. But early on in their evolution, the strings should have warped the fabric of spacetime, thereby planting the gravitational seeds of galaxies and clusters.

But cosmic string theory has fallen out of favor, partly because the mathematics has proved recalcitrant to solution and also because of observations that strongly favor the inflationary mechanism over this one. The theory hasn't gone away, however. And upcoming missions to map the cosmic microwave background radiation in unprecedented detail should offer the final word on whether cosmic strings had any role in the formation of galaxies.

In the mid-1990s, Turok began tackling inflation. As you may recall, the theory was in some trouble at the time. It predicted a universe with just enough density of matter and energy to produce a flat geometry. But observations were showing that the density of matter and energy was much lower than that required. This prompted a number of scientists, Turok among them, to see whether inflation theories could be modified to produce an open universe.

Working with Martin Bucher of Cambridge and Alfred Goldhaber of the State University of New York, Stony Brook, Turok came up with the double-dose inflation theory. With *two* bouts of inflation, they showed, one could produce a bubble universe that actually had an open geometry inside it. But the solution was like a cosmic Rube Goldberg device—it just had too many moving parts.

One day, Turok was commiserating over afternoon tea with Stephen Hawking, who suggested that it might be possible to solve the open inflation problem by joining forces. Hawking suggested that Turok try to marry some of the open bubble concepts with a theory Hawking had developed with physicist Jim Hartle in the 1980s. Such a theory, Hawking suspected, might just produce open inflation—and describe nothing less than the initial conditions of the universe itself.

"This was a genuine surprise," Turok says.[7]

In the early 1980s, Hartle and Hawking had searched for a theory that could describe the initial conditions of the universe—what it *really* was like at the very beginning—and how those conditions affected the course of cosmic evolution. Their goal was to banish, or at least sidestep, the singularity of the big bang by weaving the grainy reality of quantum mechanics into the smooth spacetime fabric of relativity.

In the 1970s, Hawking and Oxford University physicist Roger Penrose had used general relativity to show that the universe must have had a beginning. Unfortunately, though, the singularity sat at that beginning. Hawking was uncomfortable with this outcome because everything breaks down at the singularity. Thus nothing could be predicted about *what kind* of universe can come out of the singularity. As far as the hot big bang theory is concerned, as it is understood with relativity alone, *anything* could come out of it. "Thus classical relativity brings about its own downfall: it predicts that it can't predict the universe," Hawking said in a friendly debate between himself and Penrose in 1994. "Although many people welcomed this conclusion, it has always profoundly disturbed me. If the laws of physics could break down at the beginning of the universe, why couldn't they break down anywhere?"[8]

Because anything could in theory come out of the singularity at the heart of the hot big bang, then why did *our* universe come out of it?

To imagine what Hartle and Hawking devised to avoid the singularity, picture a cone. It represents a spacetime history of the universe from its origin at the tip to the present day at the top. The time dimension runs up the cone from the point. The space dimensions run *around* the cone.

To see how this cone can represent the evolution of space and time, picture a tiny insect at the very beginning of spacetime. That would be at the sharp tip of the cone. Here is the singularity, so this poor insect is skewered. There's no predictable way for it to move forward in time. So let's give it helping hand and start it moving on a time-traveling adventure from a spot just above the singularity. The insect starts crawling upward, which is forward in time. As it moves along it looks around and notices that the cone is getting bigger. In other words, as it moves forward in time, the universe is expanding. This is the expected evolution of spacetime in the standard big bang picture. Time progresses, space increases.

Eventually, the insect reaches the top and wants to head back. Crawling down the cone, it's moving backward in time now, and the space dimensions are shrinking. Eventually the insect returns to a position just shy of the tip. It would like to keep on going, but it doesn't want to get skewered by the singularity again. So it tries to figure out how to modify the point to avoid this fate. Now even though this insect is very tiny, it's exceedingly smart. And it realizes that if the point were *rounded off*, sort of like the ball at the bottom of a roller-ball pen, the time dimension would curve gently into the space direction. This would mean that it could keep crawling to the bottom of the ball and nothing bad would happen. To modify the cone in this way, the insect realizes that it has to lop off the tip and attach half a sphere—a hemisphere—in its place.

This is exactly what Hartle and Hawking did to the spacetime cone in their no-boundary theory. Keep in mind that in the quantum mechanical realm, there is no such thing as a precise point; there is an uncertainty to it. So to represent the quantum beginning of the universe in their theory, they replaced the sharp tip with a nice rounded hemisphere.

This has a funny effect, as the ant would no doubt notice if it crawled around this hemisphere. As the time dimension curves into the space dimensions in the hemisphere, it makes a right angle turn. And in quantum mechanical equations, time running at right angles to the ordinary flow of events is known as *imaginary time*. Physicists often make use of imaginary time in quantum mechanical theories to make certain calculations easier. It's an accepted technique that produces accurate results. But Hartle and Hawking used it in a different context: to devise a theory describing the origin of the universe as a round hemisphere at the bottom of a spacetime cone.

Now picture this hemisphere for a minute. Just as there is no point on the surface of a sphere like the Earth where one can say the sphere "begins," there is no distinct point on the hemispherical, rounded-off bottom of the cone where time or space begins. Within the hemisphere, there is no distinction between space and time, or even a past and future. And there simply is no starting point for either. So you might be tempted to conclude that this universe has no beginning. But time doesn't run backward eternally. The cone is self-contained. So it does have a beginning. It's just that there is no hard boundary in time, no one point where it begins. If you're still having trouble envisioning this, picture the insect again. If it kept crawling down the cone, it would simply round the hemisphere at the bottom and then head back up the cone on the other side. (But God only knows what it would be like for the insect in the realm of imaginary time — a problem that helped scotch the theory for Stanford physicist Andrei Linde, as we'll see in a bit.)

Of this model, Hawking writes, "The universe would be entirely self-contained; it wouldn't need anything outside to wind up the clockwork and set it going. Instead, everything in the universe would be determined by the laws of science and by rolls of the dice within the universe."[9]

Perhaps most famously, Hawking has said that with the no-boundary proposal, the universe "would neither be created nor destroyed. It would just BE."[10] These words are a modern echo of speculations made by philosophers centuries ago. For example, the Chinese philosopher Kuo Hsiang, who died in AD 312, put it this way: "The creating of things has no Lord; everything creates itself. Everything produces itself and does not depend on anything else. This is the normal way of the universe."[11]

And so this, in a nutshell (apologies to Hawking), is the no-boundary proposal. Spacetime has its origin in a no-boundary configuration; it inflates out of this configuration, and then experiences the big bang.

The simplest assumption in the theory is that the no-boundary hemisphere at the bottom of the spacetime cone is perfectly round. But there are actually other possibilities as well. It could be slightly flattened at the south pole for example. Or it could have other departures from perfect roundness. And each of these would evolve in a different way to produce a different kind of universe with the ordinary time we're used to. Another way of phrasing this is that there are many possible *histories* for the evolution of spacetime, depending on its initial configuration in imaginary time.

So which of the possible configurations might actually represent the origin of our universe?

You may recall from Chapter 2 a technique called *sum-over-histories* for looking at quantum mechanical behavior. Richard Feynman, who developed the technique, proposed that if you want to know the likeliest path a particle will take in moving from Point A to Point B, it's possible to calculate the answer by assuming the particle takes *every possible path* simultaneously. Each one is a history with a particular probability, and one obtains the likeliest one by summing all of them together according to a technique Feynman devised.

Feynman developed the sum-over-histories approach as a graduate student working under the direction of John Archibald Wheeler, who uses a helpful analogy in explaining it. Imagine a pitcher is throwing a baseball toward a batter. With the sum-over-histories approach, you would assume that the baseball takes every conceivable path. One goes directly over the plate, another over the outside corner, a third zooms around first base and then heads back, and others simply leave the stadium before turning back toward the batter and slamming into the catcher's mitt. Each of these "virtual" baseballs, as Wheeler describes them, has its own history. But the batter sees only one: the one path that is the result of all the probabilities of all the possible histories combined.[12]

The technique works remarkably well for determining the behavior of quantum mechanical entities like particles; it's not needed when considering macroscopic objects like baseballs, which are described perfectly well by classical physics. But Hartle and Hawking decided to apply it to the evolution of the universe as a whole. In the no-boundary proposal, there are multiple possible configurations for the universe at the beginning. Each configuration gives rise to a history in imaginary time, and each of those histories produces a history in ordinary time. Now what if you could sum over all of those possible histories to determine which one was the most probable one? You'd come up with a candidate for the initial configuration of the universe. And if its history in ordinary time looked anything like the real universe, then you could put some faith in the idea that the universe had that kind of no-boundary beginning.

This is precisely what Hartle and Hawking set out to do. But they ran into a roadblock. They could not consider all of the possible histories of the universe. They were only able to calculate histories of geometrically closed universes. So when observations began suggesting that the universe was geometrically open, the Hartle–Hawking approach was not very useful. The no-boundary proposal wasn't ruled out. But until a way could be found to use it to produce universes with open histories, it wouldn't go anywhere.

And that's why Turok's research with Bucher and Goldhaber had caught Hawking's eye. Here was an inflationary model that could produce an open geometry. So could it be welded to the no-boundary proposal?

At Hawking's urging Turok decided to take up this challenge. But his first attempt at a solution was disappointing, because the sum-over-histories approach seemed to rule out open inflation from the Hartle–Hawking no-boundary ''cone.'' The reason? Infinite amounts of energy would be produced. ''But when I brought this up with Stephen, he told me, 'Wait, you didn't include the gravitational energy,''' Turok recalls.

Since energy is equivalent to mass, energy exerts a gravitational field. And as mentioned before, gravity's energy is *negative*. So infinite positive energy would produce a *perfectly balancing* infinite negative energy. Crank this into the calculations, and the no-boundary model produces an open, inflating universe without the ridiculous result Turok first got.

But what to call the mathematically defined ''object'' that does the trick? When Turok attempted to describe it to a reporter who was working on an article for the English newspaper the *Sunday Times*, the journalist asked for a physical description. ''You must think of something we can picture,'' the reporter insisted.

''Well it's small and round, but not perfectly round, and it has little dimples—it's something like a pea,'' Turok replied. ''The reporter jumped on this and said, 'okay, it's a pea.'''

A very special pea to be sure. It's a millionth of a trillionth of a trillionth the size of a pea. But it's a lot denser than ordinary matter, so its mass actually is about that of a pea. Space and time are blended together in the pea such that its ''bottom'' half is just like the rounded-off tip of the cone in the original Hartle–Hawking no-boundary model. And when Turok and

Hawking worked out the calculations describing this remarkable object, they found to their astonishment that an open inflating universe instantly sprouted out of its "top."

The other part of the name comes from a mathematical approach, called an instanton, used to describe quantum mechanical behavior. When physicists use instantons, the calculations employ imaginary time. In the case of Turok and Hawking's pea, the instanton describes the transition from the imaginary-time portion of the universe to the ordinary-time portion.

And what about those little dimples on the surface of the pea that Turok described to the reporter? They were revealed as a likely feature by summing over many possible histories. And they actually are a physical representation of quantum fluctuations—the source of the wrinkles that inflation irons into the fabric of spacetime.

I first spoke with Turok about the pea instanton shortly after he and Hawking unveiled the idea in 1997. There was some wonderment in his voice when he described it for me. "The universe," he said, "all comes out of just one formula in a beautiful way," he said.

Turok found beauty in the theory's economy. "For sure, I think this is the simplest way for starting the universe."

But some cosmologists were quick to point to problems with the pea instanton. To begin with, critics charged that the Cambridge scientists had not eliminated the singularity entirely. It survives as a single point inside the primordial pea.

And there were other criticisms as well. John Preskill, a physicist at Caltech, emphasized that the entire field of quantum cosmology was on shaky ground. "Hawking and Turok were very pleased that they found a natural context in quantum cosmology for open universes," Preskill said. "But the whole foundation of this subject, the whole quantum approach to cosmology, is not very well founded."[13]

Andrei Linde of Stanford pointed out that the pea instanton gave rise to a universe that didn't look anything like the one astronomers were observing. This was before the discovery of cosmic acceleration so, at the time, the universe was looking very open. Even so, the density of matter was *so* low in Turok and Hawking's instanton-generated universe, that galaxies would be few and far between; even using the Hubble Space

Telescope, we wouldn't be able to see another galaxy from our own.[14]

Linde also objected to Turok and Hawking's use of an instanton to explain the origin of the universe. Traditionally, instantons can be used to determine the probability of finding a particle, like an electron, in a particular *state*, he explained. Recall that the wavelike essence of a particle can be interpreted as a *probability wave*. This "wave function" describes the probability of finding a particle in a particular place. In the case of an electron in a hydrogen atom, the particle's wave function can be used to describe the probability of finding it in its "ground state," which is the location closest to the center of the atom. "But you still do not know *exactly* where you can find it," Linde noted. "It is in the cloud *somewhere*."[15] The wave function simply describes the *chances* that it can be found closest to the center.

Now also recall that, in theory, there is a probability for the electron to be *anywhere at all in the universe*. And as Feynmann described, one must assume that it is everywhere in order to figure out where it is *most likely* to be found. In practice, this is impossible. So physicists use instantons as a calculational trick to make their lives easier. With an instanton, physicists do the actual calculations in imaginary time, and this simplifies things enormously. And it's really just a trick, Linde argued. "Fictitious imaginary time has no relation to real evolution," he said. "It's a trick that can be used to find the probability of a particular ground state." It's a well defined trick, for sure, and it works just fine. But in Linde's opinion, the equations of the instanton don't represent anything *real*. Think of them as a mathematical bridge spanning a river of complexity. They allow physicists to get to the other side — to arrive at the answer to the problem — while avoiding being swept away in the rushing waters.

But the imaginary time in Turok and Hawking's instanton is more than just a trick, according to Linde. It is used as a fundamental description of the universe in the beginning. To understand the ultimate origin of the universe using the pea instanton, "you must travel there in imaginary time," he said. "But you need to understand — what is it like *living* in imaginary time? On this question there are a lot of interpretational problems," Linde noted wryly.

And Linde had one other objection. In using an instanton in their theory, Turok and Hawking basically described the origin of the universe as a quantum-tunneling event. The theory of quantum tunneling describes how a particle or a field can move from one side of a seemingly insurmountable barrier to the other without actually passing over or through the barrier in a classical sense. (Instantons are used in these calculations.) And the way Linde saw it, Turok and Hawking's instanton describes the origin of the universe as a quantum-tunneling event *from a state of nothingness to a state of existence.*

"This is like creating a hydrogen atom from nothing. I cannot justify it. I cannot understand it," Linde lamented. "Stephen is an extremely talented person. Sometimes however—this is my interpretation—he trusts mathematics and he believes he knows the correct interpretation. But you know, it is such an esoteric science, creation from nothing, you can come with your interpretations and there is no way to check it. It is like a religion."

When I asked Turok about these criticisms, he replied that Linde and others misunderstood what he and Hawking had done. "It's *not* tunneling," he insisted. The pea instanton is not, in fact, created from nothing. 'Nothing' simply has no meaning in this context. The instanton—meaning the universe at its birth—just *is*. There is no "outside" and no "before" the pea. It is simply an object implied by the laws of physics.

"It really is a rather unique thing and I don't think this is widely understood," Turok said.

As to the surviving singularity within the instanton, it is "so mild that, like the singularity in the electric field in the center of a hydrogen atom, one can still calculate the quantum properties of the universe without any ambiguities," Turok said.

In other words, there are singularities, and then there are *singularities.* And this one wasn't so bad.

"There are singularities within black holes, yet we don't doubt that black holes exist," he said. So just because there is a singularity within the instanton doesn't mean the entire model should be tossed out. As long as the theory still works, the singularity may not be a problem. "The presence of a singularity simply says you don't know what you're doing there. It's still possible that the physics away from the singularity is fine," he said. "I like to say that what we've done is sidestep the big

177

bang singularity rather than avoid it altogether. We've found a way to go back to the beginning of time and go *around* it."

As to the charge that imaginary time does not correspond to reality, Hawking has argued that no one really knows what reality is. "I take the positivist viewpoint that a physical theory is just a mathematical model and that it is meaningless to ask whether it corresponds to reality," he said in his debate with Roger Penrose. "All that one can ask is that its predictions should be in agreement with observation."[16] But shouldn't we demand that a theory correspond to reality? No, says Hawking: "I don't demand that a theory correspond to reality because I don't know what it is. Reality is not a quality you can test with litmus paper. All I'm concerned with is that the theory should predict the results of measurements."[17]

Finally, as to the pea instanton universe being too open, Turok agreed that this was indeed a problem. "But the fact that it's wrong only by a factor of 30 is encouraging," he said, "because the calculation was done with a very simple model and yet it's not *that* far wrong." And for Turok, that was good reason to keep tinkering.

He kept at it, trying to find ways for the pea instanton approach to produce universes that look more like our own. Then, in 1998, scientists discovered cosmic acceleration, which seemed to show that the universe was flat after all. And as newer observations came in, they kept converging on a flat universe. But Turok was undeterred. If the universe turned out to be even slightly open, there would be a need for a model that could produce it through inflation.

But there were problems. The sum-over-histories calculations kept showing that the likely histories produced by the pea would be much too open. In fact, the *most* probable universe was a completely empty one, Turok now says. So if those histories were the most probable ones, why are we here?

To overcome this problem, Hawking and Turok both invoked the anthropic principle: a less probable history must have been picked out because we *are* here. In practice, this meant establishing a requirement in their calculations: all histories considered must be conducive to intelligent life. So the probable universes containing no galaxies were excluded.

Hawking has continued to use this approach to advance the no-boundary proposal. Turok, however, has ultimately found the approach unsatisfying. "I worked on Hawking's proposal for about three years, and we did some very detailed work on it," he says. "It has a nice mathematical underpinning, it's very aesthetic, and the idea that you round off the big bang in a smooth way is still very appealing. But I've convinced myself now that the state the Hartle–Hawking prescription favors is an *empty* universe. There is no wriggling you can do to get out of this conclusion. It makes a prediction, but it predicts that the universe is empty.

"We tried to get out of this with an anthropic argument," Turok continues. "But the problem is that the anthropic argument is badly defined. No two practitioners will agree on what conditions are needed in the early universe to allow the evolution of intelligence, and how to impose those conditions in calculations. In the end, it all became rather murky for me."

So Turok, in partnership with Paul Steinhardt, is now exploring something completely different. "Rather than saying that time turns into space as you go back, we are exploring the idea that time continues backward to *before* the big bang. There's nothing that actually excludes that."

This would mean that time did *not* begin with the big bang.

But what about the singularity? In the standard big bang theory, time begins at the singularity. "Hawking and Roger Penrose proved that our universe must have had a singularity in the past," Turok notes. "But all that *really* proves is that general relativity must have failed. It does not prove that time began."

So Turok and Steinhardt began looking at what it might mean if time extended before the big bang. The foundation for the theory they devised is superstring theory. As we saw before, in superstring theory, every particle consists of a bit of vibrating string. And it's the particular vibrations that endow a particle with its traits. But superstring theory requires extra dimensions to work. In M-theory, which seems to encompass the five different forms of string theory, there are 11 dimensions. Six are curled up into a very tight little ball (called a Calabi-Yau manifold) that is too tiny to be experienced in everyday life. And spacetime consists of the other five: the three dimensions of space and one dimension of time that we're familiar with, plus one higher dimension.

In this multi-dimensional spacetime, five dimensions may be bounded on opposite sides by three-dimensional parallel sheets, called branes. (Time is the fourth dimension.) These float at either end of the five-dimensional space like sheets hanging from parallel clotheslines. Our universe with all of its energy and matter is one of these boundary branes. The other is a parallel universe. The space between us and our sibling universe itself is hidden from us because particles and light cannot travel across the gap. But gravity does.

Even before Turok and Steinhardt began working on their theory, physicists were exploring how branes could have led to the origin of our universe. In 1998, for example, George Dvali and Henry Tye proposed what they called the *branefall* theory in which the parallel branes fall toward each other under the influence of gravity. And as this occurred, the gravitational field should have a dramatic effect. Wherever the three-dimensional spaces of the opposing branes are perfectly parallel, the gravitational field triggers inflationary growth; meanwhile, non-parallel regions shrink. And when the parallel regions collide, the gravitational energy is transformed into a big bang fireball.

Turok and Steinhardt have put a different twist on the idea of colliding branes. In their ekpyrotic model, the process begins with two branes, empty of matter, floating in lonely isolation. At a certain point, our sibling brane may slough off a third brane. Attracted by gravity, it slides gently across the gap toward us. Because it moves slowly, there is time for its properties to smooth out across the surface. But quantum fluctuations cause the brane to ruffle like a sheet in a gentle breeze. When it collides with our brane, the energy of the crash triggers a big bang, releasing a primordial fireball into our spacetime. Since in this picture, both branes began geometrically flat, our universe appears that way today. Since there was time for things to smooth out as the brane was drifting across the gap, our universe became homogenized, explaining why it looks smooth today. And since the colliding brane was ruffled by fluctuations, the collision happened in different places at different times, imprinting slight inhomogeneities in the spacetime fabric of our brane. These become the seeds of galaxies.

"As the branes hit," Turok says, "they make a big bang and bounce apart. And then we get the aftermath of the big bang— ripples, radiation and galaxies."

All of this adds up to an alternative to inflation's explanation for why the universe looks homogeneous and isotropic overall, and why matter is distributed in a cosmic cobweb of galaxies and clusters. But the story doesn't end there. The brane worlds, meaning our own observable universe and the parallel universe we cannot see, can be followed with theoretical calculations into the future to see what happens.

According to Turok, the cosmological constant—that repulsive energy causing the expansion of our universe to speed up—steadily dilutes matter in our brane and in our parallel brane world as well. "Eventually, it dilutes matter away to nothing," Turok says. "It empties out the branes, and we're left with two very empty, very flat, very parallel branes that are prepared for the *next* collision." Remember: gravity works across the gap. "So after some time, the branes are pulled together again. They crash and make another big bang," Turok says.

He and Steinhardt propose that the process is cyclical, perhaps eternally so—into the past and future. "This process can continue forever," Turok says. "The brane model allows a scientific universe in which the big bang happens an infinite number of times."

Given the success of inflation, there wouldn't seem to be much of a need for an alternative. But Michael Turner points out that inflation is by no means a done deal. And what happens if it fails a crucial test? "People are looking for alternatives because we don't have any back up ideas," he says.

Turok has his own reasons for pursuing colliding branes as an alternative to inflation. One is that the theory works within a framework that offers hope of uniting quantum theory with relativity. All of the fundamental forces of nature are explained within the same framework of superstrings.[*]

---

[*] But it must also be said that the calculations involved in string theory are so complicated that the equations used by theorists are just approximations. And testing string theory is incredibly difficult, because the energies involved in probing nature at the scale of strings are light-years beyond what can be accomplished in today's particle accelerators.

Turner has some grave doubts about the various brane collision theories that have been proposed. But he counts this argument as one factor in favor of the work. "Where are the frontiers in cosmology?" he asks. "One is to make some connection between cosmology and string theory. People have been trying to do that for 15 years, and they've struck out so far. But the brane world ideas are very interesting."

According to Turok, another strength of the ekpyrotic model is that it does away with the big bang singularity. Since the two branes come together with a big, flat splat, there's no point of infinite density and temperature at which all predictive power breaks down. The splat does create a kind of singular configuration, Turok admits. "But it is the mildest kind of singularity, in which one space dimension shrinks to zero for one instant. There are no infinite densities anywhere, which is the big obstacle with the standard big bang."

But as with the pea instanton model, not everyone has greeted the proposal with enthusiasm. Quoted in the online *Nature Science Update*, Andrei Linde was brutal: "It's a very bad idea popular only among journalists," he said. "It's an extremely complicated theory and simply does not work."[18]

Others are taking more of a wait and see approach. "It's a germ of an idea," Turner says. "But there's really nothing of substance there yet. The problem is getting the branes to bounce. Stringy magic may do it. But that's about all they've done so far."

Now that Turok has worked on a theory that is so different from the pea instanton, he says there are still aspects of his work with Hawking that remain attractive to him. "Hartle–Hawking faces up to a much deeper problem: why does the universe exist? Our model pushes that to one side. Instead, we say, 'Let's not explain why the universe exists but simply how it is consistent. If you look at the universe, it seems completely crazy because it has this singularity. With what we have come up with, at least the big bang is explained without a singularity. And the origin of the inhomogeneities that give rise to galaxies are explained."

But the new theory does not explain how an Ultimate Beginning to everything could have happened. "It sweeps the problem under the rug," Turok admits. "We still haven't explained how the beginning happened."

Finally, Turok concedes that he just doesn't know which of the opposite approaches he has taken may be right. "These theories are really just different ways of looking at the same problem. And I don't have any strong prejudice about which one is right. Observations will be a guide. And theoretical consistency will be another guide. Whether we will ever find what is right is open to question. But at least we can try."

So did the universe have a beginning? Or is it eternal?

These are questions that have excited people for millennia. In the fifth century BC, the Greek philosopher Heracleitus wrestled with cosmological ideas every bit as profound as the ones being addressed by scientists today. He argued for a philosophy in which the tension between opposites gives order to the world. So, for example, the tension between health and disease is what defines these opposites. In the universe overall, things seem to be divergent actually are brought together — are united in a way that makes the world coherent — by a hidden connection. That connection, that tension between opposites, brings about change in the world yet at the same time insures unity. And the connecting principle itself, Heracleitus said, is an "ever-living fire kindling in measures and being extinguished in measures."

Turok likes to think of the theory of parallel brane worlds as being a Heracleitian cosmology. In the theory, opposing branes drift apart and then come together, ever connected by gravity. When the branes collide, the energy of this hidden connection is transformed into the primordial flames of the big bang. The fires then extinguish as the branes drift apart. But inevitably they are drawn back together, beginning the cycle anew.

By contrast, Turok views the no-boundary approach as a cosmology consistent with the views of Parmenides, a philosopher who held a completely different view to Heracleitus. Parmenides argued that, at the most fundamental level, there was no change. There was only one true reality, *Eon* — an eternal, immutable and indestructible Being. Everything is a manifestation of this eternal Being, Parmenides said. As for Not-Being, meaning absolute nothingness, it cannot exist, he argued. Nor can it be expressed or even thought: "For you cannot know Not-Being, nor even say it."[19] Moreover, Being is not really something that existed in the past and continues into the future. Being neither was nor will be, Parmenides said. It simply *is*.

In the original Hartle–Hawking proposal, and in the further development of the idea by Turok and Hawking, it is technically incorrect to think of the universe as coming from nothing. In the context of the theory, it seems that there is no such thing as nothing. The concept simply makes no sense. And at the birth of the universe, the pea instanton simply *is*—the manifestation, presumably, of eternal laws of physics.

# PART 2

# PROLOGUE PART 2
## Fossils from Space

*It's not every day that Larry Pardue flies 100 miles just to have breakfast. But on the morning of June 13, 1998, the airport at Denver City, Texas was hosting a fly-in breakfast. To Pardue and three friends, bacon and eggs in Texas seemed like the perfect excuse to get out of town on a Saturday that was shaping up to be a scorcher.*

*And so a little after 7 a.m., Pardue and his friends took off from Carlsbad, New Mexico, in a Piper Cherokee. With George West at the controls, they headed east by northeast toward the town of Hobbs, a landmark on the way to Denver City. Sitting in the back seat were George's son, Damon, and brother-in-law, Robert Jesse.*

*At about 7:30, flying at an altitude of 5500 feet, Pardue gazed out at a typical, late spring cloudscape of Western skies. Puffy cumulus were off to the north, casting shadows on the buff-colored plains below.*

*Suddenly, Pardue's reverie was broken by a brilliant flash in his peripheral vision. To the north, he saw a flame shooting almost straight downward. In less than a second, it neared the horizon. Then, as quickly as it had appeared, it vanished.*

*Pardue grabbed control over the aircraft from West and pointed it toward where he had seen the flame. He carefully noted the compass heading: 012 degrees — directly toward the town of Portales, New Mexico.*[1]

On that same morning, Skip Wilson had no idea that he was about to win the cosmic lottery. His mind was occupied by more mundane affairs, such as getting to work.

By many accounts, Wilson is the Michael Jordan of his field — the best there ever was. One might not suspect this judging by the ordinary details of his life. On weekday mornings, Wilson leaves his ranch house in the flat farmlands just south of Portales, climbs into his aging, yellow Lincoln, and drives the 8 miles to his job with the New Mexico highway department.

In overalls and work boots, Wilson, in late middle-age, cuts the figure of a man accustomed to physical labor, and just maybe hard times. His rectangular face bears evidence of long days spent out on the bright plains along the border with the Texas panhandle. In summer, the sunburn extends from the tip of his nose on down, leaving his forehead and the skin around his gray-blue eyes relatively pale — protected by a "Farmers and Ranchers Realty Co." baseball cap.

Like most folks on the southern high plains, Wilson speaks with a warm, wide open accent. It's a laid back and more than slightly infectious kind of sound. Out here, "can't" becomes "cain't," "get" becomes "git," and "where" becomes "whur." For outsiders, the cowboy drawl is as colorful as the wildflowers that blanket the prairies after a wet spring. But for most of Wilson's adult life, he was blithely unselfconscious about it.

Then came an event so remarkable, it sparked national media attention. And Skip found himself right in the middle of it all.

"I was horrified because that's the first time I ever saw myself on TV," he recalls. "I thought, 'My God, you sound like a total hillbilly.'"[2]

Wilson may have seen himself this way, but his home-grown knowledge of his field is impressive. He's been building that knowledge for more than 30 years as a kind of fossil hunter. His quarry is rarer, harder to find and much more unusual than your garden variety fossil. Instead of the bony remains of long-dead creatures, Wilson hunts the stony remains of long-disintegrated planetessimals — rocky bodies, smaller than planets, born 4.6 billion years ago as our solar system was condensing out of a cloud of gas and dust.

Wilson's fossils are meteorites. And locked within them are chemical and physical traces of the origin of the Sun and planets.

Within 10 million years of the solar system's birth, almost all of the material that collectors like Skip Wilson would pick up in their careers, indeed most of the meteoritic material that has ever fallen to Earth in the past and will ever fall in the future, was bound up in an unknown number of planetessimals circling the young Sun. In the billions of years since, they have continually collided with each other, leaving a strewn field of dust, pebbles, boulders and asteroids circling the Sun between the orbits of Mars and Jupiter. This is the asteroid belt, and most meteorites ultimately derive from it.[3]

Like a giant with a slingshot, Jupiter with its massive gravitational field continually flings rocky asteroidal debris our way. Technically speaking, these are meteors. Only those bits and chunks that actually survive the intense heat of entry through our atmosphere to land on Earth are called meteorites.

Just as fossils of dinosaurs and other extinct creatures have helped paleontologists unravel the history of life, so has analysis of meteorites helped planetary scientists piece together the story of our solar system's beginnings. Writing that story is part of the bigger quest, the scientific search for our cosmic origins.

By snaring more than 200 meteorites since 1967, Skip Wilson has advanced that quest by finding more specimens than any other person alive today. He is the envy of his fellow meteorite hunters.

"Skip is a hero to all of us meteorite heads," comments Bob Haag, a renowned meteorite collector and dealer in Tucson.

Mike Farmer, another dealer in Tucson, goes further: "He's like God to us."[4]

Oddly enough, Wilson owes much of his success to the dust-bowl of the 1920s. The farming practices of the day had led to severe erosion that stripped away the topsoil from a large swath of the American interior. Parts of the landscape were left looking more like the dune fields of the Australian outback than the farmlands of the American bread-basket.

To this day, damage persists in eastern New Mexico— wounds on the landscape that are constantly picked at by strong prairie winds, preventing them from healing. In these areas, the topsoil is long gone. The underlying sand is now being scoured out, inch by inch, year by year, creating "deflation fields" depressed below the surrounding topography. Such erosion, though fatal to agriculture, is a boon to people like Skip because it continually exposes meteorites that impacted long ago.

An estimated 100 to 1000 tons of such meteoritic material falls to Earth each day, mostly in the form of fine dust; about 1% of the material is in pieces large enough to be recognized as meteorites.[5] Given that rate and Earth's size, it's clear that on a single day, any given acre of land would be lucky to receive anything more than an invisible dusting of meteoritic material.

But the dusting does build up. Run a magnet over the debris stuck in a rain gutter and it may pick up particles of metal, much of it, in fact, of meteoritic origin. And over geologic time—over thousands and millions of years—a single acre can accumulate a veritable meteoritic windfall, including some hefty chunks.

The problem, of course, is that most of this material—dust, pebbles, rocks and all—becomes buried beneath the surface before a meteorite hunter like Skip Wilson can find any of it. In eastern New Mexico, though, the ongoing erosion of areas first stripped bare during the Dust Bowl is exposing sediments long-buried for millennia. And as fine-grained sediment is blown away, the heavier stuff, including meteorites, is left behind. As a result, meteorites gradually become concentrated at the surface.

As a young man in the late 1960s, Wilson had no idea that cosmic treasure lay hidden in the sand-filled depressions scattered around town. He never really gave meteorites, or even the night sky for that matter, any thought at all. But starting at about age 21, he did develop another abiding hobby, one that would soon lead directly to meteorite hunting. Driven even then by a desire to find artifacts of lost worlds, Wilson spent much of his spare time hunting for Indian arrowheads.

Like many other people in town, he was well aware of what archeologists were digging up in an active gravel quarry near town: mammoth bones, arrowheads and tools constituting some of the oldest evidence ever found of native Americans' presence in North America. Where the quarry stood, 9000 years earlier the waters of a lake had lapped at vegetation-choked shores. Here, ancient hunters of the Clovis culture killed and butchered mammoth, buffalo and other game that had come to the lake to drink.

From the days of those long-gone people until white settlers all but exterminated native American culture, Indians had flourished around the Portales area. And they left copious evidence of their lives for those, like Wilson, with the desire and patience to search for it.

One day after work, Wilson was out searching for arrowheads when a dark object caught his attention. "I saw an old brown rock that was different from anything else around there," he says. "So I picked it up, and somehow I just knew it was unusual."

Wilson sent it off to a man by the name of Glen Huss, then owner of the American Meteorite Laboratory in Denver. A few weeks later, Huss called to say that Wilson had found a meteorite.

A new world of collecting possibilities unfolded, and Wilson began wondering whether he could be as successful hunting meteorites as he had been at finding arrowheads.

"Huss told me that nobody had ever made a successful career of huntin' meteorites," he recalls.

But Wilson was determined to try.

"It took me a couple of years to put it all together right—to learn where to hunt and how to hunt," he says. "I figured out that these rip snortin' sand storms we'd get would just peel the surface right off those deflation fields and expose new meteorites."

In the years since, Wilson has proved Huss wrong in spectacular fashion, and not just by the sheer number of his finds but also by the very precious nature of one particular specimen. It was a meteorite so rare it might have made Skip a wealthy man, had he only foreseen it's true worth before he sold it.

He found the 24-pound meteorite in 1972 on an abandoned homestead near Kenna, New Mexico, 35 miles southeast of Portales. Huss quickly agreed to cut and polish it.

About two days later, the meteorite dealer called. "This meteorite is different," he told Skip. "I can't cut it. It's just tearing my diamond blades to pieces." But Huss kept at it and eventually succeeded—at a price of three expensive diamond blades. In cross-section, the piece didn't look like much. It was mostly black. But out in the sun, Wilson says, the polished surfaces sparkled slightly, hinting at a remarkable treasure locked within.

"Wouldn't you know it," Huss told Wilson, "but this thing's got *diamond* in it."

"So Huss come luggin' it back down here," Wilson says. "And he offers me a thousand dollars for a big piece, sayin' I can keep a littler one. So I thought, 'A thousand dollars for twenty pounds?' Man, I just reached out and grabbed that

191

thousand dollars.'' Then he hid the little piece Huss gave him in a closet.

About three or four years later, Huss phoned to say that a curator with the Smithsonian's National Museum of Natural History in Washington wanted a piece of the meteorite for their collection. Huss had already sold the piece he had bought from Wilson and was wondering whether the meteorite hunter still had the little chunk in his closet.

''They're willing to pay $4000 for that piece,'' Huss said.

That sounded like a second windfall to Wilson. ''So like an idiot, boy, I just reached out and grabbed that money. Again.''

Today, a piece of what has come to be known as the Kenna Meteorite is in the national meteorite collection at the Smithsonian; another chunk is on display at the American Museum of Natural History in New York. The remainder of the 24-pound, diamond-studded rock is in university and private collections.

Skip's find is classified as an ureillite, one of the most enigmatic and rarest of meteorites. Some, like the Kenna Meteorite, contain tiny flecks of industrial grade diamond, possibly created when carbon within a parent asteroid underwent intense pressure.

Ureillites are among the most poorly understood of all meteorites. Scientists simply don't know how they formed in the early solar system.

But one thing is known for sure: they are worth far more than their weight in gold. Ureillites represent less than half of one percent of all meteorite falls. ''There are only a few of these in the world,'' says Bob Haag, the Tucson meteorite dealer. ''Collectors just have to have a piece of each type to add to their collections.'' And with so few samples of ureillites to go around, the price has skyrocketed.

The going rate for a piece of the Kenna Meteorite is about $275 a gram. That works out to just shy of $3 million for the original mass.

''I was settin' on a fortune, but I didn't have enough sense to realize it,'' Wilson laments.

Although the Kenna Meteorite didn't make him rich, it did add luster to Wilson's reputation. And by 1998, after nearly 30 years of successful hunting, there wasn't much left for him to accomplish. He had just one goal left: to collect specimens from what's known in the trade as a "fresh fall" — a meteorite that has just crashed to Earth.

At 7 a.m. on Saturday, June 13, Wilson was getting ready to head in to work at a video store he owned at the time. Summer was coming and it was turning hot. The farmers around Portales already were suffering from a severe drought, and the hotter weather heralded tough times ahead.

Wilson stepped outside to water his flowers and tomato plants.

Not far from Skip's place, Nelda Wallace, a 69-year-old widow and art teacher, was fixing coffee for her sister and brother-in-law, Fred Stafford, who were visiting from California.

A dry breeze ruffled the parched grass around her home as a few isolated cumulus clouds hovered not far off to the south — perhaps the same ones Pardue was gazing at from the passenger seat of the Piper Cherokee at the very same moment. Gray streamers of precipitation were falling from the clouds' dark bases but evaporating before hitting the ground, bringing a false promise of moisture for the parched farmlands.

Fallow fields surround Nelda's home, which is laid out in the shape of a "U." Her late husband had constructed the house by trucking two separate ranch houses to the site and linking them to one already there. The front of the unconventional complex is a facade, built by Nelda's husband to look like the storefronts of a town in an old Hollywood western.

Like the public front of her unconventional house, Nelda Wallace has a public face: polite, reserved — and inscrutable. Behind it is a tough, independent woman who knows how to take care of herself.

"You know, this is not my first rodeo," she says with just a trace of amusement wrinkling the corners of her eyes.

The coffee was ready at almost exactly 7:30. Nelda carried the steaming mugs into the living room, where her guests were waiting. "I had just sat down and kicked my shoes off when we heard this very loud boom," she recalls.

Nelda glanced anxiously over at Fred.

"Oh gee, now what was that?," she said.

Then they heard a second explosion. Nelda and Fred jumped up and started moving toward a door leading from the living

193

room to a courtyard. From a recliner, Nelda's sister protested that she wanted to come too. But having just undergone heart surgery, she was too frail.

"You cain't hit the dirt if you have to, so you stay here," Nelda ordered.

She and Fred then stepped out onto a porch overlooking the courtyard. They were met by a roaring sound.

"It's an airplane, a piece of an airplane," Wallace exclaimed. She and Fred looked for signs of a stricken plane.

Nothing.

The roaring continued, got louder, and then a third boom rumbled across the plains followed by a series of pops and crackles, much like the explosion of a July 4th rocket followed by the secondary explosions.

"It sounds like it's going to hit the house," Wallace said, now seriously alarmed.

The roaring, which had gone on unabated for what seemed like several minutes, got louder still. To Fred, it sounded like an incoming mortar round.

"I THINK IT'S GOING TO CRASH THROUGH THE ROOF," he yelled. And then he jerked his arm toward the south-west, pointing up in the sky: "Here it is, HERE IT IS!"

As Wallace followed Fred's gesture, she saw an object streaking directly toward them, smoke billowing behind it.

At about the same time, Gale Newberry was getting ready to begin his daily routine as a peanut farmer. With sandy hair, blue eyes, blonde, bushy eyebrows and a mustache that's almost but not quite a handlebar, Newberry looks younger than a man in his late fifties. His quick smile, ready laugh and easygoing manner belie the hardships that a family farmer must routinely endure.

On that hot day in late spring, his tanned face no doubt was creased with worry about the drought. Although he had plenty of well water for irrigation, in recent years the water table had been dropping precipitously. And even with irrigation, severe heat and drought can damage crops, as can the hailstorms that routinely boil up during summer.

Newberry left the house at 7:30 to move a sprinkler out in his fields when he heard the explosions.

"I heard seven or ten of them, and they were so powerful that I could feel the concussions," he says. "The sound was coming from the southwest part of town, out by this big alcohol plant. I thought maybe something had happened there. So I went back to work and half forgot about it."

Between 7:30 and 7:45, Newberry went to his brother's house. There he met his 21-year-old son, Jerrod, who reported that while outside a few minutes earlier, he had seen smoke trails in the sky and heard a loud explosion directly overhead.

The Newberrys wondered whether some fighter jocks out of nearby Clovis Air Force Base had been flying their planes a little too aggressively. Maybe they had broken the sound barrier, and those explosions simply were sonic booms. Of course, that didn't really explain the weird smoke trails. But during the growing season, farmers have more important things to do than speculate about mysterious booms and smoke trails in the sky. So Gale and Jerrod went back to work.

A few minutes before the Newberrys heard the booms and saw the smoke, Skip Wilson had finished watering his plants and returned to his house to collect his things for work.

"Just as I got inside, I heard these two sonic booms followed by this 'POP-POP-POP-POP-POP-POP.' Well that sure got my attention. So I run outside and looked up."

Wilson saw an object streaking high and fast overhead, a cloud of smoke boiling around it and a corkscrew-shaped contrail trailing it. Wilson also heard a jet, so he logically put two and two together and thought that maybe an engine on the aircraft had exploded; maybe the object he saw corkscrewing above him was a big piece falling to the ground.

But the explanation didn't quite fit with what he was seeing.

"I kept lookin' for flames," he says, "but I couldn't see any. Yet like an idiot, I kept trying to put it to that airplane."

Back at Nelda Wallace's place, a few miles southwest of Skip Wilson, she and Fred Stafford had never seen anything quite like the sight of the smoking object streaking toward the house. It passed almost right overhead, missing the house, and then impacted a little more than a hundred yards from where they stood, in stunned amazement.

195

"When it hit, it kicked up a huge cloud of dust," Wallace says.

The dust cloud quickly mushroomed to several hundred yards wide. Nelda ran back inside the house to get her shoes on, and then she and Fred headed out to the field to search for the mysterious object that had nearly slammed into her roof. At first, they couldn't find the impact site because dust and sand were swirling abrasively around them. But in a minute or so, Fred found a crater. It was about 10 inches deep and a foot in diameter. Sticking two inches out of the hole was a dark object with something like a tear-drop shape—an aerodynamic shape, Fred thought.

With booms still echoing all around them, Fred and Nelda decided to retire to the house and have breakfast before trying to handle the grayish-black object. About 45 minutes later, the dust had settled and quiet had once again descended on Portales. So they went back out to check on the strange object.

Fred pried it out of the ground. It was bigger and far heavier than a bowling ball, and its surface was charred black. It also seemed shot through with metallic veins. Some actually protruded from the surface by a half inch or more, looking all the world like they were plates of solid steel.

By about 9:30 or 10 that morning, Gale Newberry had pretty much forgotten about the booms he had heard earlier. His brother, Roy, and son were off fetching a lawn mower when Gale entered his barn.

"I went into the back room. The lights were off, so it was dark in there." But then Gale saw a bright spot of sunlight on the wall.

There are no windows in that part of the barn.

'What in the world is that?,' Gale thought to himself. As he followed the shaft of light leading up from the spot on the wall to its point of origin, he saw an 8-inch-by-5-inch hole in the roof.

His first guess as to what had happened was none too charitable: perhaps Jerrod, his son, had been using a forklift in the barn and had mistakenly punched a hole in the roof. But given the height and angle of the roof, Gale realized this was physically impossible.

At about this point, Jerrod and Roy entered the barn.

"We got to lookin' down near that spot of light on the wall, and we saw a hole in the insulation," Gale says. "Then Jerrod just reached in behind the insulation, and wouldn't you know it, he pulled out a rock."

Up in the Piper Cherokee, Larry Pardue was astonished by the intensity of the flaming ball he had seen streaking toward the ground.

"I've seen fireballs at night, but certainly nothing during the daytime near this bright," he says. "On a bright morning, to see it like that, it must have been incredibly bright."

His conclusion? A big meteorite had just fallen to Earth. And he had a hunch that it must have exploded above the ground, scattering many pieces around the countryside.

Larry Pardue wouldn't know it for two days, but he was exactly right.

At Nelda Wallace's place, she and Fred Stafford had also concluded that the astonishing event they had just witnessed was a meteorite fall. What other explanation could there be? Fred wondered.

The two of them joked for a few minutes; Nelda laughingly invoked her late husband as a possible explanation. But a part of her refused to believe that a meteorite had fallen on her property. It was just too incredible. And so she called Canon Air Force Base to check whether there had been any fighter activity that morning. The answer was no. Then Wallace tried calling the local newspaper. No answer. She decided to sit tight and see what developed.

For the Newberrys, standing in Gale's barn, the explanation for the potato-sized rock Jerrod was holding in his hand also seemed clear. It had to have come from space.

"We figured out pretty quickly that it must be a meteorite," Gale says. "Now, I know Skip fairly well—we went to school together. I knew he was a meteorite hound, and that he had been doing this for years. So I called him up and I said, 'Skip, are you very busy? I got something to show you.' And he said, 'Does it have anything to do with those explosions?'"

Gale told him to just come on over to the barn.

A few minutes earlier, an extraordinary thought had hit Wilson like a jolt of adrenaline. Was his fervent wish to collect meteorites from a fresh fall coming true, literally right before his eyes? As soon as the question came to mind, he pushed it away. Having a meteorite land almost in his own back yard was just too bizarre for Wilson to believe.

To be sure, stranger episodes involving meteorite falls have happened. In 1954, for example, a meteorite falling over Sylacauga, Alabama, severely bruised Elizabeth Hodges' side. And in 1992, a young Ugandan boy escaped serious harm when a meteorite that had been slowed down by the leaves of a banana plant struck him in the head.

But these are the only modern records of people actually being struck by meteorites. And even though meteorites impact the Earth every day, it's rare that anyone ever gets to see it happen, let alone a famous meteorite hunter.

So Wilson was determined not to jump to conclusions. He would believe that he had witnessed a meteorite fall only when he was holding a fragment of the thing in his own hands. And so he rushed over to Newberry's place.

When he arrived, Newberry didn't waste any time.

''Hey Skip,'' he said, ''I'm gonna' make your day.''

''And boy, did he ever,'' Wilson says.


According to scientific reconstructions, the Portales Valley Meteorite, as it has since officially been named, likely streaked into the atmosphere at 20 000 miles per hour. Roughly a yard across and weighing more than 200 pounds, it was crushed by increasing air pressure as it fell through the increasingly dense atmosphere. As it neared Portales, the mounting pressure finally caused it to explode — just as Pardue up in the airplane had thought — throwing off three large pieces. Moments later, as the largest of these pieces continued streaking toward the northeast, there was a second explosion, which showered Portales with many smaller fragments in a strewn field about six miles long by a mile or so wide.[6]

Now that he was holding one of these fragments in his hand — a dark gray spud of a rock weighing about a pound and scorched black in places — Wilson finally acknowledged what he had known in his heart all along.

Gale Newberry called the local newspaper, and this time, someone was home. A skeptical editor and photographer came

198

over to have a look. The top story on page 1 of the *Portales News-Tribune* the next day, Sunday, June 14, ran under a journalistically cautious headline:

**"Farmer: A meteorite crashed into my barn."**

Now the news was out. A collecting frenzy was about to ensue, and Nelda Wallace, the Newberrys and Skip Wilson were about to get their requisite five minutes of fame.

As Skip Wilson puts it, "Man, this thing just went crazy."

By Monday morning, Mike Farmer from Tucson had arrived at Gale Newberry's door, identifying himself as being from the University of Arizona. Although he was not the scientist Gale thought him to be, technically speaking, Farmer had told the truth: he was a student at the U of A. But his interest in the meteorite was commercial, not academic. Farmer was determined to buy pieces of what he thought would be a valuable meteorite.

Within minutes of seeing the rock that had punched through the roof of Gale Newberry's barn, Farmer offered to buy it for $1500.

Gale turned him down flat.

At about noon, Gale says he was working out in his fields when he saw a Humvee—a hulking sport utility vehicle originally designed as a replacement for the venerable Army Jeep. It was approaching his house, and strapped to the top was an ultralight aircraft. At the wheel of the Humvee was a man with shoulder-length hair and dark sunglasses: Robert Haag, perhaps the most colorful meteorite dealer in the world. He had arrived to make an offer for Newberry's meteorite. But as he had done with Farmer, Gale said no.

By this point, newspapers and television stations were carrying stories of the Portales meteorite fall. And Skip Wilson had made the appearance on television news that horrified him so.

If all that were not enough to work folks up into a frenzy, Haag got on a local Portales radio station to say that he had thousands of dollars to spend on pieces of the meteorite.

"That sure got folks to huntin'," Gale Newberry says with a grin.

People began crawling all over the Portales area. Determined not to let anything stand in their way of striking paydirt, including the law, many trespassed on private land in their search for fragments, Wilson recounts.

199

At this point, Gale Newberry told Jerrod that he ought to bicycle around looking for some pieces. He was really joking—he didn't take the idea all that seriously. But Jerrod did.

"Well, he just hopped on his bicycle and tore off down the road," Gale says. "I went to plowin' and it weren't 15 minutes later that he come up carryin' a piece of that meteorite. He had found it lying off to the side of the road."

The fragment had hit the pavement, where it left a small crater, and bounced. Later, a meteorite hunter would literally cut a chunk of asphalt containing the crater right out of the road.

Meanwhile, Nelda Wallace had remained mum about what had happened out at her place. But by Tuesday, she had told her story to the *Portales News-Tribune*. And the next day, a photo of a beaming Nelda Wallace with her whopping 37-pound fragment graced the front page.

It wasn't long before Haag drove up in his Humvee. He offered Wallace $500 just to *look* at the meteorite. But she wasn't in the mood to bargain with any of the dealers, whom she suspected of seeking to take advantage of her. So in so many words, she told him to get lost.

Gale Newberry decided to hold on to his piece, but his son sold the chunk he had found to Haag for $5000.

Over the next few weeks, Wilson scoured the Portales area and, all told, found six fragments of the meteorite totaling a little more than 20 pounds. It was, to say the least, a dream come true, "almost a fairy tale for me," he says.

"Of course it's always goin' through your mind, you're out huntin', and you think, 'Man, it'd be neat if I could see this sucker come screamin' in.' The chances are it's not gonna happen. But yet it did. It was close enough to where I heard it, I seen the visual effects of it, and then I got to go out and pick up fresh meteorites from it."

"These others I find, when I pick one up, I know I got somethin' in my hand that's old, old, old, and come a long, long ways. Very few other people get to do that. But this, now *this* was the biggest highlight of my meteorite huntin' experience by far."

Skip Wilson finally got to collect specimens from a fresh fall, and that was gratifying enough. But there was even more reason for Skip to be thrilled. As it turned out, scientific analysis has revealed that the Portales Valley Meteorite is quite unusual—in some respects, unlike any meteorite in its class ever found. So once again, Wilson was involved in finding a precious meteorite.

It may not allow scientists to make a quantum leap forward in their understanding of the origins of our solar system. But the Portales meteorite has already yielded new details about events that occurred on a long-lost mini-planet—a planetesimal—that circled the young Sun in the very earliest days of the solar system, probably before the planets as we know them today had taken shape.

Perhaps more important, the chunks of rock that rained down on Portales that sultry summer morning represent a new, precious addition to the world's collection of early solar system fossils. Without this collection, the birthing story of the Sun and planets might consist of little more than a broad outline based on guesswork, theoretical modeling and inferences about what happened 4.6 billion years ago drawn from how the bodies of the solar system look today. With the knowledge gleaned from the collection, the story now has rich texture and even drama.

And one other thing: the hard, direct, supporting evidence—or what scientists call "ground truth"—that only preserved samples of the early solar system itself can provide.

Scientific analysis of the Portales meteorite began not long after it's spectacular Earthfall. Within a few weeks, Nelda Wallace donated an 11-gram piece of her chunk to the University of New Mexico's Institute of Meteoritics in Albuquerque. (Wallace would later sell the lion's share of her chunk to a syndicate of universities, led by UCLA. When UNM meteoriticist Adrian Brearley first saw it, he was struck by the thick and finely delineated metal veins crisscrossing the sample.

"I was quite taken aback by it, really—quite astonished by its appearance," he says. "It was unlike any meteorite I had ever seen."[7]

Experts like Brearley have many tools for teasing out the composition and structure of a meteorite—and in so doing, its history. The simplest technique is to cut a section from the meteorite and polish it. Metals, oxide compounds and other materials can be seen with light reflected off such a sample. "Most meteorites can be identified this way," Brearley says.

201

But for a more in-depth analysis, other techniques are needed. In one common technique, a sliver about half as thick as a human hair is cut from the sample. This is so thin that light can pass right through the solid rock. When such a thin section is viewed through a microscope with polarizing filters, a montage of forms—rectangles, squares, broken amoeboids, spheres and rods—explodes through the eyepiece in a vivid rainbow of colors. If it weren't a creation of nature, one might guess that it was a work of abstract expressionism. To a trained eye, the colors and shapes reveal the crystalline structure of the sample, as well as information about its composition.

When Brearley looked at a thin section from the sample donated by Nelda Wallace, he could clearly make out chondrules, millimeter-sized spheres consisting of olivine and pyroxine crystals embedded within a surrounding silicate matrix. Chondrules are thought to have originated early on during the solar system's birthing process. According to the conventional theory, tiny balls of dust orbiting around the young Sun were heated and thereby melted, forming little droplets of molten material. As the droplets floated amidst dust and gas of the early solar system, they cooled and solidified. Eventually, they accreted with lots of other orbiting material to form planetesimals. These mini-planets—toddlers on the planetary growth scale—are thought to be the parent bodies of meteorites.

By examining the textures of bar-shaped crystals of olivine within chondrules, scientists can get a good idea of how hot the original dust balls got and how quickly they cooled. This in turn yields clues to conditions within the early solar system when the chondrules formed. For example, the longer cooling takes, the more time the crystals have to grow. Thus, larger olivine crystals indicate slow cooling, which in turn suggests that the chondrules formed during a phase in the young Sun's development when it heated up and then remained hot for awhile. Smaller crystals, by contrast, indicate a rapid rise and then rapid drop in temperature. This suggests that the crystals formed during a solar outburst that came and went relatively quickly—like a young child's temper tantrum.

Based on what he saw through the polarizing microscope, Brearley quickly concluded that Nelda Wallace's sample had come from a chondrite, the only kind of meteorite that contains chondrules. Chondrites typically consist mostly of stone with some metal typically distributed in little flecks, specks and patches. The Portales sample had lots more metal than that, but it was almost perfectly segregated in the shiny, steel-like veins.

"There are a few ordinary chondrites like this," Brearley says. "But the Portales meteorite is an extreme example. This is something we've never seen before to this degree."

What accounted for the unusual nature of the Portales meteorite? Brearley was eager to know, but his wife was expecting a baby, and he also had pressing projects to work on. So an explanation would have to await work by other scientists.

In Tucson, Mike Farmer and Bob Haag had donated 16 specimens of the Portales meteorite to the Lunar and Planetary Institute of the University of Arizona. There, meteoriticist David Kring and colleagues began analyzing the samples.

Like Brearley, Kring prepared thin sections and looked at them under a polarizing microscope, which revealed the chondrules. And like Brearley, he was struck by the segregation of the metal into thick veins.

Next, Kring's group used a device called an electron microprobe to tease out detailed information about the meteorite's composition. First, they cut a slice from the meteorite 100 times thinner than a human hair. In the microprobe, they bombarded this ultra-thin section with a beam of electrons. This caused atoms within the section to emit X-rays. By analyzing the energies and wavelengths of those X-rays, Kring's group determined the precise chemical composition of the minerals in the sample.

With this and other analyses, Kring and his colleagues decided that they were working with a type of meteorite known as an H6-chondrite. All H-chondrites are believed to have come from a single, parent planetesimal that formed 4.6 billion years ago. The metallic veins in the rock enabled Kring to flesh out the story of this tiny planet's early evolution.

"Analyzing meteorites like the one that hit Portales gives us an opportunity to teach people what we're learning about the geologic evolution of the solar system," he says. "If we simply stood up and said this is how it happened, no one would listen."[8]

The story that resolves from the analyses by Kring's group is one of great violence. Based on the shape and size of barred olivine and other crystals within the chondrules, he concluded that the little dust balls from which they formed were heated in a blinding flash. The sudden outburst of intense energy required for such an event is not surprising, given that our Sun probably was prone to great, eruptive tantrums during its infancy as dust and other materials were drawn by gravity onto the solar surface.

Not long after they became molten, the dust balls began to cool. Once solidified, these chondrules combined with each other, residues of evaporated compounds, and new dust, which had condensed like snow that forms spontaneously in air. Eventually, enough of this material accreted to form the planetesimal, a mini-planet that orbited around the young sun somewhere between the current orbits of Mars and Jupiter.

By analyzing the nature and size of minerals from meteorites like Portales Valley that came from this parent body, scientists have learned that material within the planetesimal was heated to high enough temperatures to become metamorphosed. And for this to have occurred, the planetesimal had to have been of sufficient size. That's because relatively small bodies radiate their internal heat (generated by impacts and radioactive elements) more quickly than bigger bodies. The reason? The blanket of rock insulating a small body's interior is skimpier than that of a bigger body. So smaller bodies just can't get hot enough for rocks to be metamorphosed.

If a body is too big, though, it will retain so much heat that it will not simply metamorphose; it will melt through and through. When this happens, metal within the body melts and thereby becomes mobile. And move it does, sinking to the center and forming an iron core. (Earth's core is believed to have formed in the same way.)

Accordingly, pure iron meteorites are thought to have originated as material from the cores of large parent planetesimals. Similarly, stony-iron meteorites, consisting of roughly equal measures of the two materials, are believed to have formed at the lumpy boundary between the iron core and stony exterior in such planetesimals.

The Portales meteorite, as well as its siblings—all the other meteorites that came from the same parent planetesimal—

shows no evidence of complete melting. Instead, its mineral grains indicate a degree of metamorphosis consistent with a parent body about 125 miles in diameter, about the distance between Los Angeles and San Diego.

If that were the end of the story, the Portales meteorite would not have become shot through with metal veins. To David Kring and other meteoriticists, those veins suggest that at some point an asteroid smashed into the planetesimal, gouging out a large crater and creating mayhem in the mini-planet's interior. The scenario goes like this. Metal once was distributed in flecks and specks throughout the stony parts of what was to become the Portales meteorite. Then it was heated to the melting point, enabling the metal to move and collect in cracks. A large impact on the parent body would have provided the requisite heat and also cracked the surrounding rock. And the high pressure from the impact would have jetted the now molten metal into the newly opened fractures.

Based on the crystalline structure of the metal in the veins, Kring concludes that it cooled and solidified extremely slowly—perhaps only a few degrees per million years.[9] For such slow cooling to have occurred, the planetesimal should have been very young at the time of the impact—perhaps only 100 million years old. That's because its interior had to have retained much of the intense heat of its youth.[10] Moreover, the veins must have been buried deep within the asteroid, where the overlying rock would have effectively prevented interior heat from escaping.

Kring's best estimate is that the metal veins so sensationally evident on pieces of the Portales meteorite came from at least four miles deep within the parent planetesimal. Given that depth, he calculates that the crater gouged out by the impact should have been as large as 23 miles across—enough to accommodate most of the City of New York and some of the bordering New Jersey suburbs. For a present-day analogue of such a crater, one need only look to the asteroid belt, Kring says. Here, the asteroid Mathilde (itself probably a fragment of a parent planetesimal) is scarred by five craters with diameters in this size range.

Of course, the story doesn't end there, because somehow, veined rock sitting 4 miles deep within a mini-planet had to find its way to Earth on June 13, 1998. To make that possible, big chunks of the planetesimal first had to break off, including a piece carrying the future meteorite. And then that piece had to become fragmented, and the fragments fragmented until a metal-veined boulder about a yard across was created. Since the asteroid belt where the planetesimals were orbiting was littered with all manner of debris and mini-planets, a simple game of cosmic bumper cars over billions of years — driven by Jupiter's great gravitational energy — could easily have done the trick.

By chance, the boulder's ride in the asteroid belt ended when Jupiter's gravitational field interacted with its orbit in such a way as to swing it into an Earth-crossing path. And simply by chance, the boulder found itself streaking toward the very town where Skip Wilson, one of the greatest meteorite hunters of all time, happened to live.

If it seems astonishing that scientists like Brearley and Kring can glean so much information from a sliver of rock a hundred times thinner than a human hair, consider this: meteorites have even deeper and older stories to tell. In fact, by studying faint residues of isotopes trapped within a specific class of meteorites, some scientists believe they can discern events that happened *before* the solar system formed.

The meteorites in question are flecked with inclusions of aluminum and calcium. Examination of the materials in these meteorites has revealed that they formed at extremely high temperature, suggesting that they date to the very earliest events in the birth of the solar system, when temperatures must have been hottest. Radiometric dating puts these meteorites at 4.559 billion years old, the very oldest that have been found.

Within the aluminum–calcium inclusions in these meteorites are tiny amounts of magnesium 27. This isotope is like a fingerprint found at a crime scene that identifies the crime's perpetrator. Many scientists believe that its presence in the very oldest meteorites reveals a sort of cosmological triggerman: the perpetrator in an event of great violence — an event that led directly to the formation of our solar system.

Magnesium 27 is produced through the decay of another isotope, aluminum 26. None of this parent isotope remains today because it has an extremely short half life. But the magnesium 27, its decay product, reveals that the aluminum 26 must have been present when this class of meteorites formed 4.559 billion years ago. This is important because one very likely source for aluminum 26 is a massive explosion of a star in its death throes.

In other words, a supernova.

"The Holy Grail for some of us," Brearley says, "is understanding what sorts of processes were happening before the birth of the solar system." He notes that astrophysicists have put forward many theories about those processes. "But meteorites are the only ground truth for these astrophysical models."

According to these models, solar systems begin to form when giant interstellar clouds of molecular hydrogen laced with dust and other elements begin to collapse. But what triggers the collapse? One possible mechanism is the blast wave emanating from the explosion of a nearby supernova.

And so chemical fingerprints gleaned from ancient meteorites suggest to some scientists that the formation of our own solar system was triggered when a star exploded at the end of its life. Of course, it's also possible that supernova explosions simply seeded the interstellar medium in our vicinity with aluminum 26 and that something else actually triggered the collapse of our parent molecular cloud.

In the chapters that follow, we'll investigate these and other issues concerning the birth of stars and planets.

*   *   *

Regardless of the precise sequence of events in our own solar system, it's plainly clear that the origin of stars and planets is part of an overarching cycle of stellar life and death. And according to University of Colorado astrophysicist John Bally, supernova explosions and other events of unimaginable violence play key roles in that cycle.

Given just how violent these events can be, the formation of a planet like ours may well be a miracle—as we are about to see.

# 8

# INTERSTELLAR ECOLOGY

## *Cycles of Birth, Death & Rebirth*

*What seest thou else in the dark-backward abysm of time?*
— Prospero in William Shakespeare's,
*The Tempest* [1]

*We had the sky, up there, all speckled with stars, and we used to lay on our backs and look up at them, and discuss about whether they was made, or only just happened...*
— Huckleberry Finn [2]

With the Sun heading for the horizon, John Bally bounds up a slope of volcanic cinders covered in snow, pursuing the last photons of daylight. This is no easy feat, as he is climbing in thin air. His goal? The tip of the summit cone of Mauna Kea, the high point on the Big Island of Hawaii and the tallest mountain in the Pacific basin.

Constructed of lava flows, Mauna Kea's broadly sloping bulk tops out at 13 796 feet. At this elevation, the oxygen available to fuel exertion is 60% that of sea level. But hypoxia seems of no consequence to the University of Colorado astrophysicist, who is moving enthusiastically toward the roof of the Pacific.

The summit cone, Pu' u Wekiu, is a 300-foot-high blister of cinders erupted in a final moment of molten exuberance. It rises from an austere plateau of red, brown and black lava flows. There are no trees or bushes up here. No grasses. Only cinder and rock. As far as I can tell, there aren't even any lichens on the rock. There simply is no visible trace of life at all.

But humans have staked a claim in this almost alien environment. The distinctly Mars-like summit of Mauna Kea is studded with nine optical/infra-red telescopes, and two instruments designed to observe in the submillimeter range of the electro-magnetic spectrum. All are dedicated to capturing photons from much farther away than the Sun. The Canada–France–Hawaii Telescope, housed in a classic white dome just below us, is what brought Bally across three time zones and 2700 miles of Pacific Ocean. For the past several nights, he's been using the CFHT to explore the Orion Nebula, a region where stars—and possibly planets—are being born.

We drove up here in the early evening on a breathtakingly steep and winding dirt road. As the sun began to set, Bally suggested that a dash up Pu' u Wekiu would be just the thing to convince our bodies that as night falls, it's time to wake up, not get ready for bed. The only problem is that now we're dashing over snow that seems decidedly glacierlike. The white stuff fell earlier in the week during a two-day blizzard that arrived on 100 mile per hour winds. Drifting and white-out conditions required the temporary evacuation of the summit observatories. Today, though, it's just cold, windy—and impossibly clear. And the snow that a day or so ago was as close as Hawaii gets to powder has since hardened into a crusty sheet.

"This stuff would be absolutely perfect for crampons," notes Bally, who has done a bit of ice climbing back at home in Colorado.[3]

In just tennis shoes today, Bally wins the race with the photons as we reach the summit. With the Sun dipping behind some clouds, the last rays of daylight are painting the snow orange. From this elevation, we are looking *down* at the clouds, as well as the Big Island's notorious "vog," volcanic smog from active vents on the flanks of Mauna Loa, a volcano to the south. The soupy mix sits about 7000 feet below us.

Above the western horizon, the sky has just turned a color best described as neon green. And to the east, Mauna Kea has somehow etched its shadow onto the atmosphere itself, forming a perfect triangle of darkness in the purple eastern sky.

For protection against near-freezing temperatures and a stiff breeze, I've donned a down parka. Bally wears a thin fleece jacket. Slightly balding but still pony-tailed, he bounces up and down to ward off the chill—an activity that makes me feel faint after just two or three jumps.

We linger for as long as possible, keenly aware that standing on the summit of a snow-tipped peak on a remote tropical island is a treat that comes along infrequently at best. But as night falls, the shutters on the dome of the Canada–France–Hawaii Telescope are opening, signaling that it's time for us to go. Before we do, Bally sets the tone for the coming night's activities.

"There's a romanticism to the notion that to understand our origins, we climb mountains and look to the sky. Observatories like this are almost like temples to the universe."

With stars blinking on in the east, we head back down the slope to join two of Bally's colleagues in the observatory's control room. We have to hurry: observing time on this and the other telescopes dotting Mauna Kea's summit — including the twin Kecks, the world's largest optical telescopes — is precious. Orion will be visible for about 8 hours tonight. But there is a daunting amount of work to do if Bally is to accomplish his goal of constructing what he calls a "data cube" of the Orion Nebula, a celestial spawning ground for thousands of stars.

The cube will be a three-dimensional map of sorts, charting how raw materials and byproducts of star formation move through the nebula. With this information, Bally can work backwards to reconstruct the recent history of events in Orion as stars have been born from collapsing clouds of gas and dust. The result, he hopes, will be a clearer understanding of the origin of solar systems.

"The ultimate reason why we are doing this work," Bally told me before we left for Hawaii, "is that we want to learn how we got here."

For carrying out ground-based observations to do just that, Mauna Kea is among the best sites in the world. The summit sits above 97% of the atmosphere's water vapor, placing it far above the marine inversion layer that holds obscuring moisture (and comforting warmth) close to the humid lowlands of Hawaii. As a result, when clouds appear, they're usually well below the summit, providing typically clear skies and a Saharan dearth of precipitation. (On average, only 6 inches of precipitation falls here each year, compared with nearly 25 inches in Chicago.) Given that the relative humidity at the top of the CFHT dome hovered at an astonishingly low 2% last night, my first night on the mountain, it's a miracle that any rain or snow falls at all.

Adding to the astronomical attractiveness of this site is the relative lack of turbulence in the air above the observatories. Turbulence comes from temperature differences and anything that disturbs the free movement of air through the atmosphere. To passing light waves, the swirling, rippling and jiggling air currents present differences in density that are the gaseous equivalent of tiny lenses. As the light waves streak through the atmosphere, the "lenses" refract, or bend, the waves slightly, this way and that.

For casual observers looking at the night sky, the result is the lovely twinkling of stars. For an astronomer, the result is anything but. When a star is observed through a telescope, its image will appear indistinct, as one minute some light waves enter the telescope along one path, and the next minute waves coming from the same spot on the star come in shifted ever so slightly.

Because the summit of Mauna Kea rises in the middle of the Pacific, higher in the atmosphere than anything else for thousands of miles around, air flow is steady here, and with minimal differences in temperature from place to place. So atmospheric turbulence is kept to a minimum.

The combination of clear, dry conditions and the steadiness of the air over Mauna Kea produces some of the best astronomical "seeing" in the world. This translates into telescopic images of the highest quality. No matter how large the telescope, there are few sites on Earth where the seeing is good enough to clearly image objects smaller than 1 arcsecond — a unit of measure on the sky that is 1800 times tinier than the diameter of the Moon as seen from Earth.* Atop Mauna Kea, average seeing is about 0.40 arcseconds, and sometimes as good as 0.25. (With a new technology called adaptive optics that compensates for atmospheric turbulence, some ground-based telescopes are doing even better than this.)

Of course, one way to do better is to do what NASA did: spend $2.1 billion to put a telescope in space. That telescope, of course, is the Hubble Space Telescope.

---

* Astronomers measure objects using degrees, arcminutes and arcseconds. The sky is 360° around. Each degree consists of 60 arcminutes. And each arcminute consists of 60 arcseconds. The Moon is 1800 arcseconds across. Very distant galaxies can measure just 1 arcsecond.

The space telescope's advantage, of course, is its perch high above the atmosphere, where there's no moisture, no pollution, no clouds and no turbulence to ruin things. But the Canada–France–Hawaii Telescope does have one advantage: it's primary mirror is considerably bigger: 3.6 meters across compared to 2.4 meters for Hubble. With such a large area for collecting photons, the CFHT can hold its own when imaging faint objects.

So if all goes well tonight, the seeing will live up to usual Mauna Kea expectations. And Bally and his two colleagues, Doug Johnstone of the University of Toronto, and Gilles Joncasse of Laval University in Quebec, will catch some revealing photons from the very heart of Orion.

By constructing their data cube, Bally and Johnstone are trying to work out the details of what John calls the "ecology" of the Orion Nebula. (Joncasse is working on another project.) As in biological systems, cycles rule interstellar ecosystems in places like the celestial spawning grounds of Orion. The raw materials of such ecosystems are predominantly hydrogen and helium, with a tiny amount of heavier elements, such as oxygen and carbon, and a sprinkling of interstellar dust. These raw materials are transformed into new stars, are blown back out into space when stars mature and die (with some exploding as supernovae), and then are recycled as a new generation of stars is born.

Orion happens to be a great place to study these processes because the nebula is "only" 1500 light-years away. While that's 8700 trillion miles distant, it's right in our galactic back yard on the cosmic scale of things—close enough to be explored in detail.

"The reason I'm interested in the Orion Nebula in particular is that it's one of the nearest rich nurseries of young stars," Bally says. "Our solar system was formed by precisely the kind of processes visible in Orion right now. We're literally seeing right in front of our eyes the birth of other planetary systems."

Some of the questions astrophysicists like Bally are trying to answer include these. Why do clouds of the raw materials of stars and planets begin to collapse and coalesce? Precisely how does this process unfold? And how does the material swirling in disks encircling young stars evolve to give birth to planets?

To close the ecological circle, Bally and others also are tracing the movement of gas and dust blown out by supernovae and other processes, and are trying to learn what makes this stuff fragment into new clouds of raw materials ready to form stars and planets all over again.

"Star formation is *the* fundamental process in the universe," Bally emphasizes. "It determines the fate of baryonic matter—in other words, ordinary neutrons, protons, electrons and atoms. So the key to understanding the evolution of the universe is understanding how star formation works."

But this research may yield even more than that. Bally believes that by exploring places like Orion, and their starforming regions, astrophysicists may learn whether the birth of solar systems similar to our own—ones with planets capable of sustaining life—is common or rare. And already, Orion is yielding clues that solar systems like our own, with planets conducive to complex life, including intelligent species, may be the exception, not the rule. The interstellar environment in places like Orion is hellaceous—Bally's word. It's filled with blistering radiation and harsh stellar winds. Based on evidence like this from his study of Orion, as well as research by others, Bally says the embryos of planets may simply get blown apart by these onslaughts.

Aside perhaps from the Big Dipper, the constellation of Orion, of which the Orion Nebula is a part, may be the most widely recognized of astronomical landmarks. Orion consists of more than 20 stars and nebulae arranged in the sparkling outline of a hunter with a pin-sized head (well, actually, a star-sized head), a drawn archer's bow, and a sword.

At Orion's waist are the most obvious objects in the constellation: three very bright, bluish-white stars that form his cinched belt. When looking south at Orion in the night sky of the Northern Hemisphere, Mintaka is the westernmost star (to the right) in the belt. Appropriately enough, its name comes from the Arabic word for *belt*. The center star is Alnilam, which means *a belt of pearls*. And the easternmost star (to the left) is Alnitak, which means *girdle*. These three stars, which are about 1500 light-years away, are prominent because they are supergiants. With 20 times the mass of the Sun, each shines with 20 000 to 40 000 times the brightness of the Sun.[4] They formed about 10 million years ago, but they won't last much longer.

"All of them will go kaboom as supernovae in about 1 million years," Bally says.

Meissa, Orion's pin-head, is much like the belt stars in that it's a bluish-white super-giant thousands of times brighter than the Sun. Meissa, which derives from the Arabic word for *white spot* or *shining*, is also 1500 light-years away and was born about 10 million years ago. Saiph, the star at Orion's right knee (on the *left* side of the body when you look up at Orion), has a similar profile. The name comes from the Arabic word for *sword* or *hilt*.

It's probably no coincidence that all these stars are about the same distance from us and reside in the same general region. They were born, astronomers believe, from the same parent cloud of gas and dust.

At Orion's right shoulder is a different sort of star, the distinctly orange-red Betelgeuse (in the United States, pronounced "beetlejuice"). The amusing name is derived from the Arabic *bayt al jauza*, meaning *house of the twins*. (There may be some confusion as to the derivation of the name. One source says it is a corruption of an Arabic phrase meaning *armpit of the giant*,[5] which would be appropriate in the sense that this is just about where the star is located in Orion). Betelgeuse is a red super-giant that's 20 times more massive and shines 10 000 times more brightly than the Sun. With a Jabba-the-Hut-like girth of 700 million miles, this star would bulge beyond the orbit of Mars if it were the center of our solar system, and we'd most certainly be cosmic toast. Luckily, Betelgeuse is 600 light-years away — significantly closer than the others, but comfortably far from us.

Finally, we come to Rigel, which marks Orion's left knee. In the night sky it appears nearly as bright as Betelgeuse, and so it looks to be the same distance away. But in reality, it is at least 300 light-years more distant from us. The reason Rigel seems almost as bright is that it radiates as much as six times as much light.

To the naked eye, Orion looks like a flat object. But as the difference in distance to Betelgeuse and the other stars in the constellation reveals, the hunter's seeming two-dimensional outline is deceptive. Orion is three-dimensional, reaching hundreds of light-years *into* space as well as across it. (The Big Dipper is also three-dimensional.)

And that's the thing about Orion. There's much more than meets the eye — more, even, than mere stars. In fact, the most important member of the constellation is invisible: an elongated cloud of gas and dust stretching from near Betelgeuse at Orion's right shoulder to Rigel at his left knee, to which most of the constellation's stars can trace their ancestry.[6] The cloud is so big that even though it is 1500 light-years away, it takes up more of the sky than you can cover with your outstretched hand. In 10 million years, cores of concentrated material within this cloud have spawned thousands of stars.

Within the disk of the Milky Way are billions of stars and a patchy distribution of gas and dust from which they all formed.[7] The gas and dust comprise what's known as the interstellar medium, the space between stars. In some places, the ISM is extremely tenuous. Within a cubic centimeter, you'd be lucky to find 10 particles of matter (atoms, electrons, etc.), which is thousands of times emptier than the best vacuum attainable by machines on Earth. But in some regions, the density is more than 10 times that amount, giving rise to a discrete cloud. At these densities, atoms can join together chemically to form molecules, which is why such structures are called giant *molecular* clouds.

These clouds can extend across 300 light-years of space or more, making them the largest known objects in the Galaxy. With enough gas and dust to build something like a million solar systems, they are the spawning grounds for stars and planets. Several thousand are scattered across the spiral arms of the Milky Way. Based on estimates of the efficiency of star formation within giant molecular clouds, astronomers estimate that roughly two or three solar mass stars are born in the galaxy each year.[8]

Giant molecular clouds are frigid — as cold as 5° above absolute zero, making these objects the coldest things in the cosmos. They contain dust — about 1% by mass — that plays an important role in the origin of planets, and more than 120 varieties of molecules. These include a rich stew of organic chemicals that may seed forming planets with the materials needed for the origin of life. (More about this in Chapter 12.) But the most common constituent of giant molecular clouds by far is molecular hydrogen gas, or $H_2$: more than 99% of the molecules in a GMC are hydrogen.[9]

The Sun is also made mostly of hydrogen (although not in molecular form, because it's too hot, and not quite as much hydrogen as a molecular cloud).[10] This is no coincidence, since giant molecular clouds are what give birth to stars like the Sun and the ones we see glittering in Orion. So why, then, is the Sun hot and these clouds so cold? The answer is simple. The matter in the center of the Sun, consisting mostly of electrons and protons, has been compressed so much that the temperature is immense: some 15 million K. That's hot enough for protons to overcome their mutual electrostatic repulsion and to join in the process of thermonuclear fusion, which releases great amounts of heat. Without fusion, the Sun would have run out of energy in just 50 million years. Because of it, the Sun has been shining for 4.5 billion years.[11]

For any star to reach temperatures hot enough for fusion fires to ignite, something has to trigger a gravitational collapse of small, faint wisps of material within a giant molecular into a super-dense ball of firey gas. Candidates include a density wave passing through the galaxy, a shock wave from a supernova, or a wall of dense material blown out by particularly hot and active stars. Any of these would squeeze the cloud from the outside, causing clumps of gas deep inside to contract. Contraction raises the density of gas within these clumps. And with more mass packed together into a tighter space, gravity gets stronger — strong enough, perhaps, to begin drawing gas together quickly, triggering a collapse.

It's also possible that clumps of gas within a giant molecular cloud may be poised at the threshold of gravitational collapse all on their own. In this case, all it may take for a collapse to begin is for a small amount of gas to seep into the center; theorists believe this happens naturally along magnetic field lines. According to this idea, pioneered by University of California theorist Frank Shu, collapse happens from the inside out.

Whatever causes the collapse, once it gets going gravity ensures that it keeps going, pulling material together into denser and denser cores embedded within the extended giant molecular cloud. And it is within these cores that stars are born, most typically in clusters.

Astronomers wanting to study the overall structure of giant molecular clouds and how stars form within them are faced with a substantial problem. The cold hydrogen and helium that make up the bulk of clouds are invisible. Astronomers try to get around this in a number of ways. One is by using radio telescopes, which can detect emissions of light from trace gases within the cloud, such as carbon monoxide. By seeing what these gases are doing, astronomers try to get a sense of what the overall cloud is doing. But analyzing this kind of data can be difficult, and interpreting it unequivocally isn't always possible.[12]

Dust also poses a problem. Although it makes up only a small portion of molecular clouds, like black soot billowing from a fire, it is extremely dark and obscuring. Just how much so has been demonstrated by astronomers affiliated with the European Southern Observatory. They made use of the fact that the dust in clouds does allow some invisible infra-red light from background stars to pass through. Depending on the density of the dust, the light passing through is reddened by a specific amount. (The denser the dust, the redder the light has to be in order for it to get through.) Using a detector called SOFIA on the New Technology Telescope in Chile, they measured the reddening of infra-red light from stars behind a cloud called Barnard 68 located about 500 light-years away toward the southern constellation Ophiuchus, the "Serpentholder." From these data, they calculated the density of the dust, and what they found is astonishing. If a sheet of dark particles with the same light-blocking power were placed in front of the Sun, there would be eternal darkness here on Earth. The Sun would be 15 times too faint to be observable by the naked eye![13]

The ability of infra-red light to penetrate dense veils of dust also has allowed Hubble Space Telescope astronomers to see what's going on inside a molecular cloud in the constellation of Orion. Using the NICMOS infra-red camera on HST, they imaged beautiful details of the cloud's interior, including tendrils and arcs of dust glowing in the infra-red—evidence of massive outflows of gas from young stars—as well as a number of forming stars that were previously hidden by the dust.

By lifting the veil that previously hid star-forming activity within molecular clouds, infra-red astronomy has proved very valuable to astronomers. But one significant limitation of this approach is that only a small portion of a cloud can be probed. In addition to close up views of small regions, astronomers would like to get an overview of the internal structure of an entire cloud. And for this, they use radio telescopes designed to see light emitted by dust in the *submillimeter* part of the spectrum, which has a longer wavelength than infra-red light.

In 1998, Bally and Doug Johnstone trained the James Clerk Maxwell Telescope atop Mauna Kea on Orion's giant molecular cloud. The telescope consists of a dish 15 meters in diameter that gathers submillimeter photons, making it the world's largest instrument designed to operate in that part of the spectrum. (Attached to the front of the dish is the world's largest piece of Gore-Tex, for protection from the elements.)

To take pictures of the Orion cloud, Bally and Johnstone used a camera, called the Submillimeter Common User Bolometer Array, or SCUBA, with detectors cooled to a tenth of a degree above absolute zero ($-273°$ C). This minimizes background noise and allows the instrument to measure tiny emissions from freezing cold dust particles. In SCUBA images, the elongated Orion cloud looks a bit like a neuron, with a cell-like bulge roughly in the center and two axon-like appendages, one running north, the other south. (Astrophysicists describe it somewhat differently, preferring to see the integral symbol of calculus, $\int$, in the outline of the cloud.) Along the length of the cloud are filaments and fingers, as well as loops and bulges. And along its central axis, yellow clumps mark spots where the dust is densest. These are stellar pregnancies, wombs within the cloud that have collapsed and begun to spawn new stars. One of the brightest such clumps lies in the central, bulging area. This is the site of hyperactive star formation in the Orion Nebula.[14]

To the naked eye, the overall Orion Molecular Cloud is mostly invisible. But young clusters of stars it has spawned in the Orion Nebula within the cloud have heated some of the gas to high enough temperatures so that it glows—and you don't need an instrument like SCUBA to see it. Through binoculars, glowing portions of the Orion Nebula can be seen quite clearly as a fuzzy patch of light in the middle of Orion's sword. (*Nebula* actually derives from the Latin word for mist or cloud.) The glow comes from a bubble of mostly hot, ionized gas that astronomers call an *HII* region.

218

The discovery of the nebula is attributed to Nicholas-Claude Fabri de Peiresc, a French lawyer and scholar who trained his telescope on the glowing patch in 1610. Peiresc was the quintessential Renaissance man — an expert in optics, astronomy, biology and law, and an acquaintance of Galileo, as well as the Flemish painter Peter Paul Rubens. In the late eighteenth century, famed astronomer William Herschel studied the nebula with a telescope he built himself. (Herschel was born in Germany, but moved to England as a young man.) He anticipated the modern view of Orion as a stellar nursery in 1789, describing the glowing gas as ''an unformed fiery mist, the chaotic material of future suns.'' Almost a century later, the Orion Nebula became the first such object ever to be photographed successfully. The photographer was Henry Draper, and he took Orion's portrait in 1880.[15]

From Peiresc to Herschel and then to Draper, the geography of Orion's nebulosity proved an irresistible target for telescopic exploration. Today, the lay of the land is known well, even to backyard astronomers with modest scopes.

The most conspicuous landmarks in the nebula are four large, hot and bright stars that make up the Trapezium. (With imaginary lines drawn between the stars, the resulting shape is a *trapezium*: a figure with four sides, no two of which are parallel.) Like the other super-giants in the Orion constellation (Betelgeuse excepted), the Trapezium members are classified by astronomers as 'O' stars and were born 10 million years ago — the progeny of collapsed clumps of dust and gas within the Orion giant molecular cloud. They are no longer cloaked by surrounding dust because they've literally burned and blown it all away.

The Trapezium's O-stars are not alone, of course. All together, 700 stars reside in the same neighborhood. Together, the big stars and their smaller neighbors comprise what astronomers call the *Trapezium Cluster*. Neighborhoods such as this are the Calcuttas of the cosmos, with as many as 1500 stars packed into each cubic light-year — almost 500 000 times the average star density in the vicinity of our own solar system.[16]

Just like their cousins elsewhere in the constellation, the Trapezium's O-stars are many times more massive than our own sun, and each one blazes with tens of thousands of times the intensity. Some of the massive stars have set temperature records. Using the orbiting Chandra X-ray Observatory, scientists

219

have calculated that these stars are as hot as 108 million °F, breaking the previous record for hottest massive star known by more than 60 million degrees.[17]

As newborns, O stars like the Trapezium members are particularly obstreperous. Working with data from Chandra, Bally and colleagues have shown that within 10 000 years of their birth, O stars can emit X-ray flares that are a *billion* times as powerful as those produced by our Sun.[18]

In short, O stars are the Jim Morrisons of the cosmos: they burn bright, live fast, and die young. The most intense ones blow up as supernovae in as little as 1 million years. (Thank goodness there aren't a lot of them around. Out of 4 billion stars in the entire Milky Way, there are an estimated 20 000 O stars.[19])

The Trapezium's four O stars are the most conspicuous objects in the Orion Nebula. But they are not the only O stars in the region. All told, nine such objects now reside in the nebula, some of which are twins, or binaries. They are joined by objects just a little less hot and massive known as B stars, forming what astronomers call an OB association. As the bad boys of the nebula, the O and B stars have disturbed everything else in the neighborhood.

"Put such a star in a sea of hydrogen, and it's like a giant ultraviolet bulb," Bally says. The star produces high-energy UV photons. These slam into the hydrogen atoms, knocking electrons from their orbitals and thereby ionizing the gas. The freed electrons eventually collide with others, lose energy and then return to their former orbits, emitting the light we see as the lovely nebulosity coming from places like Orion. Regions of ionized gas like this are *HII* regions. Clouds of mostly non-ionized gas, like those comprising giant molecular clouds, are called *HI* regions.

Although some 90% of the ionized gas is hydrogen, smaller amounts of ionized helium, oxygen, nitrogen and other elements are present as well. Emissions from these gases contribute many of the lovely colors visible in telescopic images of the Orion nebula.

In the process of ionizing the gases, the UV radiation heats them as well. The result is a *bubble* of ionized gas inflating within the nebula. "As the bubble expands," Bally explains, "it rams into its surroundings, sweeping up material into a dense shell that marks the outer envelope of the bubble." In the bubble's interior, dust is swept clean, causing a region around the stars to be evacuated of obscuring material. Meanwhile, the expanding bubble wall causes compression of gas and dust within the surrounding molecular cloud, triggering gravitational collapse of knots of material and thereby giving birth to new stars.

This simple picture of an inflating bubble is fine as far as it goes. But as observations of nebulae like Orion have revealed more and more details of the action within, scientists have realized nature is messier than the simple picture suggests. For one thing, the gas and dust into which the bubble expands is not uniform, so a bubble can be quite irregular. For another, "people realized that stars are more than mere blobs of light," Bally says. Massive stars also produce fierce winds of atomic particles. Bally calls them "stellar hurricanes," and scientists recognized that their energy should help drive the inflation of a bubble and excavate a clear region inside.

In fact, observations have shown that one O star in Orion produces winds 10 million times stronger, in terms of mass and speed, than the wind produced by our own Sun, according to Bally. The amount of material thrown off by such a star is immense. In its short, 30-million-year lifetime, it will lose 10 to 20% of its monstrously bloated mass—much more than the entire mass of the Sun—to the wind. (If only dieting were this easy.)

One additional complication to the simple picture is this: as a bubble gets bigger and bigger, it might manage to reach the very edge of the giant molecular cloud itself, where the profound vacuum of empty interstellar space begins. To predict what might happen in such an event, scientists came up with the "Champagne flow" theory. In this picture, when the bubble's shell reaches the edge of the cloud, it breaks and vents its contents out into the vacuum at supersonic speeds.

What's left from all this violent action is a kind of cavity exca-vated in the side of the molecular cloud. Using images from Hubble and ground-based telescopes, C. Robert O'Dell and Zheng Wen of Rice University have created three-dimensional renderings of the cavity excavated by the Trapezium stars. The bubble these stars blew has broken through the edge of the molecular cloud, just as the Champagne model predicts. Now that this part of the bubble wall is gone, we can get a clear view of what's inside the cavity. In O'Dell and Zheng's renderings, the wrinkled, glowing inner surface of the cavity forms something like an oblong valley with the Trapezium stars roughly in the center.

In the constellation Cassiopeia, 7100 light-years from Earth, Hubble images have revealed what a more intact bubble looks like. Appropriately named the "Bubble Nebula," the structure is remarkably spherical. A blue, diaphanous veil of glowing material makes up the shell. Inside, a massive star sits off-center, its stellar winds clocking in at 4 million miles per hour. The glowing bubble wall marks the gust front where the wind slows as it plows into the denser material of the cloud.

Here in the Bubble Nebula, at least, nature is not so messy.[*]

During the past 10 million years, the Orion molecular cloud has spawned at least several dozen O stars, each of which has blown its own bubble. Many of these stars have already exploded as supernovae, according to Bally. "The key idea here is that stars are never born in isolation," he says. "They are always born from these molecular clouds, which are localized. And that means that when the stars die, their supernovae are very clustered. They occur in one region of space, like the Orion cloud. And Orion supernovae—all 30, 40, 50 or 100 of them—all of them cluster in a region that is no more than 100 light-years across."

In other words, many massive stars have blown bubbles and exploded within the Orion molecular cloud, even before the Trapezium came on the scene. The net effect of all this celestial hyperactivity is the mother of all bubbles—a *superbubble* encompassing all of the smaller ones.

Imagine the first massive stars lighting up in the giant molecular cloud. They clear out dust and blow local bubbles of ionized hydrogen gas. Then one dies in a supernova, creating an even bigger, hotter bubble. "The next supernova goes off more or less in the debris left behind in that bubble," Bally says. Then

---

[*] A Hubble image of the Bubble Nebula was the starting point for the photo illustration that graces the cover of this book.

more pop off, and eventually dozens of supernovae explode, creating bubbles within bubbles within bubbles.

"The collective effect is something akin to a stellar wind that blows from the center of the first association of O stars," Bally says.[20]

Just one supernova remnant—the shell of debris it throws out—may be up to 100 light-years across, Bally notes. "If you put dozens or even hundreds of these things into one region of space, as happens in a molecular cloud like Orion, the collective bubble they blow can be a *thousand* light-years across."

In Orion, the first supernovae probably went off about 12 million years ago, just below Bellatrix (the star at Orion's left shoulder). According to Bally, that explosion blew a small bubble that reached the belt stars about 6 million years ago, causing compression that triggered a burst of new star formation there. O stars then went supernova, pushing out a new bubble. About 4 million years ago, its shell reached the sword, compressing the Orion Molecular Cloud, where the Orion Nebula is located. More stars blinked on, and more supernovae popped off. Today, the superbubble from these and succeeding supernovae stretches across $40°$ of the sky. On a winter night, with Orion roughly overhead, the superbubble would extend half way to the horizon. "It's big. *Darn* big," Bally says.

If Orion were the only star-forming environment in the universe where such tempestuous events occurred, it would be a mere curiosity. But it is not the only place. Massive stars blowing bubbles in their parent molecular clouds, and superbubbles resulting from generations of supernovae explosions, have been observed elsewhere in the Milky Way—and beyond our own galaxy as well. And some of these are much bigger than the Orion superbubble.

In the Tarantula Nebula 150 000 light-years from us in the Large Magellanic Cloud (actually a sister galaxy of the Milky Way), a superbubble has grown to 3000 light-years in diameter, according to Bally. "If the nebula were at the same distance from us as Orion is, it would fill half the sky and would be as bright as the Moon," he says. The reason? Whereas Orion has a few dozen high-mass stars, the Tarantula Nebula has several *hundred* of them.

"The message here is that interstellar space and star forma-tion is incredibly violent," Bally says. "There are lots of bangs and booms."

On the plane trip across the Pacific heading toward the Big Island, I had tried to imagine what it would be like to use a telescope as large as the CFHT. The biggest telescope I had ever looked through was a 10-inch reflector set up by an amateur astronomer on a street corner in New Orleans. The guy had erected several scopes right next to Café Du Monde along the Mississippi, and he was offering passers-by free looks at Jupiter. So I stepped up to the eyepiece to look at one of the wonders of the cosmos. In my naïveté (before writing this book, mind you), I had once imagined astronomers doing the same thing, just with bigger telescopes.

In fact, it wasn't long ago that astronomers using telescopes like the CFHT had to be hoisted up in a crane to the focus point. Here, often in the freezing cold of an open observatory dome high on a mountain, they'd spend the night crammed into a small, dark capsule, taking pictures of distant nebulae and galaxies using a traditional camera to expose photographic plates. Today, the CFHT's old observing capsule lies near the base of the telescope, an antique. Now, astronomers do their work in a comfy control room, watching flickering images on computer monitors.[21] While at work, they listen to their favorite CDs on a stereo system. And at the CFHT, they take occasional breaks in room 403, otherwise known as Café du Mont Blanc, a lounge complete with an oxygen bottle and a goodie-stuffed refrigerator for those suffering from hypoxia-induced light-headedness or the munchies (or both).

Most important, in place of the traditional cameras, astrono-mers at the CFHT and other major observatories use charged-coupled devices, or CCDs. This is a huge improvement over photographic plates. For every 100 photons brought by a telescope into a traditional camera, the photographic plate will capture only five, according to Doug Johnstone. The human eye is even less efficient. It must collect 200 photons before a person can even begin to sense any light. (Which is one reason why astronomers actually spent most of their time imaging astronom-ical objects on photographic plates, in the days before CCDs, rather than peering through eyepieces.) With a CCD, however, for every 100 photons that fall, 95 actually are captured. This sensitivity combined with time exposures, which can last for hours, explains why modern astronomical images can be so lavishly bright, saturated and vivid.

Tonight, however, astronomical imaging is not the main goal for Bally, Johnstone and Joncasse. To construct the data cube of what's moving where and how fast within the Orion Nebula, the scientists need to take *spectra*. By measuring the spectra of light coming from features in Orion, they can tell which ones are blueshifted, meaning moving toward us, and which ones are redshifted, meaning moving away. Not only that, the spectra can reveal how much material is moving and with what velocity.

To accomplish this, Bally and his colleagues have attached a device called a Fabry–Perot spectrometer to the telescope. It is designed so that the scientists can look at features by the light they emit in very specific wavelengths. For example, Bally and his colleagues will be using the device to image light emitted by ionized hydrogen in Orion. Known as hydrogen-alpha emission, this radiation is common (because hydrogen is the most abundant element in the universe) and uncommonly bright. The hydrogen itself is not the point. Its ubiquity and brightness make it a convenient *tracer* of the movement of gas overall. Similarly, the scientists will use the Fabry–Perot to image emissions of sulfur and oxygen, also convenient tracers.

Once the telescope is aligned precisely with the portion of Orion Bally and his colleagues want to study tonight, photons will be gathered by the CFHT's giant mirror and directed into the spectrometer. Within the Fabry–Perot are two partially reflective plates. Some of the light goes right through the plates while some bounces back and forth between them. By changing the distance between the two plates, the scientists can dictate which specific wavelengths of light will build up and which ones will not. Thus, they can tune the spectrometer to ''look'' for the specific wavelength emitted by, say, ionized hydrogen.

The Fabry–Perot is an exquisitely sensitive instrument. It can be tuned to capture minute variations in a particular element's wavelength. Since such variations typically are caused by Doppler redshifting or blueshifting, the sensitivity of the device enables scientists to detect movement of light-emitting matter. For example, ionized hydrogen is known to be a component of fast-moving flows of gas throughout the nebula. With the Fabry–Perot set up to detect the hydrogen alpha light emissions, the device can also detect the subtle shifts in the element's wavelength caused by its movement within the overall flow. With these data, scientists can determine the density, mass and velocity of the material in the flow.

But first, light built up between the plates must be transmitted out of the back of the instrument to the CCD. When photons of the light land on the device's light-sensitive surface, electrical signals are generated in the CCD and then processed by a computer to construct an image on a monitor.

A typical Fabry–Perot image consists of a series of concentric circles, like the bull's-eye pattern of a dart board. Objects within the telescope's view will appear on the dart board if they are emitting light in the specific wavelength the scientists are interested in. And if an object's light in that wavelength is slightly redshifted, meaning it's moving away from us, it will appear on the outside of a ring (toward the periphery of the dart board). If it's blueshifted, it will appear on the inside (toward the center).

By slightly changing the distance between the plates in the spectrometer in succession, Bally and Johnstone will move through what they call the ''velocity space'' of the nebula, successively charting material that's moving at different speeds and directions. By building up a series of such observations in different wavelengths (hydrogen, sulfur, oxygen, etc.), they hope to accomplish their goal of mapping the movement of material within the nebula.

After watching sunset on Mauna Kea's summit, Bally and I rode the observatory's elevator to the fourth floor and got settled in the control room. Gilles Joncasse, who is working on another project with the Fabry–Perot for the first few hours of the evening, will get a crack at the telescope first. So we while away the time, waiting for Gilles to finish and for Orion to rise higher in the sky.

Over snacks in the Café Mont Blanc, I ask Bally why he became an astronomer. Like many astronomers I've interviewed, he can trace his interest to a vivid childhood memory:

''When I was about 9 years old,'' he says, ''I saw a picture of the Pleiades. Not long afterward, I looked up in the sky, and I saw it there. I was able to identify it! That had a big impact—it stayed with me."

By 14, Bally had built his own telescope. He didn't just put the pieces together out of a kit. He literally ground the glass for his own mirror. This dual interest in both astronomical science and technology influenced his choices as his education and, later, his career unfolded.

In graduate school, he started out in particle physics—a field known not just for the complex science but for its huge machines. After awhile, though, Bally says he felt "like a small cog in a very big machine." He soon found his way back to astronomy.

After finishing his Ph.D., Bally was hired in 1980 at Bell Labs to work under Arno Penzias, the scientist who along with Robert Wilson discovered the cosmic microwave background radiation—the echo of the big bang. Bally notes with wonder just what an amazing place Bell Labs was for a research scientist. The budget was huge—$400 million a year. And even though this was essentially *the* phone company (Bell Labs was owned by ATT when it still held a monopoly in telephone service), there were few strings attached.

"It was an incredible environment for basic research," Bally says. "There were no limits on funding." ATT's philosophy "was to get 100 Ph.D.s together and enable them to do whatever research they desired. It was recognized that every five years or so, someone would invent the laser or something similar."

Or discover the echo from the big bang.

While at Bell Labs, Bally did basic research, gathering spectra of astronomical objects using instruments tuned to the sub-millimeter range of the spectrum. Because this was the phone company, he of course did applied research as well. But for the most part, he and his colleagues were free to pursue whatever interested them.

But then the ATT monopoly was broken up as part of the historic settlement that created the so-called "baby Bells" (the local phone companies). "Suddenly, the bottom line became much more important," Bally says. "It was clear that the remarkable environment at Bell Labs wouldn't last forever."

He lasted there until 1991, when Bally's wife, Kim, got a job working for the phone company in the Denver area. By coincidence, a position was open for an astronomer at the University of Colorado, and Bally was the right guy for the job.

He estimates that between 1980 and 1991, when he left Bell Labs, he collected something like 1 million spectra of astronomical objects. Thanks to improved technology, in the three nights total that Bally will spend observing Orion with the CFHT and the Fabry–Perot spectrometer, he expects to collect *10 million spectra*.

"That's kind of sobering," he says. "What was I doing with my life before?"

If things keep going the way they are tonight, however, Bally may not reach his goal. Some high thin clouds have slid over the Big Island, much to Joncasse's dismay. (No observing site is perfect.) Ken, the CFHT telescope operator, or T.O., calls a meteorologist, who says a weak atmospheric trough is moving through. Things should improve, but he's not making any promises.

By 10:45, it begins to look like the meteorologist called it right. The clouds are dissipating a bit, and Orion has risen into position. It's time for Bally and Johnstone to get to work.

No one other than the T.O. gets to move or otherwise mess with the multi-million-dollar telescope. So at Bally's behest, Ken taps commands into his keyboard, slewing the motorized telescope toward the nebula. When he gets it precisely aligned, Ken sends a command through the computer, and the telescope begins to track Orion automatically. The observatory dome moves with it, keeping the telescope window in the correct position.

Within a few minutes, the first image from the Fabry–Perot appears on Bally's computer monitor. I must admit that I came to Hawaii with romantic notions of what this moment would be like. I pictured lavishly detailed color images materializing onscreen—images that would look much like the beautiful cosmic landscapes painted by the HST. But this picture is nothing of the sort. There is some hint of billowing cloud filling the spaces between the dartboard rings of the spectrometer. But it's faint and rendered in shades of gray.

Projecting inside the innermost ring is a feature of interest to the two astronomers: a short dark line pointing slightly away from the center. Bally reminds me that we're looking at hydrogen alpha emission. So it appears that this tiny, nondescript feature is a narrow flow of gas containing ionized hydrogen. It's blue-shifted slightly, so the gas is moving roughly toward us. At one end of the line is a slight bulge. This, Bally says, is a young star still in its formative years. And so the nature of the enigmatic line becomes clear: it is an image of a highly collumated *jet* of gas spewing from the star. Such jets, the focus of tonight's explorations, are believed to play a crucial role in the origin of stars and planets. Without them, in fact, we probably wouldn't be here.

Evidence of their presence first came in the 1950s, when George Herbig and Guillermo Haro discovered spectrographic signs in star-forming regions of fast-moving gas that changed velocity over time. But the origin of these gaseous features, which came to be known Herbig–Haro objects, was a mystery. Later, astronomers came to believe that these features were shock waves of some sort. Such waves are visible because atoms can't move out of their way fast enough. As a result, atoms pile up, temperatures rise, and gas ionizes, causing it to glow.

In 1983, Bo Reipurth, now at the University of Hawaii, discovered the source of the shocks: giant flows of hot plasma spewing from forming stars, or what astronomers call young stellar objects. Where these jets were plowing into surrounding gas, Reipurth found the tell-tale glow of Herbig–Haro objects. In the years since, these features of star-forming regions have become common sights through large telescopes. The shocks appear in images from HST and other telescopes as glowing knots and arcs of ionized gas. The jets are a bit more difficult to detect. The flowing gas within a jet does not emit much of its own light. It may give itself away when it stimulates surrounding material to emit radiation. And sometimes a jet is lit up by ultra-violet light from surrounding stars. Such irradiated jets reveal themselves as tendril-like trails, which are classified as Herbig–Haro objects as well.[22] A number of these features are the object of Bally and Johnstone's attentions tonight. They may provide some insight into whether embryonic planets can survive and grow in the nasty environment of Orion.

Stellar outflows are unusual in that they seem to spew from nozzles at the north and south poles of a protostar. Scientists now believe these jets are actually part of bigger outflows from across the face of a protostar. According to current theory, magnetic fields gather material from these outflows and funnel it into jets. (More about this in the next chapter.) In this way, jets can spew trillions of tons of gas into space at up to half a million miles per hour. And they can stretch across many light-years of space, punching right out of a molecular cloud and into the surrounding interstellar medium.[23]

Observations with Hubble and other telescopes reveal that the outflows are not so much like water spewing continuously from the nozzle of a fire hose, as rounds shot out, one after another, from a machine gun. The history of these ratattat outflows is recorded in chains of Herbig–Haro knots and bow shocks, with each object marking a burst from the stellar gun.

But what triggers the outflows to begin with? Astronomers believe they actually are a star's way of transforming itself from a cold wispy cloud to a dense ball of burning gas without self-destructing in the process.

To help me understand how, Bally first pulls up images taken by Hubble of tiny nuggets of dark material in Orion — tear-shaped blobs floating against a drapery of glowing gas. The pointed tails of these objects point uniformly away from the Trapezium stars; their rounded heads, which glow brightly, point toward them. Like a bullet train, these objects are *aerodynamic*. They've been sculpted by the wind emanating from the O stars. And the glow is ignited by ultraviolet radiation striking the objects head-on.

These objects have been dubbed *proplyds*. Consisting of hydrogen gas and dust, they are the amniotic sacks from which forming stars draw their sustaining supplies of gas and dust. We can see them in the Orion Nebula because the O stars have burnt and blown away the molecular cloud's surrounding dust; the proplyds originated as cores of denser material within the larger giant molecular cloud. They hold together because of their gravity. But not for long. Even as the star and perhaps planets are forming within, the cocoon's edges are being eroded away — *photoevaporated*, as astronomers put it — by radiation from the Trapezium stars.

In one particularly stunning image of a proplyd, captured by the HST, it's possible to see signs of the birthing process within. From head to tail, the tear-shaped object is about 1500 times the distance between the Sun and Earth, or 1500 astronomical units. Embedded near the center of this womb and sitting perpendicular to its axis is a bluish Frisbee-like disk of dark material seen edge-on. The disk's diameter measures about 250 astronomical units—solar system sized. (An astronomical unit, or AU, is equal to the distance between Earth and the Sun.) The material in the disk, Bally says, is dust and gas originally from the parent molecular cloud that's now rotating and condensing to form a protostar at the center. (The star is not yet visible because it is still shrouded in obscuring dust.)

In other HST images Bally shows me, circumstellar disks are seen not edge-on but more or less looking directly down on them. In these pictures, a hot spot at the center betrays the existence of a protostar hidden within. And in still other HST images, tendril-like trails point in opposite directions from a central point within a disk. These mark the paths of jetting gas. At the head of each jet is an arc-shaped shroud of a material: a bow shock. And in many cases, a series of shocks is aligned along the axis of the jets.

"In star-forming regions like Orion, jets crisscross all over the place," Bally says. "It's clear that they're very important."

According to the current theory of star formation, gas and dust in a circumstellar disk steadily spirals inward and accretes to the spinning protostar. The physics of the situation demands that the protostar find a way to rid itself of the angular momentum it gains when it gobbles cosmic manna from the disk.[24] Otherwise, it would fly apart. According to the most widely accepted theory, the star does this through the bipolar jets. Even as gas falls onto the star from the circumstellar disk, the star sheds just enough material to keep itself from committing suicide. Meanwhile, stuff in the disk that's not swallowed by the star or flung out in jets may coagulate into lumps, which grow in size and eventually form planets.

In very broad strokes, that is the modern picture of how stars, and possibly entire solar systems, are born. Of course, there are many details left out, such as how magnetic fields funnel material into jets, and what role they play in the formation of planets. But in its broad outlines, this theory has been spectacularly confirmed by images of proplyds, disks and jets.

The idea that a solar system forms through the construction services of a circumstellar disk is not new. Remarkably, it was first proposed in 1755 by the German philosopher Immanuel Kant, not long after Benjamin Franklin did his famous kite experiments with electricity. Kant anticipated the revelations of modern astronomy by more than 200 years, devising a theory of star and planet formation that still stands today.

Rather than invoking direct intervention by God, Kant devised a theory relying simply on Newton's laws of motion to make a solar system from raw materials. As Kant put it, he was proposing a way for the solar system "to evolve itself out of the crude state of chaos." It was, to say the least, a revolutionary approach. And Kant saw this well, as the preface to his work, *Universal Natural History and Theory of the Heavens*, clearly shows:

> *"To discover the system which binds together the great members of the creation in the whole extent of infinitude, and to derive the formation of the heavenly bodies themselves, and the origin of their movements, from the primitive state of nature by mechanical laws, seems to go far beyond the power of human reason."*[25]

Not surprisingly, Kant felt intimidated by the task. "I have ventured, on the basis of a slight conjecture," he wrote, "to undertake a dangerous expedition." The danger lay in his challenge to the dominant Christian view that God created the solar system — a view championed by no less a giant of science than Isaac Newton. Writing in his monumental work *Principia*, Newton insisted that natural causes alone were woefully insufficient for explaining the wonders of the solar system:

> *"This most beautiful system of the sun, planets, and comets, could only proceed from the counsel and dominion of an intelligent and powerful being."*[26]

Contrary to what many students are taught about the scientific method, pure observation was not the starting point for Kant's investigations. No astronomical imagery pointed the way specifically toward the theory he devised. It came simply from his understanding of the laws of nature, and probably a good measure of intuition.

In the beginning, according to Kant's theory, the universe had no stars or planets. "I assume," he wrote, "that all the material of which the globes belonging to our solar system — all the planets and comets—consist, at the beginning of all things was decomposed into its primary elements, and filled the whole of space of the universe in which the bodies formed out of it now revolve... At that time nothing had yet been formed."[27]

And then, Kant proposed, gravity drew cosmic matter together into a swirling, flattened disk. At the center, where the density was greatest, the star was born. In the disk around it, planets arose.

"...we see a region of space extending from the center of the sun to unknown distances," he wrote, "contained between two planes not far distant from each other." Within this disk "the attraction of the elementary matters for each other" draws particles together, forming the Sun and the planets in orbit around it.[28] So where Newton saw the hand of God at work in making the planets all orbit in the same direction, Kant presciently saw a primordial cloud rotating and conferring that motion on the bodies that condensed from it.

Kant's idea was taken up later in the century by the French astronomer Pierre-Simon Laplace. But with no empirical evidence to give credence to their ideas, it faded into obscurity.[29]

Then, in the 1970s, astronomers found that very young stars, called T Tauri stars, emitted more ultraviolet and infra-red radiation than had been expected. Donald Lynden-Bell and Jim Pringle of Cambridge University suggested this excess could be explained if the stars had circumstellar disks. In the region between the inner edge of the disk and the star, accreting material would heat up dramatically and radiate ultraviolet light. Meanwhile, in the outer part of the disk, matter spiraling slowly inward would radiate in the infra-red.[30]

In 1983, astronomers using NASA's Infra-red Astronomical Satellite, or IRAS, discovered indirect evidence of a dust disk surrounding the star Beta Pictoris, about 63 light-years away. The next year, astronomers Brad Smith and Rich Terrile imaged the faint disk in optical wavelengths. Then in the 1990s, early pictures taken by the Hubble Space Telescope clearly revealed that the matter surrounding many young stars was confined to disks.[31] In fact, out of more than a hundred stars surveyed in Orion, more than half have turned out to harbor disks.[32]

233

"The Kant–Laplace hypothesis forms the basis of all modern work on planet formation," comments Alan Boss, an astrophysicist at the Carnegie Institution of Washington. "It has been spectacularly proven likely through astronomical observations of the existence of suspected protoplanetary disks. We need only to detect the presence of protoplanets in these disks before finishing the proof."[33]

Bally echoes Boss's sentiments:

"What amazes me is that 200 years ago, Kant and Laplace described the formation of the solar system from a disk, and they more or less got it right. If they were alive today, they would be astounded."

So, as undistinguished as the little jet on the Fabry–Perot image looks, it is incredibly significant. With the discovery and study of such features, scientists like Bally are filling in the details of an overarching theory first proposed in Benjamin Franklin's day.

As we examine the jet on the monitor, Bally estimates that the plasma flowing off the star is moving out into the nebula at 150 kilometers per second. He refers to another image, made two years ago using a telescope at Kitt Peak, to see if he can find it there. That observing run was a reconnaissance mission designed to identify targets for closer scrutiny with the Fabry–Perot here. And the jet we're seeing now is not there.

"That's kind of amazing," Bally says. "We've made a discovery. Sir Gilles, it looks like you've been party to the discovery of a new jet."

In a few minutes, another jet appears on the screen. "That's phenomenal," Bally says, really into it now. "That's a new one."

All told, astronomers have discovered about 500 Herbig–Haro objects. We've just added two tonight.

Over the course of the night, Bally and Johnstone move their way through the nebula's velocity space. Starting with light emitted by ionized hydrogen, they take one exposure after another, each time issuing computer commands to move the plates of the Fabry–Perot slightly to capture red and blueshifts. Then they switch from ionized hydrogen to sulfur, imaging the same portion of space in different wavelengths to fill up their data cube.

According to Bally, the data contain a wealth of information that will take a year or more to "reduce." For example, by analyzing the ratio of intensity of two particular wavelengths of sulfur, he and Johnstone will be able to calculate the density of the moving gas. From that, they can calculate the velocity and mass of material in the jet. "And if I have the velocity and the mass of material in jets, I can determine how much material has been ejected at different points in the past," Bally says.

Looking at the movement of plasma shooting from young stars is like examining the geology of the Grand Canyon, he explains. By charting how this material has moved over time, he hopes to decipher the region's history, much like geologists have reconstructed the geological history of the Grand Canyon by charting and interpreting the exposed rock layers. More specifically, Bally hopes to trace the loss of material from forming stars through their jets. "Once we understand that," he says, "we can understand better how the stars formed. The idea is to decipher how stars assemble themselves."

By 3 a.m., it's time for a break. I suggest to Bally that he and I test a Mauna Kea urban legend. At this high altitude, taking a hit of pure oxygen can turn the Milky Way from a visual spectacle into a religious experience. So the experiment will go like this. First, we'll go into the lounge, turn all the lights out to get acclimated to the dark, and head outside to look at the sky. Next we'll come back inside, take a hit or two from the oxygen bottle, and see if there's a difference in the way we perceive the stars.

We bundle up in our winter gear, acclimate to the dark, and exit through a hatch onto a catwalk suspended from the side of the observatory dome. Even though we took time to get used to the dark, I can't see a thing. This is just a tad unsettling, since I'm about 200 feet up and a black void looms before me. But there is a guard rail—if only I could find it.

In about a minute, our pupils dilate enough for the spectacle of the night sky to unfold before for us. Shivering as near freezing temperatures and a brisk wind conspire to whisk away our body heat, we gaze upward. To our naked eyes, the Milky Way looks much as it does in astro-photographs: a bright, uneven arc of white on black, as the light of countless stars within the Milky Way's disk blends together to make a glowing splash. This is pretty incredible, considering that our eyes are much less efficient at capturing photons than film is, *and*, unlike astro-photographers, we don't have the benefit of a time exposure here.

Next, we head back inside for some oxygen. We each take a few deep breaths, return to the catwalk, and gaze up at the Milky Way.

"Pretty spectacular," I note, noncommitally. "What do you think?"

"Yes, quite amazing," he replies.

"Better? Yeah, I think it's better."

"Maybe it looks better because we *think* it should look better," Bally ventures.

"Yes, maybe. No, it's better, I'm sure."

With that issue settled, Bally pulls out a pair of binoculars. I scan the shoulder of Mauna Loa, the other almost-14 000-foot-high mountain on the Big Island, looking for signs of active volcanism. In a minute, I find it—a faint orange glow. Lava flows on the south side of the mountain are illuminating the vog above volcanic vents.

Then I aim at the Orion Nebula. The binocular's lenses, only an inch or so across, gather enough light so that we can make out a small, glowing smudge. Bally comments that we're seeing the nebula as it looked not long after the fall of the Roman Empire, for it has taken its photons 1500 years to travel to us tonight. I squint, trying to make out detail in a smudge that's trillions of miles across. And I marvel at the capacity of the huge mirror in the telescope behind me to concentrate photons from features much tinier than the smudge.

As the evening wears on, good seeing comes and goes. On most other mountains, tonight's conditions would be considered great. But compared to the best conditions Mauna Kea has to offer, the seeing is a bit disappointing. While we wait for things to clear up, I ask Bally and Johnstone what causes portions of a molecular cloud to collapse to produce young stars and their jets.

According to Johnstone, there are two basic possibilities. Either the self-gravity of a patch within a cloud causes gas and dust to contract and eventually collapse on its own, forming a dense region in the center that spins down into a star; or pressure from the outside, perhaps from a supernova explosion or even the collision of two galaxies, forces gas within a cloud to contract and then collapse. Either way, gravity is the key ingredient because once material in a cloud has come (or has been pushed) together in a dense enough clump, gravitational collapse is inevitable.

"All this is at a very theoretical phase now," Johnstone emphasizes. "That's why we need instruments like SCUBA—to map the insides of clouds and see how accurate our theories are."

Whatever the trigger may be—and several mechanisms may work, depending on the circumstances—it takes about 100 000 years for a patch of material within a molecular cloud to collapse into a star. During this time, the faint wisps of gas with a smattering of dust increase in density by a factor of 10 million, and temperatures go from 15 Kelvin to more than 11 million Kelvin. The result? Ignition of thermonuclear fires, and a ball of burning gas that we call a star.[34]

By 4:30, Bally and Johnstone quit chasing photons and begin getting ready to head down the mountain to Hale Pohaku, the lodge at 9300 feet where Mauna Kea astronomers sleep. But during the night's discussions, one key piece of the puzzle has been left out. How do giant molecular clouds like Orion form in the first place? According to Bally, astronomers really aren't sure. But it's clear that he has strong views on the subject— ideas not shared universally by astronomers.

He and others in his camp believe that superbubbles play a key role in the formation of new molecular clouds, as well as in triggering new waves of star formation. As a superbubble expands, Bally explains, it may blow out of the top and bottom of the galactic disk. But within the disk, the bubble's expansion is retarded as it tries to force its way through the denser gas there. "Therefore, you get this intersection of the superbubble and the disk of the galaxy," Bally says. "This makes a *super-ring*, which expands slowly—at tens of kilometers per second at most."

As the last supernovae in a cluster like the Trapezium die, the superbubble loses its source of inflation. Since there isn't much to retard the bubble walls, they keep expanding but at a much more leisurely pace. Accordingly, within the galactic disk the super-ring just coasts along, like a wave expanding outward when a stone is tossed into a pond. As it expands, the super-ring sweeps up material in its path—gas and dust from molecular clouds, heavy elements from supernovae remnants, etc.

"As it sweeps up more and more material," Bally continues, "the ring slows down. After about 30 million years, it's moving only at a few kilometers per second and has swept up material in a ring with a radius of up to a thousand light-years. The amount of mass it can sweep up just from the debris of molecular clouds is a *million* solar masses or more, or the equivalent of the mass contained by 10 giant molecular clouds." Then as the coasting super-ring of gas and dust cools, individual clouds begin to condense from this material under the influence of their own gravity. Each cloud should contain mass equivalent to that of 100 000 Suns. "And it's my personal view that this is the origin of molecular clouds," Bally argues. "That's how they're born. They are the fossilized super-remains of supernova explosions, fragmenting under the force of their own gravity," he says.

"Of course, there are people who might differ," Johnstone interjects. The conventional view, he says, is that spiral density waves periodically sweep through the arms of galaxies like the Milky Way, leaving new molecular clouds in their wake. According to this theory, the waves gather gas into giant blobs, and the blobs evolve into giant molecular clouds.

When Johnstone finishes sketching out the conventional theory, Bally picks up the thread of his argument: "So the question is this. What dominates the formation of giant molecular clouds? Is it spiral density waves? Or is it this kind of process?—super-rings sweeping up gas and dust, which fragment into molecular clouds?"

"I appeal to the observations," Bally says. "Look around the Sun and what do we see? Gould's Belt is all around us, which is 50 million years old." This feature, he argues, is the super-ring that gave birth to the Orion cloud, among others.

Gould's Belt is a ring of young stars and star-forming regions. It's visible in the night sky as a band of bright stars forming a circle that extends between the constellations Orion and Scorpius. Earth is located inside and near the edge of the ring. In the center, Bally notes, is a cluster of stars known as Alpha Persei. He believes that a battery of supernova explosions in the Alpha Persei cluster some 40 to 50 million years ago inflated a super bubble. As the superbubble's wall expanded outward in the galactic disk, it swept up millions of solar masses of material, creating the super-ring.

When the supernova activity stopped, a cluster of stars remained in the heart of Perseus, and the material in the super-ring fragmented into a chain of molecular clouds. Contraction of gas and dust in one of those clouds ignited star formation in the Scorpius–Centaurus region. Giant O stars here have in turn blown bubbles and gone supernova, sending forth a new super-bubble. Within the past 250 000 years, our solar system has moved into a flow of hydrogen gas emanating from this star-forming region, Bally says, citing research by University of Chicago astronomer Priscilla Frisch.[35]

Elsewhere in Gould's Belt, the Orion cloud took shape. Contraction of patches of gas and dust began there about 12 million years ago, Bally says. The expanding bubble from the first resulting supernovae has since triggered compression of gas and dust, triggering more star formation elsewhere in the cloud — and in stepping-stone fashion, down to Orion's belt, into the sword near the Horsehead Nebula, and finally to what is now the Orion Nebula.[36]

''So we can see two generations of this process,'' Bally says. The first generation began with the super-ring expanding outward from Alpha Persei. The clouds that formed within that coasting ring, have now given rise to a second generation of super-rings, including Orion's.

''So even if another mechanism modulates this process, it's an observational fact that these superbubbles *are* forming. What may be debatable is whether the process leads to the formation of super-rings and giant molecular clouds from them. But if it doesn't, then what happens to all that swept up gas? Where does it go? I mean you sweep it up. We see it around us.''[37]

To help me understand his point, Bally scrawls a diagram on a piece of paper of a super-ring expanding within a galaxy's spiral arm. Looking on, Johnstone does not seem entirely convinced by Bally's argument. At one point, he grabs the pencil and modifies the picture. As I watch, he and Bally begin debating both the geometric details and the overall picture of molecular clouds forming from supernova remnants within super-rings. The debate gets rather heated. And watching the two of them go at it at five in the morning, I wonder whether I've misunderstood how new scientific knowledge is created. Maybe the traditionalists who emphasize a neat, orderly method consisting of experiment, observation and testing are missing an intangible but more important dimension — a *personal* dimension. It's clear that Bally and Johnstone hold to fervently held beliefs, and that these ideas about how nature works forcefully shape what they choose to investigate — and even how they interpret what they find. The traditional picture of cool, detached observers drawing pure conclusions from raw data just doesn't fit what I'm seeing here in the control room of the Canada–France–Hawaii Telescope. Instead, it seems that new scientific knowledge may eventually spring from this kind of heated argument.

In short order, the debate makes my presence irrelevant. The two astronomers go at it on their own for half an hour, while I retire to the Café Mont Blanc for a sugary pick-me-up. We're all exhausted, our eyes are bloodshot, and our brains are addled by jet lag, sleep deprivation and lack of oxygen. Gilles Joncasse and I are desperate to get to bed. Finally, he makes an announcement: "The bus is leaving for Hale Pohaku." In this case, the "bus" is the CFHT's Suburban that Gilles is about to climb into, with or without his two colleagues.

The prospect of descending 3000 feet on foot through a Martian landscape of lava flows finally does the trick. Bally and Johnstone join Gilles and me in the elevator and we descend to ground level. That doesn't stop the argument, however. It continues in the Suburban on the twisting drive down the mountain, and even over breakfast in the lodge's cafeteria at 6 a.m. But before we all drag ourselves off to bed, Johnstone wants to make sure that I don't get the wrong idea. "This doesn't mean I will never write a paper with John again," he says. And then Bally jumps in: "In fact, it means we probably *will* write a paper together, about this very issue."

\* \* \*

The observing run on Mauna Kea in December of 1999 was a bonanza for the two astronomers (as well as for Joncasse on his independent project). In several days, they filled up their data cube with many gigabytes of information. And as Bally had predicted, the cornucopia of data took more than a year to analyze.

240

In their work on the mountain, Bally and Johnstone had targeted the jets of young, low-mass stars that were lit up brightly by radiation. Some 20 or 30 examples of such irradiated jets had been discovered previously in Hubble images of Orion. By using the Fabry–Perot to look at emissions of sulfur in two wavelengths, the two astronomers calculated the density of material in a number of irradiated jets. And by determining their red and blue-shifts, they were able to calculate how much material was moving and with what velocity. What they found might seem counter-intuitive, given the violence so common in Orion.

"These flows are really *wimpy*," Bally reports. The jets of young, low-mass stars in the nebula are flowing at lower speeds and carrying less material than expected. Like a pum-meled boxer on the ropes, each one of these stars can't seem to muster much fight.

The villains, not surprisingly, are the heavyweights of the nebula, the massive O stars. "I think the jets are wimpy because they have suffered a lot of damage," Bally says. Wind and radia-tion are rapidly stripping away the circumstellar disks from the lower-mass stars, slashing the accretion rate of material onto the stars and thereby starving the jets of their fuel.

For Bally, this is precious new evidence for one of his most fer-vently held views: namely, the very act of *star* formation is a hazard to *planet* formation. With the circumstellar disks being stripped off, Bally says, "what you're seeing is the deaths of these failed planetary systems—and I think they *are* failed planetary systems."

Because of the O stars in places like Orion, it's very difficult for solar systems to form there, especially ones like our own, he argues. Not only do the massive stars' radiation and winds strip dust from circumstellar disks before rocky planets can coalesce, they also drive off hydrogen and helium, depriving the system of the gases needed to construct large gaseous planets like Jupiter and Saturn. If solar systems like our own are to form in such environments, dust in circumstellar disks would have to coalesce quickly to form rocky planets like Earth. And huge amounts of gas would have to condense into Jupiters and Saturns much more rapidly than the conventional planet-formation model suggests is possible.

Even if rocky planets managed to form quickly, the absence of gaseous giants could doom such solar systems to an existence with-out complex life forms like our own. According to this theory, first proposed in 1992 by George Wetherill of the Carnegie Institution of Washington, the huge gravitational fields of planets like Jupiter and Saturn vacuum up flotsam and jetsam in the early solar system (and fling some out of the solar system) that would otherwise form swarms of comets.[38] Impacts of comets and asteroids have profoundly shaped the evolution of life on Earth, causing mass extinctions like the one that did in the dinosaurs and countless other species 65 million years ago. Mammals survived that extinction and went on to thrive in the ecological niches opened by the demise of the dinosaurs. So as a species, we may well owe our existence to that impact. But if Jupiter and Saturn had not reduced the population of comets, and if they did not continue to act as *gravitational guardians* today, the impact rate on Earth might be 1000 times greater,[39] making it very difficult for species like our own to thrive for very long.

All over the Orion molecular cloud, O stars are already bringing star and planet formation to an end, Bally says. Even more dramatically, the stars are poised to destroy the mother cloud itself.

''We're nearing the end, and statistically, that makes sense,'' he says. ''If you look at giant molecular clouds throughout the galaxy, once you form O stars, the clouds seem to dissolve in a few million years.''

But even as the violence in Orion brings this generation of star formation to a halt, it is sowing the seeds of new generations. The O stars in the nebula will go supernova soon, Bally points out. As they do, they will spew heavy elements such as oxygen, phosphorous and nitrogen into the interstellar medium. These elements will be added to the material ejected by previous supernovae. In time, hydrogen and these heavy elements will find themselves in new molecular clouds, where they will be available to be recycled into new stars.[40]

So what is the bottom line? In Orion, the destruction wrought by O stars is balanced by a holistic vision of interstellar ecology, replete with cycles of life, death and rebirth. Is interstellar ecology as nurturing of life as our Earthly ecology is?

The answers depend to a great degree on the details of how stars and planets form. Those details determine whether solar systems like our own can form quickly enough in Orion-like environments to beat the O stars to the punch. Or whether they are common in calmer, more placid environments, where O stars are unknown.

We turn now to those details. The next chapter focuses on how a profoundly cold and diffuse patch of gas within a molecular cloud contracts down into a hot, dense ball of burning gas—in other words, a star.

# 9

# A STAR IS BORN

## *Prelude to Planets*

*There are more stars in the visible universe than grains of sand on all the beaches on Earth.*
— Seth Shostak[1]

*A frog poses a more daunting scientific challenge than a star.*
— Sir Martin Rees[2]

''If you want to understand how solar systems form you really need to talk to Frank Shu,'' John Bally told me toward the end of our observing run at the Canada–France–Hawaii Telescope.

By that point, I had begun to grasp the big picture of interstellar ecology. The next step was to drill down to the details of how a molecular cloud assembles itself into a star. Bally and Johnstone told me that Shu is renowned among astronomers for his theories on star formation, galactic structure and other topics. I later learned that he is legendary for less conventionally scientific pursuits as well. He's something of a pool shark, for one thing. And he is said to be as adept at poker as he is at solving complex astrophysical problems. Shu is reported to have skipped out of an astronomy conference in Vegas to spend most of his time at the poker tables.[3]

When you think about it, these extra-curricular interests make perfect sense. After all, in poker as in astrophysical theory-making, the player makes predictions about future behavior based on his model of the problem at hand. In astrophysics, the model describes the behavior of nature in terms of mathematical algorithms stored in a computer. In poker, the model is in the gambler's head—it's his approach to assessing the probability of certain cards being thrown. In a similar vein, pool is a game of dynamical interactions between spheres. And in many ways, so is the formation of a solar system—a complex theoretical problem Shu has tackled with characteristic enthusiasm, producing major breakthroughs in understanding.

Shu's enthusiasm is grounded in math, an interest first nurtured by his father's career as a mathematics professor at Purdue and other Midwestern universities. In 1943, the father's path toward a mathematics career led him from Kunming, China, to graduate school at Brown University in the United States. Shu was two months old when his father left home for America. Five years later, when Mao Tse Tung and the Communist Party were about to take power in China, Shu's father decided it was time to extract his family. Moving first to Taiwan, Shu, his mother and three siblings later joined the father in the United States.

Like his father, Shu had a talent for mathematics. At the age of 16, he skipped his last year in high school and entered the Massachusetts Institute of Technology to pursue a degree in physics. His mentor there was a renowned astrophysical theorist named C. C. Lin, who helped steer Shu toward astronomy. Enrolling at Harvard for his Ph.D., Shu focused his dissertation on a vexing theoretical question: why do galaxies like our own have spiral arms? He proposed that density waves sweeping periodically through the galaxy were responsible for the observed structure — a theory that has long since been the textbook explanation. (The same density waves also are believed by many scientists to gather gas and dust into giant molecular clouds.)

Since then, Shu has worked on many other problems. In the 1970s, for example, he and a Ph.D. student proposed a theory explaining how one star cannibalizes another in what astronomers call compact binary systems. In the 1980s, Shu took a theoretical detour into planetary science, figuring out why Saturn's rings consist of ringlets and gaps reminiscent of the grooves and ridges of an old-fashioned phonograph record. Saturn's moons and many moonlets gravitationally herd the material in Saturn's rings, producing the complex structure, Shu proposed.

But he is probably best known for his theoretical modeling of star formation. In 1977, he came up with a model for how material within a molecular cloud collapses to form a protostar complete with a girdling disk of material. According to Ray Jayawardhana of the University of Michigan, the model described the progression of the collapse and was the foundation for later, more

245

detailed work on the origin of stars. "Frank is one of the most influential theorists in more than a generation," Jayawardhana says. "You could talk to anyone in the field, and even if they disagreed with the content of his work, they would acknowledge that he's had a tremendous influence. It's hard to overstate it."[4]

And so on a foggy summer morning, I entered a nondescript building on the University of California, Berkeley campus and found Shu in his office.* Papers and books were piled on the floor, on tables, on every available surface that would hold them without slipping, as well as on others that wouldn't. There didn't seem to be any hidden order to the mess, which looked as if it could have been the result of a violent wind blowing formerly organized office contents into heaps and scraps. But like the stellar winds that Shu believes helped bring order to the forming solar system, the wind at work here—his intellect—is bringing order to our understanding of our cosmic origins.

Right from the outset of our discussion, Shu betrayed a philosophical streak. "What is our purpose as a civilization and as a thinking species?" he asked. "Exploration is the only one that is truly worthwhile. Not just physical exploration, but *intellectual* exploration. In that sense, scientists and artists have similar desires and yearnings. The scientist explores the external world, and the artist seeks internal truth."[5]

As a theorist, Shu's explorations of the external world have not taken place through big telescopes. (He has observed with a telescope only twice in his career.) His exploration has taken place through his *mind*. But Shu's ideas are known for being closely tied to what observers are finding through their telescopes. His M.O. is to make detailed mathematical models of phenomena documented by observers, and then to use the models to make specific predictions that observers can check against reality.

When I asked Shu to describe some of the more important recent developments in research on the origin of solar systems, I thought he might focus on theoretical advances first, observational ones second. I got it backwards.

---

* Early in 2002, Shu left U.C. Berkeley to become president of National Tsing Hua University in Taiwan.

"Thanks to high-resolution observations in the infra-red and millimeter parts of the electromagnetic spectrum, we have examined the earliest stages of the formation of stars, as well as the birth pangs of what we believe are planetary systems," he said. "There has also been a correspondingly healthy development of theory, which in some cases actually *preceded* the observations, and in others was motivated by the observations. So, there have been confirmations of old ideas as well as new surprises."

And what of those surprises? "The biggest one, I think, is that the formation of stars and planetary systems generally is much more violent than anyone imagined," Shu commented. And chief among the causes of that tumult, he said, are stellar jets.

As evidence has mounted that jets are as fundamental to the origin of a star as contractions are to the birth of a child, Shu has been motivated to explain why this is so. He began work on the problem in 1988, and he has been refining a comprehensive theory ever since. Shu's model elegantly accounts for stellar jets and the chondrules found in certain meteorites, explaining both as the result of a single process. As you may recall from the Portales meteorite story, chondrules are tiny blobs of material present in chondritic meteorites that date to the earliest moments of the solar system. They are believed to be among the basic building blocks of planets. Shu's model ties seemingly disparate phenomena together into one theoretical framework, which in turn provides fundamental insight into the origins of stars *and* planets.

The model is not yet universally accepted. But according to Alan Boss, a theorist at the Carnegie Institution of Washington who is sometimes at theoretical odds with Shu, it is one of the most promising explanations for the early stages of star formation. "There are other ways to do just about everything that Shu's model proposes to do—in some ways better ways. But the model has the benefit of being based on observed outflows, which are known to be common in young stars. So to that extent, one has to take seriously whatever consequences strong outflows might have on the entire system."[6]

Shu has emphasized just what a remarkable feat it is for a star to form from a core of gas and dust within a molecular cloud. It requires a transformation so profound that it has tied theorists in knots for several decades. The problem, he says, is that cores are wispy, cold and huge, and they rotate like a pirouetting Sumo wrestler. By contrast, stars are dense, hot and compact. And they *spin*, like an Olympic figure skater.

If you could travel in a spaceship to a core buried in a giant molecular cloud, you'd hardly know when you'd arrived. That's because on average, each cubic centimeter of a cloud core contains just 10 000 particles (mostly hydrogen molecules). That might sound like a lot, but it's more than a billion times more rarefied than the air we breathe on Earth at sea level.[7]

Meanwhile, your spaceship's climate control system would need to be in good working order because the temperature in the core would be some tens of degrees Kelvin — pretty close to as cold as cold can be.

If you wanted to take a jaunt across the core, even at light speed it would take you longer than an average summer vacation. That's because a typical core is about 2 light-months across.

And if you were of a scientific bent and wanted to know the speed of the core's rotation, you'd be ill advised simply to wait and see how fast your position changed relative to background stars. That's because typical cores spin at a leisurely pace, completing just one turn every million years.[8] So you'd die long before noticing any appreciable change (assuming you didn't use any sophisticated equipment or put yourself into suspended animation).

Contrast these initial conditions with the finished product, a star like our Sun. Assuming you were in a spaceship capable of making the journey safely, on your way toward the center, you'd find something like $10^{24}$ particles, mostly protons, packed into every cubic centimeter. That density is nearly 20 orders of magnitude greater (much more than a trillion trillion times) than the starting density in the cloud core. (You'd obviously know then that you'd arrived.)

Depending on how deep you got, you would measure temperatures in the order of 10 million Kelvin, which is more than a million times hotter than the cold raw material. On the other hand, it wouldn't take you much time to burrow all the way through the Sun because it is more than a million times more compact than its parent cloud core. And upon your exit, if you lingered on the surface of the Sun to measure its rotation, you'd only have to wait a few weeks for it to complete one turn.

To characterize the difference between the starting cloud and the finished star, Frank Shu borrows a term used by audiophiles to describe the difference between the quietest and loudest sounds in a piece of recorded music: *dynamic range*. In a stereo system, too much dynamic range produces problems for speakers not designed to handle it. In the case of star formation, the huge dynamic range causes a host of problems—which is another way of saying opportunities—for theorists.

One of the most significant of such problems is best understood with an analogy. Imagine you're watching a figure-skater, and it's her finale. She leaps into a spin with her arms outstretched. As the music builds to a crescendo, she draws her arms in. Anyone who has watched the Winter Olympics knows what happens: she spins faster. The reason? Conservation of angular momentum.

Angular momentum is the rotational cousin to linear momentum, or the momentum of an object moving in a straight line. Linear momentum is equal to the object's mass multiplied by its velocity. But you don't have to know the formula to feel intuitively that the faster an object moves in a straight line, the more momentum it carries. Similarly, the more *mass* an object has as it moves in a straight line, the greater is its linear momentum.

In the case of a spinning figure-skater, her angular momentum is equal to her *radial velocity* multiplied by something called the *moment of inertia*. The laws of physics dictate that the product of these two quantities must equal the same number before and after the skater draws her arms inward. Another way of saying this is that her angular momentum must be conserved. So, if her moment of inertia *decreases*, her radial velocity must *increase* to conserve her angular momentum. And that's exactly what happens when she brings her arms in. The reason? Moment of inertia is a function of the mass of a spinning object and its radius squared. When the skater brings her arms in, she reduces her radius, dropping her moment of inertia. As a result, she must spin faster.[9]

A forming star is not unlike the spinning figure-skater. It contracts as it transforms from a gigantic core to a compact ball of gas, thereby experiencing a decrease in its moment of inertia. To conserve its angular momentum, the star must speed up. And here is where one theoretical problem lies. Calculations

249

show that given its enormous contraction, the star would have to speed up so much that it would rotate once every second. "In fact," Shu says, "it's *impossible* for the Sun to turn around once every second, because it would be spinning faster than the speed of light."[10] Forming stars can't ignore the laws of physics any more than a figure-skater can, and those laws state that *nothing* can travel faster than light speed.

One way around this seeming impasse is to change the rules of the game. In the case of the figure-skater, she could avoid spinning faster as she drew in her arms by shedding some mass, thereby giving up some of her angular momentum. The total momentum in the system would stay the same — it would be conserved — because the skater would impart some of hers to whatever mass she threw off. In reality, of course, this is unlikely. The television network censors would never put up with the skater shedding the only mass available to her. But a *star* might be able to strip off some of its mass as it contracted.

Nature has obviously figured out a way to solve the angular momentum problem. (Otherwise, we wouldn't be here wondering about it.) And according to Shu's model, two of the key consequences of the solution turn out to be bipolar jets and — lucky for us — planets.

So let's follow the evolution of gas and dust within a molecular cloud to the formation of a protostar complete with jets and a planet-forming protostellar disk. To do it, a virtual tour of the Milky Way's spiral arms seems in order. The vast distances involved are no object now. We've installed a warp drive . . .

We point our virtual spacecraft toward Saiph, the star that marks Orion's right knee; it's a landmark that will help us find the southern end of Orion's giant molecular cloud. As we cruise from the profoundly empty reaches of interstellar space into the cloud, we use an infra-red telescope to see what's up ahead in the Orion nebula. (In the infra-red, we can peer through the obscuring dust). Stars seem to be very highly clustered in the nebula. Within just a few parsecs,* there are something like 1000 stars. This is a seething, stellar tenement district.

---

* 1 parsec = 3.26 light-years = 19 trillion miles (or 31 trillion kilometers)

Cruising through the Orion cloud toward the nebula, we must use imaging devices sensitive to submillimeter radiation to see the cloudscape passing by our portholes. That's because the thin wisps of hydrogen gas that make up most of the cloud are invisible to the naked eye, and the dust here is profoundly dark. In the submillimeter, we can see the dust, which traces out the overall structure of molecular gas. A rich cloud topography is revealed on our monitors. The scene is reminiscent of a radar image of a line of thunderstorms. Bulbous blobs and slender streamers of gas and dust comprise an elongated cloud structure. Buried within the blobs and streamers are still denser structures with similar shapes.

Astronomers have seen this hierarchical pattern, and they have adopted the terms *cloud*, *clump* and *core* to describe it. Dense cores are embedded within clumps which, in turn, are embedded in the overall cloud.

But some astronomers argue that these descriptions are artificial. They are the result, they say, of our need to bring order to things by making up discrete categories even for objects that are continuous in form. They say the interstellar medium, of which molecular clouds are a part, is fractal in nature.[11] This means the same shapes are repeated over and over again on different scales.

A watershed is a good example of a fractal structure. From the biggest to the smallest scales, the shapes are the same. So, from the river to its tributaries, from the tributaries to their streams, from the streams to their creeks, and finally from the creeks to the tiny, erosional rivulets that run into them, the shapes of all the channels are geometrically indistinguishable. And each of these features, although discrete in our terminology, is not really a discrete entity in nature. One thing flows (literally) into the next, forming a continuous whole—a watershed.

Whether or not molecular clouds are truly fractal in a rigorous mathematical sense, it is clear that they exhibit what scientists call *self-similar* structure. But this description goes only so far, because it says nothing about the details of how a star forms. And by definition, to form a star, nature constructs something—a ball of gas—that looks nothing like the self-similar structures of core, clump and cloud.[12]

To figure out how nature departs from self-similarity to transform a cold, wispy cloud core into a hot, compact star—

251

or a cluster of stars, in the case of Orion-like environments—scientists first have to address the question of how the clumps and cores form in the first place.

The collective gravitational pull of the gas and dust in the cloud must be the key. So from our virtual galactic cruiser, we use a gravity meter to take some measurements. The instrument reveals that the self-gravity of the molecules comprising the cloud ought to be enough to collapse the cloud to high density.[13] Yet here we are in the Orion cloud and it doesn't appear to be collapsing everywhere. Moreover, based on the best evidence, astronomers believe that GMCs survive for millions of years. So something must be supporting the cloud in the face of gravity.

There are two good candidates. The first is magnetism. With our detectors, we find that the cloud is laced with lines of magnetic force. Astronomers have theorized that such magnetic field lines act something like metal reinforcing rods within concrete: they gird the cloud against gravitational collapse.

Our detectors also pick up evidence of turbulence in the cloud. This is no surprise, since wherever we look, the jets and winds of forming stars are pumping huge amounts of material into the cloud. This is keeping things stirred up quite nicely. Theorists say the pressure from turbulent movement of gas within a cloud pushes outward even as gravity tries to pull things inward.

We also note that even though the cloud is frightfully cold, it retains a hint of heat in the form of thermal movement of molecules. This could also help support the cloud. (But many astronomers say it is not nearly as important as the other two.)

Scientists aren't so sure which is more important, magnetic fields or turbulence. Ask that question at a conference on star formation and you're likely to get quite a lively debate going. Suffice it to say that one or the other (or both) probably keeps gravity at bay. But gravity certainly has to win at some point. The density of gas within the clumps and cores inside the Orion cloud is greater than outside those structures. Something must have enabled the material in these features to pull together. Moreover, from our spaceship we can see that, in each clump and core, molecules are tending to move toward the center—sedately, to be sure, but inexorably nonetheless. *Something* must be allowing this to happen.

And there's one other reason to conclude that at some point gravity must win over magnetism and turbulence: there most definitely are stars all around us. To form them, some parts of the cloud must have drawn together and collapsed gravitationally into dense balls of hot gas.

So how does that happen? Scientists have identified two possibilities. A patch within a molecular cloud can be jostled by some *outside* force, causing it first to contract and then collapse gravitationally. Or conditions within the patch itself can lead automatically to contraction and collapse, *from the inside out*.

The inside-out scenario depends on mass. Theory suggests that there is a critical mass at which a patch of matter within a cloud is perfectly supported against its own gravity by magnetic fields alone. If a patch has just a little more mass than this critical amount, molecules within it will drift inward toward a central point due to their self gravity. They won't accelerate, at least not at first, because the magnetic fields are still providing considerable resistance. Such gentle inward movement (scientists call it *quasi-static* movement; go figure) probably is one way that clumps and cores form within clouds.

Eventually, theorists say, a threshold is passed. Enough mass gathers together in a clump so that gravity suddenly becomes overwhelming, and gas and dust accelerate inward in a catastrophic swoosh. This is a gravitational collapse, and it's called an inside-out collapse because it happens in the center of a core first and works its way outward. Frank Shu played an instrumental role in figuring out the details of how this may work.

Another alternative is that a patch within a cloud has *less* mass than is needed for gravity to win over the magnetic fields. Such a patch is said to be *subcritical*. In this case, even external pressure can't force the cloud to collapse, because the magnetic fields provide more than adequate reinforcement. If stars are going to form in such a patch, something has to give.

Considering these alternatives, one could ask whether most real clouds out there are subcritical or supercritical. Given the current state of knowledge, astronomers just aren't sure. Perhaps future observations will provide an answer.

Finally, how big a role does turbulence play in supporting a cloud? The answer seems to be that in supercritical clouds, in which magnetism alone is not enough to forestall gravity, turbulence is crucial. In computer simulations, when turbulence is allowed to decay, gas readily drifts together in denser and denser structures. By contrast, in subcritical clouds, nothing happens if the turbulence is allowed to decay, because the magnetic fields by themselves provide enough reinforcement.

So, at least according to theory, whether structures form in a molecular cloud depends on the luck of the draw. If there are patches with enough mass to overcome the supporting forces of magnetism and turbulence, clumps and cores will form easily — indeed, automatically — and gravitational collapse of gas and dust into stars is inevitable. But what about patches that aren't endowed with enough mass for gravity to overcome the supporting forces? In this case, stars can still form from the inside out because magnetic support can slowly weaken over time. The culprits are charged particles, or ions, within molecular clouds. Because ions are charged, they "feel" the magnetism and lock to the field lines and therefore resist gravity's pull toward the center. The neutral molecules in the cloud, on the other hand, don't feel the field directly and are therefore free to move.

"The neutrals are attracted by gravity to try to come to the center of the cloud," Shu explains. "When they try, very frequently they collide with an ion and bounce back. But over time, they steadily drift toward the center of the cloud."[14] At the same time, the collisions of neutrals with ions weakens the magnetic field. Meanwhile, gravity strengthens as the neutrals drift toward the center. With gravity increasing and magnetic support weakening, lumps and cores can form.

But the process doesn't stop there. Simulations show that at first the density at the center increases slowly. In time, however, this portion of the cloud becomes more and more centrally condensed. Eventually a threshold is crossed and material whooshes inward in a sudden gravitational collapse.

Scientists call this slow leakage of neutral molecules toward the center of a patch within a molecular cloud *ambipolar diffusion*, and many believe it is the basic process by which isolated stars like our own Sun form.

Do these theoretical mechanisms work as advertised to produce the kinds of structures actually seen in molecular clouds? In pursuit of an answer, Shu and a colleague completed computer simulations using what they call a "toy model" — a set of equations that represent the behavior of molecules within a cloud. For the sake of the simulation, mathematical representations of particles stood in for molecules and were arranged in a grid of cells. Four such particles were distributed randomly in each simulation cell, with each such cell corresponding to a region of space measuring a tenth of a parsec on a side. The mathematical equations were designed to take into account the effects of gravity, magnetic fields, turbulence and ambipolar diffusion on the particles.

After 2.5 million years of simulated time, particles begin to gather into clumps. After 15 million years elapse, small cores form; these in turn give birth to dense balls of particularly concentrated matter. These are protostars. Moreover, the protostars begin to spit out material in jet-like outflows, which replenish the turbulence within the simulated cloud.

After 28 million years of simulated time, long filaments studded with dense cores develop. ''The overall impression is not qualitatively dissimilar to observational maps of the Taurus molecular cloud complex,'' Shu and his colleague wrote in a description of their work.[15]

Finally, after 50 million years of simulated time, 40% of the cloud is converted into stars. And their outflows have carved out a cavity in the cloud.

Shu admits that simplifications in the model limit its application to the real world.[16] For example, the simulation is done only in two dimensions. Expanding to three would obviously be more realistic. But that would make the computations more difficult. Even so, the scientists believe their model lends support to the overall theoretical picture.

Simulations such as this support the idea that portions of a cloud can contract and then collapse from the inside out, forming the clumps and cores—and stars as well—seen within molecular clouds. Most scientists who study star formation find this theory appealing because it relies, elegantly, on physics alone—the physics of magnetic fields, turbulence and gravity. There's no need to invoke some external bolt from the blue.

But according to the Carnegie Institution's Alan Boss, clues sifted from 4.559-billion-year-old meteorites—the oldest ever found—hint that our own solar system owes its existence to just such a bolt. The clues consist of faint residues of various radioactive isotopes trapped within tiny inclusions in the rock. One of the radioactive isotopes is magnesium-27. It is like a fingerprint at a crime scene that points to the triggerman. In this case, Boss and other scientists say, the isotopic fingerprint points to the culprit in the collapse of the cloud core that gave birth to our solar system.

As described earlier, magnesium-27 is produced through the decay of another isotope, aluminum-26. None of the parent aluminum remains today because its half life—740 000 years—is very short compared to the longevity of the solar system overall. In just 740 000 years, half of the aluminum-26 decays into magnesium; in another 740 000 years, another half decays, and so on until it's all gone and all that remains is much longer-lived magnesium. Finding it in meteorites that formed from the primordial material of the solar system shows that aluminum-26 must have been present at the outset. So how did it, along with a number of other tell-tale isotopes, get there?

One way, Boss says, is through furious irradiation of precursor chemicals by a star that's going supernova. So he and other scientists believe that a supernova exploded near the core that would later collapse to form our solar system. When the blast wave reached the core, it compressed gas and dust within it, perhaps accelerating movement of molecules inward and ultimately triggering a gravitational collapse. At the same time, the isotopes created in the explosion of the star, including aluminum-26, were injected by the blast into the cloud core and then incorporated into the forming solar system. And 4.559 billion years later, meteorites containing the decay products of these short-lived isotopes, including magnesium-27, fell to Earth, giving us evidence of the original crime.

''There is good observational evidence that regions of star formation occur on the boundary of old supernovae remnants,'' Boss says. ''So we pretty much know that supernova shocks can sweep up a lot of gas and dust and compress some of it enough to form a new generation of stars.''[17]

But a supernova explosion is not the only way that gas and dust within a molecular cloud can be triggered from the outside into a gravitational collapse. Images from the Hubble Space Telescope reveal another possibility: collisions between galaxies.

In 1997, for example, Hubble revealed 1000 star clusters bursting to life as two galaxies smashed head on. They are known as the Antennae galaxies, because their calamitous gravitational inter-action has pulled two long antenna-like tails from them. Wrapped around the centers of these profoundly disrupted galaxies are necklaces of glowing gas studded with the bright, young star clusters. Dark filaments of dust are woven among the clusters — silhouetted traces of the molecular clouds from which the stars were born. According to Brad Whitmore of the Space Telescope Science Institute, the collision between the galaxies has heated the gas surrounding the molecular clouds, causing it to expand. This has driven shock waves into the clouds, triggering collapse of the cloud and spawning the forma-tion of the star clusters.[18]

As a graduate student at the University of Pennsylvania, Sarah Gallagher used the Hubble Space Telescope in 2001 to capture a similar burst of star-cluster formation in a group of five galaxies known as Stephan's Quintet. At least two of these galaxies have smashed into each other; like hit-and-run drivers, they are now speeding away. The collision seems to have ripped huge streamers of gas out of the galaxies, which now reside as clouds thousands of light-years from their parent galaxies. According to Gallagher, the galaxies are still interacting in a gravitational wrestling match. This has compressed the gas in the orphaned clouds, triggering intense bursts of star for-mation.[19]

Galactic pileups such as these seem to be responsible for enormous numbers of stellar births. "Galaxy collisions are believed to be very significant in the history of star formation in the universe," Gallagher says. "They result in a very different rate of star formation than you currently see in most galaxies locally. For example, the Milky Way is converting about 1 solar mass of gas into stars each year. By comparison, as a result of galaxy interactions, you can get star formation rates of 50 to 1000-plus solar masses per year. So you can see that much of the star formation over the history of a galaxy can happen during a relatively brief epoch of a galaxy interaction."[20]

Star formation is so rapid in galactic collisions that almost all of the gas in molecular clouds is converted into stars. Globular clusters — groups of up to a million stars all bound together by their collective gravity — are what's left once things settle down.

The Antennae and Stephan's Quintet are close enough for Hubble to image in reasonable detail. Astronomers are lucky to have them so close by because galaxy collisions are relatively few and far between at this point in the history of the universe. But such events appear to have been much more common in the past, when the universe was smaller and galaxies were therefore closer together. Astronomers believe that galaxy collisions may have played a crucial role in the origin of stars earlier in the history of the universe.

"The current star formation rate in the local Universe is puny compared to what it was at redshifts of 1 to 2, or about halfway back in the history of the Universe," Gallagher notes. "Galaxy interactions and mergers were also more common then. Coincidence? Probably not. The two phenomena, increased rate of star formation and galaxy interactions, are quite likely to be related."[21]

Whatever causes material within a molecular cloud to contract, once gravity wins out over the supporting forces of magnetic fields, turbulence and thermal pressure, the formation of a protostar is almost inevitable. That process unfolds somewhat differently depending on the cloud. In places like Orion, cloud cores are believed to fragment into smaller, clustered chunks, spawning dense star clusters like the one that has taken shape in the Orion Nebula.

Although the vast majority of star formation happens in such cosmically urban settings, our solar system is believed to have formed in a different kind of cloud: a quiet, boring suburb — a not-so-giant molecular cloud where stars were born individually and in isolation, rather than together in dense clusters. So for insights into how our own solar system formed, we must leave Orion.

At warp speed, we head for a complex of clouds in the constellation Taurus. We call up galactic topographic maps on the computer and note that the complex is an elongated archipelago of molecular clouds. From east to west, the archipelago is 25 parsecs long — considerably smaller than the Orion molecular cloud. Astronomers have labeled the Taurus complex a *dark cloud* — dark because of the obscuring nature of its dust. We can see it out of our portholes only because it is blocking the light from stars behind it, so it appears as splotches of black against the sparkled background.

As we sail in and among the islands of gas and dust, peering through the dust with infra-red instruments, stars seem to be few and far between. Each one stands on its own, with no neighbors for a few parsecs around. And there's also none of Orion's O-star violence.

In short order, we find a core of gas and dust that seems to have grown quite dense. Molecules are streaming toward the center, where the concentration of gas is getting higher and higher. We don't know it, but contraction has been occurring for about 10 million years, which is just about the amount of time it takes for a cloud core to form, contract and begin to collapse catastrophically.

If Frank Shu were on board, he'd probably tell us to get the hell out of here. At a certain point, the density at the center of a contracting core like the one we're in now actually approaches *infinity*, raising the specter of a singularity just like the ones found at the centers of black holes. Shu calls this the "pivotal state," and it's at this point that catastrophic gravitational collapse takes over.[22]

Unbeknownst to us, our cloud core has already reached the pivotal state and, as a result, material is free falling toward the center. The collapse is happening from the inside out, and the collapse zone hasn't expanded to reach us yet. So as far as we can tell from the view out of our portholes, nothing unusual is happening. But make no mistake about it, a tidal wave marking the edge of the expanding collapse zone is approaching us. Interior to this wave, material is zooming toward the center. As the wave reaches ever larger portions of the cloud core, matter is moving inward to fill the hole left by the stuff that started free falling just before it.

Before we know what hits us, the massive wave rolls over us—and like a swimmer grabbed by a riptide, we are sucked irresistibly inward. Everything around us, all the gas and all the dust, is moving along with us at supersonic speed. We are in free fall toward the very center of the cloud.[23]

Eventually we fall onto the outer reaches of a whirling hurricane of gas and dust circulating around a bulbous central object. From the north and south poles of this monstrosity, huge collimated fountains of gas are spewing into the surrounding cloud. We've landed in a disk from which planets may eventually form.

But right now the blob is hungrily gobbling stuff from the disk. As the gas and dust reach the blob's surface, it slows down catastrophically. This material's energy of motion can't simply disappear. Energy must be conserved. So the kinetic energy of the infalling material is released in bursts of heat and light. Eventually, temperatures within the blob will increase to the point where atomic nuclei can fuse, which will release still more energy. This fusion energy will power a Sun-like star for billions of years. But it's interesting to note that the energy responsible for igniting the fires of thermonuclear fusion ultimately comes from gravity, because that is what draws material onto the blob in the first place.

Our blob is steadily heating up. But it's not a fully formed star yet. It hasn't reached fusion temperatures. Technically speaking, our blob is a *protostar*—a stellar embryo still attached to its placenta, the circumstellar disk, and still nestled deep within the womb of the ever-collapsing cloud core.

To form a protostar like our's takes about 100 000 years of gravitational collapse. Then heat within the blob increases to a point where thermal pressure almost (but not quite) stops its contraction.[24] When the thermal pressure just about counteracts gravity, the blob is said to be in hydrostatic equilibrium. In other words, it has become a stable protostar about four or five times the radius of the Sun. Accretion will continue until all the material in the disk and surrounding cloud core has either been gobbled up or blown away. At this point, the protostar will have made the transition to a full-fledged star.

From our position in the mid-plane of the disk, we can't see the stellar embryo at the center. The stuff all around us is just too thick. So we can't tell exactly how far along it has come in its gestation. One thing is clear, however: it's quite hungry and seems bent on having us for lunch. With its gravity, the ravenous monstrosity is drawing us inward through the placental disk.

But why a disk? Why didn't we and everything else simply fall onto the protostar itself?

Formation of a disk makes sense on a gut level. Remember, the process started with a spinning cloud that was somewhat elongated and flattened to begin with. It seems natural that the whole thing should flatten further as it spins, like pizza dough spun up into the air. More specifically, though, centrifugal forces in the cloud should tend to resist the inward pull of gravity toward the axis of rotation. As a result, material spiraling inward in the cloud core should form a disklike shape as it reaches the center, a shape supported by centrifugal force.

That explains the disk. But as Frank Shu says, "At the center, you've got to build a sphere, not a disk. *How do spheres come out of disks?*"[25]

Balls of pizza dough don't do *that*.

According to Alan Boss, since the physics of the situation dictate that most of the cloud core's mass will fall onto the disk, "the problem then is to get the disk's mass transported *inward* onto the protostar. And if it's going to do this, the disk must lose nearly all of its angular momentum."[26]

Put another way, if you're going to build a star, you need to move mass inward and angular momentum outward through the disk. Simultaneously. And that's no easy trick. According to Scott Kenyon of the Smithsonian Astrophysical Observatory in Cambridge, MA, the details of the process are unknown. Different mechanisms may work at different times. But in very simple terms, according to Kenyon, the process works something like this.[27]

For the sake of simplicity, Kenyon asks us to imagine two narrow rings of mass orbiting around a protostar, one just inside the other. There's no gap between the rings; they butt up against each other. Material in one ring is moving somewhat faster than the stuff in the other ring. As a result, faster material rubs against slower material. The rubbing causes friction, which heats the material in the rings. Another way of looking at this is that some energy of motion of the material (it's *kinetic* energy) is converted into heat. With less energy of motion, molecules and dust slow down. And when they slow down, they are drawn inward in the disk. This is how material is thought to move toward the star, where it will eventually be accreted.

But then there's the angular momentum problem to solve. As mass moves inward in the disk, its angular momentum would increase. Because the laws of physics state that angular momentum must be conserved, that's just not allowed without some compensation. If you'll recall the spinning Olympic figure-skater, when she drew her arms in, she compensated by speeding up. In the case of material orbiting a protostar, angular momentum is conserved in a different way, according to Kenyon. Even as some material moves inward toward the star, other material expands *outward* in compensation. The result? Material in the rings *spreads*. This allows mass to be transported inward while conserving angular momentum.[28]

Of course, in a real disk, material probably doesn't move in simple, discrete rings, as in this cartoon explanation. Movement within a real disk may involve all sorts of complex phenomena, such as convection, shocks, magnetic stresses, sound waves, density waves, tidal forces and gravitational instabilities. That being said, the result seems to be the same.

But there is still another problem for nature to solve. She may have licked the angular momentum problem for the *disk*, but the *protostar's* problem remains. As it has evolved to this stage, it has contracted enormously. But it can't spin as fast as it needs to in order for angular momentum to be conserved. So what gives?

In this case, Shu says, nature has not hidden her answer.

''For some reason, when you have the combination of a compact object which is trying to accrete matter from a rapidly rotating disk, nature likes to solve the problem by shooting stuff out,'' he says. ''This is a great puzzle with star formation. Theory says the way to form stars is by gravitational collapse— infall. But observations say it's very hard to see infall. Everywhere we look, things are coming *out*. Why is that?''[29]

The outflowing stuff he is referring to is in the bipolar jets. And these outflows, he argues, provide one way for nature to handle the protostar's angular momentum problem. Just as the disk moves some stuff out at the same time it moves other stuff in, so does the protostar eject some stuff even as it eats other stuff. In both cases, the yin and yang movement of mass is nature's way of conserving angular momentum.

''Well, what's wrong with this picture?'' Shu asks rhetorically. ''*Naïve theory*!'' he says. ''You have a star, you have a disk, you accrete onto the star and *magic happens*. Two jets leave in these two opposite directions. Why is that naïve? Well, what collimates these jets like this?'' The conventional answer is that magnetic fields do the job. ''But magnetic fields are not magic,'' Shu says. ''They have to satisfy equations.'' And to satisfy those equations, Shu says, material should be spewing out *all over the place*, not just in jets.[30]

Well, in nature, it *does* spew out all over the place, Shu continues. We just don't see it. The reason? For gas in stellar outflows to emit radiation—in other words to glow so that astronomers can see it—enough matter must be crammed together so that it will be excited to emit photons. But with the exception of the jets, the gas in outflows from forming stars just isn't dense enough for this to happen. So it's invisible.[31]

What, then, is producing all of these outflows? That's where Shu's model comes in. But it doesn't just explain the mechanism for producing jets and other outflows. It also neatly explains how nature makes a sphere come out of a disk while also satisfying the laws of nature. And the fact that the model also explains where chondrules came from, thereby helping fill in details about the origin of planets, is icing on the cake.

But perhaps the best part of Shu's model is the name he has given it: the *X-wind*.

*X* stands for *extraordinary*, and to understand how this wind works, it helps to see material from the disk actually accreting to the protostar. Trouble is, we're locked in this slow motion death spiral in the outer reaches of the disk. At the rate we're moving inward, it'll take thousands of years to reach the accretion zone. So it's obviously time to see whether this spacecraft can defy the gravity that's drawing us inward by hovering like a helicopter above the disk.

Ascending from our groove in the circumstellar LP, we move to a vantage point that offers a clear view of the accretion zone. Deploying spectrographic equipment, we observe hot ultraviolet radiation pouring from this area. Meanwhile, farther out in the disk, infra-red radiation seems to dominate.

As you may recall from the previous chapter, Donald Lynden-Bell of Cambridge University and Jim Pringle of Cambridge University proposed in 1974 that this was the result of accretion of material from a disk to a protostar.[32] The infra-red radiation, they said, was the energy emitted by gas and dust as it spiraled inward through the disk. The ultraviolet light was coming from gas as it actually accreted to the protostar.[33] As Shu puts it, ''All this stuff is falling down onto the central star,'' he says. ''BANG! It slams into the star at a few hundred kilometers per second. A lot of radiation is generated.''[34]

To observe the goings on in the accretion zone, we're wearing special sunglasses. But they're not just to protect us from frying our eyeballs with ultraviolet light. They've been designed by Frank Shu himself to allow the user to see magnetic field lines and streams of gas arcing from disk to star. Some theorists who aren't so sure about his X-wind theory might complain that we're seeing the world through Frank Shu glasses. So keep in mind that the theory is just one possibility — albeit one accepted by many experts.

What do we see? Because it has been contracting, the protostar should be spinning at enormous speed — in fact, just shy of the speed at which centrifugal forces should tear it apart. But our protostar is spinning much more slowly than that. As we look closer we note that the protostar's magnetic field has truncated the disk, creating a gap between the star and the inner edge of the disk. Cables of magnetic force connect from the equatorial region of the protostar to the inner edge of the disk. The two are essentially locked together and are therefore rotating at about the same speed.[35] That's what is slowing the star down.

The magnetic field lines are also having another important effect: they have formed funnels that are channeling gas and dust from the disk onto the protostar. We're watching accretion as it's happening.

But nature still has a physics problem to solve here, right? Like the spinning ice-skater who drew her arms inward, the spinning protostar has become more and more compact as it has contracted down from a much more expansive cloud core. And like the skater, it wants to spin more rapidly to conserve its angular momentum. But it can't because it's locked into co-rotation with the disk's slower-spinning inner edge. In other words, the star has gained a huge amount of angular momentum, and the laws of physics dictate that it must get rid of it. As Shu said, it's accomplishing this feat by blowing stuff back out. But how does it do this?

Viewing the protostar through Frank Shu glasses, we see that as the protostar tries desperately to speed up, it strains mightily on its magnetic shackles. This transfers a torque through the confining cables back to their attachment points on the inner edge of the disk. The resulting forces are whipping up gas and dust in the disk and spewing it outward in an extraordinary wind—an X-wind.

As they say in politics, if you want to know how things really work, follow the money. But since we're talking physics here, we need to follow the momentum—which is precisely what we just did. We watched as the star transferred excess angular momentum onto the inner edge of the disk through back torques on the magnetic field lines. This, in turn, whipped up material in the disk and propelled it outward. And as this material spun off in all directions, like batter spit out by an egg beater, it ultimately carried away the protostar's excess angular momentum.

But what about those massive jets of gas spewing from the top and bottom of the protostar? According to Shu, the complex interplay of forces between the protostar and inner disk cause some magnetic field lines to form channels that curve up and down toward the north and south poles of the protostar. High above each of the poles, these channels narrow and point out into space. About 50% of the material sprayed outward by the X-wind flows into these channels and is funneled toward the poles. Like water spewing from a nozzle at the end of a fire-fighter's hose, the material flowing through the magnetic channels jets outward into space, attaining outflow speeds of 100 miles per second or more.[36]

Ultimately, according to Shu, roughly a third of the material that accretes to the star from the disk is thrown back out by the X-wind. The laws of physics are satisfied. And so is Shu, because his theory neatly explains much more than the physical mechanism behind bipolar jets and how stars use them to get rid of their excess angular momentum.[37]

Material that jets from the poles of the protostar ultimately is lost to space. (Well, not really "lost," since it becomes available for later generations of star formation.) But remember that this accounts for only half of the outflows produced by the X-wind's egg beater effect. What happens to the other 50%? According to Shu, this material is lofted out of the plane of the disk and sprayed out over it, like water from a lawn sprinkler. The gas forms a wind flowing outward in all directions from the proto-star; ultimately, it helps clear out the remaining material from the cloud core. But a considerable amount of dust is lofted from the disk and sprayed around as well. When they are within the plane of the disk, the dust particles are protected from the heat of the protostar. (The dust is so dense there that the protostar's radiation can't penetrate very far.) But once the dust is lofted out of the disk, it basks in the full glare of the protostar's radiation and is quickly melted.

"In this way," Shu says, "the dust is turned into a spray of molten rock that rained over the entire solar system."[38] These tiny molten droplets, measuring a millimeter in size, are the chondrules that comprise something like 80% of chondritic meteorites such as the Portales Valley meteorite. Why is this important? As we've seen, chondrules are believed to be one of the basic building blocks of planetesimals. And planetesimals, in turn, were mini-planets that were brought together by their mutual gravity to form the Earth, the other rocky planets and, perhaps, the rocky cores of the giant, gaseous planets like Jupiter. But Shu goes even further, arguing that the spray of chondrules ultimately *triggered* planet formation in our solar system. In other words, the formation of chondrules by the X-wind was a pivotal moment in the cosmic chain of events that led to us.

But this is getting ahead of our story, since the origin of planets is the subject of Chapter 11. We still have to observe the birth of our star.

From our vantage point above the disk, we can see molten material spraying outward in every direction. We've also got a good view of the protostar itself, as well as its greater surround-ings. Based on its size, it seems as if it's destined to become a star about the same size as our Sun. Moreover, its winds are blowing away the veils of dust and gas from the original cloud core than once completely enveloped it. In fact, we can now see out of the core into the region of space beyond. We're in a calm, Taurus-like molecular cloud, so we don't see any giant O stars around. Way off in the distance, a few light-years away, we can make out another protostar about to emerge from its womb.

If any astronomers back in our solar system had been looking in our direction earlier in the evolution of our protostar, they might have seen knots of shocked gas strung out in a line from the cloud core. Based on their observations of these Herbig–Haro objects, they would have concluded that a protostar with bipolar jets was embedded within the cloud core. But they would not have been able to see the protostar itself because of the shroud of dust around it. Now, however, any astronomers looking our way can see our protostar revealed for the first time—a bright ball of hot gas surrounded by a disk of remnant cloud material.

The protostar has reached the end of its gestation. No longer a stellar embryo, it has finally been born. By conventional astronomical classification, it is now a T Tauri star: a colicky star baby prone to nasty eruptions of hot gas from its surface. Unlike a human infant, the star will remain attached to its placenta—the disk—for awhile. And it is from the remaining gas and dust in this disk that the final creative act in the origin of a solar system will take place: the fashioning of planets.

# 10

## SOLAR SYSTEMS

### *From Rococo Clockwork to Extrasolar Planets*

*And so there are innumerable suns, and likewise an infinite number of earths circling about those suns . . .*

— Geordano Bruno[1]

*For I will prove that the earth does have motion, that it surpasses the moon in brightness, and that it is not the sump where the universe's filth and ephemera collect.*

— Galileo Galilei[2]

*It is not improbable, I must point out, that there are inhabitants not only on the moon but on Jupiter too . . .*

— Johannes Kepler[3]

Nearly 2000 years before Christopher Columbus set off on his journey to reach land by sailing west across the Atlantic, the ancient Greeks had already used their sophisticated grasp of geometry to correctly calculate the size and shape of the Earth. From the shadow cast by the Earth on the moon during a lunar eclipse, the Greeks figured out that our planet was a sphere. And by measuring the angles made by solar rays on summer's first day in two parts of Egypt, Erastothenes calculated the size of the Earth.

But despite these impressive intellectual feats, the Greeks did get one thing wrong about the Earth. In their cosmic maps, they placed it at the center of the universe. In the fourth century BC, Aristotle and the astronomer Eudoxus proposed that the stars were affixed to invisible, concentric spheres orbiting the Earth. This made sense, given that the vast majority of stars appeared to move en masse across the vault of the sky each evening, remaining motionless with respect to each other, year in and year out. But even they recognized that there was one little problem: a handful of objects seemed to wander relative to the fixed stars. Noticing this itinerant behavior, the Greeks called these objects *planetes—wanderers* in English. At the time, seven such objects were known: the Sun, the Moon, Mercury, Venus, Mars, Jupiter and Saturn. (They were all grouped as planets because the Greeks didn't know that the Sun and Moon were different kinds of objects.)

Although they wandered too and fro, overall the planets progressed around the Earth, just like the other celestial objects. So the geocentric model still seemed logical, and the planets' departure from perfect orbital circularity simply remained an annoying anomaly.

In the second century AD, Ptolemy attempted a fix that retained an unmoving Earth at the center of the cosmos while explaining the wandering of the planets. In his model, he kept the non-wandering stars pinned to their own orbiting sphere enclosing the universe. Meanwhile, he accommodated the motions of the planets by ascribing them to epicycles, a complex system of circular motions within circular motions. While a planet traced out a circular path around the Earth, Ptolemy proposed that it also had a secondary, *epicyclic* orbit centered on the path itself. (In some cases, he even had to resort to additional epicycles.)

Like the internal gearing of a mechanical clock, Ptolemy's model was elaborately complex. But that didn't make it accurate. Although the system seemed to explain the odd wanderings of the planets, ancient mariners found that it failed to predict the movement of celestial bodies with the precision needed for navigation. Some were lost at sea.

Despite its flaws, the Ptolemaic system was the best thing going in the Christian world. And by medieval times, the Christian cosmic map consisted of 10 rotating spheres concentric to the Earth at the center. Ptolemaic epicycles within two spheres explained the movements of the Sun and Moon. Other epicycles within the spheres of Mercury, Venus, Mars, Jupiter and Saturn explained their movements. Two spheres contained the stars. And enclosing the entire rococo mechanism was an outermost sphere, the *Primum Mobile*. This marked the physical boundary of the universe. Beyond it was a realm outside of space and time itself—the realm of God.

But the medieval picture of the cosmos went even deeper than Ptolemy's celestial clockwork. Medievals believed reality consisted not just of the physical world but of a *metaphysical* one as well. The metaphysical order consisted of a "Great Chain of Being," with God at the top, according to Margaret Wertheim, author of *The Pearly Gates of Cyberspace: A History of Space from Dante to the Internet*. Next came the angelic beings, followed by humans. Beneath us in the chain were the animals, plants and, lastly, non-living things. Earth was the realm of physical space. Everything else—the other planets, the Moon, the Sun and the other stars—were seen as perfect, unchanging bodies in the heavenly, metaphysical realm of angels.

We bring a mechanistic bias to our understanding of medieval cosmology. When we read that people of the medieval period believed humanity was at the center of the universe, we assume this meant the *physical* universe, which we believe to be all there is. But medievals believed humans were at the center of the metaphysical cosmos as well. Within this space, Wertheim says, humans were halfway between the heavenly "ethereal beings" and the earthly "material things." And as such, we occupied a *special* position in the metaphysical Chain of Being.[4]

Today, of course, we know we inhabit a vastly different kind of solar system, and we no longer think of our relationship to the cosmos in the same way. Today, we know that the Moon and planets are different than stars, and that they are *physical* entities that *evolve* over time, just like Earth. From the point of view of science, there is no "soul-space," just the physical universe (or the multiverse, if it exists).

Most important, once we realized that space beyond Earth was physical and could be explored, we used telescopes and spacecraft to discover the gargantuan scale of the universe. We've felt insignificant and inconsequential ever since.

Whereas medievals believed that each position in the physical cosmos was special in some way, and that our distinction was to be at the center, today astronomy tells us that *no* position in the cosmos is special. And whereas medievals believed we alone among living things enjoyed the special quality of intellect, today many of us assume our species is but one tiny clan in an extended community of intelligent creatures populating the universe—and that in all likelihood, we're cretins on the cosmic IQ scale.

In short, today we think of ourselves as an average species living on an ordinary chunk of rock orbiting a typical star in a typical solar system within a spiral galaxy that's just like billions and billions and billions of others. After 500 years of astounding intellectual achievement, science has left us with this terrible inferiority complex. Despite our amazing success at deciphering the origin and evolutionary history of the universe—and despite our remarkable ability just to *ask* where we fit in the cosmic scheme of things—many of us believe that the most profound lesson of science is that we are truly nothing special.

Many of the astronomers who have helped shape our modern view of the cosmos do not share this view, as will become clear in this chapter and those to follow. The truth is, we just don't know whether Earthlike planets are common or rare. The most incisive theoretical models on the origin of solar systems, as well as the very best observational evidence about solar systems other than our own, do not yet allow us to conclude that our solar system is typical. And while there is good reason to believe that life is common in the universe, the same cannot be said of intelligent life.

In fact, based on the limited evidence gathered so far, it is equally possible that our planet and our solar system as a whole are quite *atypical*, and that the conditions needed to nurture the origin and sustenance of complex life, including intelligent life, may be quite rare. But even if the newly flourishing field of astrobiology should eventually show otherwise, our very ability to arrive at any conclusion argues that we've gone too far in demoting ourselves from Wertheim's ''soul-space'' linchpin to the cosmic equivalent of a bacterium on an elephant's shoulder.

With this in mind, the next two chapters will highlight not only what we're learning about solar systems and the origin of planets, but also what these findings are revealing about our place in the universe.

Before scientists could begin to tease apart the details of how planets form, they had to know that there was such a thing as a physical solar system to study. And for that to happen, the medieval notion that celestial bodies were not physical entities had to give way to our modern view of the cosmos.

271

The first cracks in the medieval edifice appeared in the 1400s, when Nicholas Cusanus, or Nicholas of Cusa, a Catholic Cardinal, proposed an idea that today is a cornerstone of modern cosmology: *The universe has no center.*

Cusa arrived at this conclusion from a religious perspective. As Margaret Wertheim tells the story, Cusa believed that God alone was absolute, which meant the universe had to be boundless. Otherwise, it would have an absolute end. Now, if the universe had no boundary—if it were infinite—it could not have a center, and every point would be equivalent to every other point. "With one blow, then," Wertheim writes, "the cardinal from Kues shattered the medieval 'world-bubble' and released the cosmos from the crystalline prison of its celestial 'spheres.'" [5]

Cusa repudiated the idea of a physical hierarchy in the cosmos. And if the stars occupy no more special a place than we do, then they are no closer to God than we are. In fact, Cusa even proposed that our planet was just like all the other celestial objects, with its own light and heat. What a remarkable idea for a person who lived centuries before the modern era of remote sensing of Earth from space. But Cusa actually went further, proposing that we humans were not alone in the cosmos. The universe, he said, had many worlds, each of which was inhabited.

And so in one master-stroke, Cusa anticipated not only modern cosmology but *astrobiology* as well.

The abandonment of the medieval geocentric cosmology has been held up as the start of humankind's demotion. But according to Owen Gingerich, an astronomer and historian of science at Harvard, some scholars argue that in the medieval scheme of things, "the center of the universe was the sink-hole of all the dross. It was not the preferred place." So Cusa and others who championed similar views actually *elevated* our status, making Earth's position more noble, not less so. [6]

If you think about this as Cusa did, but with our current understanding of the physical universe, our solar system is no less exalted than a galaxy. Indeed, if you go further and think about it in John Bally's ecological terms, each part of the galactic system has its own special role to play. When they explode as supernovae, massive stars are the primary producers, supplying the raw materials for construction of new stars and planets. The expanding shells of supernovae remnants are the galaxy's recycling agents, responsible for sweeping up raw materials and, possibly, forming them into molecular clouds. Molecular clouds, in turn, are the spawning grounds for solar systems.

And us? Maybe because of the cosmic luck of the draw — being in the right place at the right time as a species — it is our special role to figure all of this out. In other words, to provide a way for the universe to become aware of itself.

In any case, Cusa was too far ahead of the curve. He died 150 years before the telescope was even invented. So there was no way for anyone to test his radical cosmology. Moreover, printing with moveable type had not yet been invented, so the dissemination of ideas like Cusa's was very limited.

Even so, Cusa did have an important influence on other thinkers, including Giordano Bruno, an Italian philosopher born in Nola, Italy, in 1548. Like Cusa, Bruno proposed that the universe was infinite and contained an infinite number of worlds:

*"All things are one: the heavens, the immensity of space, our mother earth, the encompassing universe, the ethereal region through which all things move and continue on their way. Herein our senses may perceive innumerable heavenly bodies, stars, spheres, suns and earths; and reason may deduce an infinitude of them. The universe, immense and infinite, is the sum total of all that space and all the bodies it contains."*[7]

Bruno became a candidate in 1591 for the chair of mathematics at the University of Padua. But his main competition, Galileo, got the job. Bruno's ideas increasingly came to be seen as heretical, and in 1600 the Roman Inquisition burned him alive.

But even before Bruno was sent to the flames, the crack in the Earth-centered, clockwork cosmos first pried opened by Cusa was already getting wider, thanks to a practical problem: ships getting lost at sea because of the inaccuracies in the Ptolemaic system. And then, Nicholas Copernicus came along with a crowbar.

By the early 1500s, according to Owen Gingerich, "there were criticisms of the Ptolemaic system in the air." By 1514, Copernicus began to develop an alternative. But it took about 20 years of waiting for observations "to check the parameters" before he really started writing his book in earnest.

273

In 1530, Copernicus presented a tentative outline of a model he hoped would solve some of the problems of the Ptolemaic system. Much to the shock of some of his contemporaries, he placed the Sun at the center of the universe and set the formerly stationary Earth in motion around it. Thirteen years later he published a full description of the theory in his book *On the Revolutions of Heavenly Spheres* (*De revolutionibus orbium coelestium*).

Copernicus, a Pole who studied canon law, mathematics and astronomy and then became a physician, didn't invent the heliocentric idea. It had been considered by the ancient Greeks, among others. But in much more painstaking detail, he worked out a system in which the Earth and the "wandering stars" revolved around the Sun. Like Ptolemy, Copernicus held that the "sphere of the fixed stars" bounded the universe; but he broke ranks by holding that it was "immovable." They only appeared to move because of the motions of the Earth around the Sun.

Among the justifications he gave for putting the Sun at the center of the universe was this one:

*"For who would place this lamp of a beautiful temple in another or better place than this wherefrom it can illuminate everything at the same time?"*[8]

The heliocentric idea went against a thousand years of belief, so Copernicus knew it wouldn't exactly be adopted by acclamation. And in the very first line of his book, he anticipated his critics:

*"Since the newness of the hypotheses of this work – which sets the earth in motion and puts an immovable sun at the centre of the universe – has already received a great deal of publicity, I have no doubt that certain of the savants have taken grave offense and think it wrong to raise any disturbance among liberal disciplines which have had the right set up for a long time now. If, however, they are willing to weigh the matter scrupulously, they will find that the author of this work has done nothing which merits blame."*[9]

But there was one thing for which Copernicus merited blame: his model turned out to be just as clunky as Ptolemy's. And it remained inaccurate in predicting the motions of celestial bodies. This is because Copernicus clung to the idea that planets orbited in *circular* orbits, as opposed to the slightly elliptical orbits we now know are true.

"Copernicus did not increase precision," Gingerich says. "He created the framework on which a more accurate system could be built."

Even so, for young scholars of the day, the heliocentric scheme actually offered one advantage. By setting the Earth and all the other planets in motion around the Sun, Copernicus enlarged the cosmos. In the Ptolemaic system, the sphere of the stars had to be quite close. Otherwise it would have to orbit at enormous speed to complete one full circle around the Earth in a day — so fast that the sphere might disintegrate. In the Copernican system, if the stars are as close by as in the Ptolemaic system, you would expect to see them moving back and forth as the Earth circles the Sun. The fact that they don't means the universe has to be big so the stars can be far away. To understand why, imagine looking out the window of a moving car. After you've driven just a short distance, telephone poles in the foreground have moved across your field of view. But the apparent position of a mountain off on the horizon hasn't seemed to move at all. The reason, of course, is that the mountain is very far away.

Copernicus was not really a revolutionary. As the canon of the Frauenburg cathedral, he was a professional expert in the laws of the Catholic Church. He also maintained friends in high Catholic places, including Tiedeman Giese, Bishop of Culm, "a man," Copernicus wrote, "filled with the greatest zeal for the divine and liberal arts."[10] Most telling, though, was the preface to his book, which is written in the form of a letter to Pope Paul III. In the preface, which dedicates the book to the Pope, Copernicus writes in self-deprecating terms that "the scorn which I had to fear on account of the newness and absurdity of my opinion almost drove me to abandon a work already undertaken." But learned men of the church urged him to go public with "a work which I had kept hidden among my things for not merely nine years, but for almost four times nine years."

According to Gingerich, Copernicus shows here that he is "justifiably worried about the reception his book will have because he knows it will alarm a lot of people." In the preface, he seeks to protect his book from criticism. He implies that he is, in Gingerich's description, "only an astronomer trying to predict the positions of the planets." In fact, Copernicus makes no claim that his system is a real description of the physical world. Instead, he presents it merely as a "mathematical trick," Gingerich says. "He is just trying to fiddle the geometric figures until he can predict the seemingly capricious motions of the planets in the heavens."

In this way, Copernicus holds fast to the basic medieval notion of a cosmic hierarchy. The spheres of the planets, other than Earth, remain in the metaphysical realm. These are not *places* that anyone can *visit*. And beyond the planets is the most exalted of the spheres: "...the first and highest of all is the sphere of the fixed stars," Copernicus writes, "which comprehends itself and all things, and is accordingly immovable."[11]

Copernicus barely got to see his work published. On May 24, 1543, he lay on his deathbed in Frauenburg, having been stricken by dysentery and then partly paralyzed by a stroke. An advance copy of his book was brought to him before he passed away.

Cusa's ideas may have been more revolutionary in some respects, but it was the Copernican notion of a Sun-centered solar system that ultimately changed the world. By placing the Earth in motion around the Sun, Copernicus began the dismantling of the medieval edifice. It would take other men, including Johannes Kepler, Galileo Galilei and Isaac Newton, to bring it down completely. This development in human thought has come to be called the Copernican revolution.

"There was a certain kind of change in the air, and Copernicus crystallized it and gave it focus," Gingerich says. "But the change didn't end there. Kepler forged it—it was Kepler's work that won the technical battle. And Gallileo made it intellectually acceptable—his book won the psychological battle."

Why did the Copernican revolution begin at that particular moment and not earlier? Observations of the heavenly bodies did not make the difference. "There were no particular observations that Copernicus could make that someone could not have made 500 years earlier," Gingerich says. So other factors had to be important.

One was the invention of moveable type. This gave Copernicus access to research materials that made it easier for him to devise his system. And it also insured that his ideas would be well distributed in the form of a book. "His idea was out there, and people could find it," Gingerich says.

Another factor was the discovery of America. "That's important, and it may take you by surprise," he says. The reason? It undermined Ptolemy's reputation. In the 1480s, he was considered the world's greatest geographer. "But suddenly it was realized that Ptolemy didn't know it all," Gingerich says. "Suddenly, there was a whole new hemisphere that Ptolemy was oblivious to. This created a level of openness, allowing a challenge to the received wisdom."

One of the challengers was Kepler, whose great contribution to the unfolding revolution was to see the cosmos beyond Earth as a physical place. And the same laws that operated on Earth were at work there as well, he argued. "Kepler took the basic notions of Copernicus to their logical conclusion," Gingerich says. "He treated the Earth as just one of the other planets."

Since Kepler believed that heavenly bodies like the Moon actually had a physical reality just like our own planet, he also was the first person to imagine that they could be visited. He is even credited with authoring the first ever work of science fiction: *Somnium*, or *The Dream*, in which he writes of a journey to the moon, which is inhabited by lizard-like creatures.

But Kepler was not just the first science fiction writer. He also is credited with connecting the heliocentric model of the universe to physics. And that, according to Margaret Wertheim, makes him the first *astrophysicist*.

Working with observational data gathered by the renowned Danish astronomer Tycho Brahe, Kepler discovered the three laws of planetary motion. The first law rescued the heliocentric model from epicycles. It states that the planets move not in circles but in *elliptical* orbits, with the Sun at one focus of each orbit. In working through the details of how such a system would work, Kepler vanquished the inaccuracies of the Ptolemaic and Copernican models.

In his second law, Kepler removed the hand of God as animator of the day-to-day motion of the planets. They continue to move, he said, because of a physical force coming from the Sun. Finally, the third law provided a formula for calculating the size of a planet's orbit relative to those of other planets. With this tool, astronomers could begin to *map* the new-found physical space of the solar system.

"Kepler became the architect of the Copernican system," Gingerich says. In so doing, he blew away the metaphysical mists obscuring humankind's view of the purely physical solar system.

Writes Wertheim: "From here on in, celestial space will ring not with the songs of cherubim and seraphim, but with the roar of rockets and the woosh of warp drives."[12]

Kepler was the first in a long line of astrophysical theorists whose most prominent practitioners today include the likes of Frank Shu. Like Shu, Kepler is said to have used a telescope only once or twice.[13] As in the case of Shu's theoretical work, Kepler's theories were tied closely to observations. And just as the astrophysical theories of scientists like Shu are put to observational tests, so was Kepler's radical notion that the heavenly bodies were physical places, just like the Earth.

The person to do it was Galileo. When he pointed his telescope at the Moon, he found a rugged landscape of mountains and valleys. Pointing it toward the Sun, Galileo discovered spots moving across its surface. Here was clear evidence that Kepler was right. These were real, physical, ever *changing* worlds.

Galileo also made another critical contribution. This time, his tool was not a telescope. Instead, he pushed wooden blocks across a table to test traditional ideas about inertia. Aristotle had said that bodies at rest tend to stay at rest. And that was that. But by polishing his table to reduce friction and pushing his blocks again, Galileo found that they moved farther. From these experiments he was able to add the key second part to our understanding of inertia. Not only do bodies at rest tend to remain at rest, but *bodies in motion tend to remain in motion.*

This eliminated one of the prime objections to the idea that the Earth moved around the Sun. The Greeks had rejected the idea for this reason: if the Earth moved, they wondered, why doesn't a person who jumps in the air land away from his take off point? Shouldn't the Earth simply move out from under him? Galileo's revised concept of inertia provided the answer. As a person begins a jump into the air, he is already moving with the Earth. Because an object in motion tends to remain in motion, he retains that movement throughout his leap and therefore lands in the same place.

"Kepler's insistence that astronomy had physical causes was radical and revolutionary," Gingerich says. Unlike Copernicus, "he was not simply using geometrical devices as fictional elements." So Copernicus could get away with championing a system that placed the stars at large distances from Earth because it was just a mathematical trick. It didn't really threaten the metaphysical cosmos, "which was very cozy with Earth in the middle and these spheres of the planets and the fixed stars with heaven just beyond," Gingerich says. "There was an easily envisioned position for the heavens, and everything was more or less on a human scale." But then Galileo came along, claiming that we really *are* in a vast cosmos. "Now, you're not sure exactly where heaven is. And this causes a tremendous psychological upheaval."

It didn't help Galileo's cause that he was pushy. He urged the Church to recognize that the Ptolemaic system was obsolete.

"Galileo was too good a propagandizer," Gingerich says. "And he was stealing the thunder from the clerics, who thought they were guardians and adjudicators of truth." Already threatened by the rise of Protestantism, the Vatican just could not tolerate anyone "rocking the boat." So Galileo was forced to recant his ideas, and he was placed under house arrest for the rest of his life.

But the edifice of medieval cosmology was crumbling nonetheless. All that was needed to complete the demolition was a quantitative description of the force responsible for the planets' revolutions about the Sun. Sir Isaac Newton, born in 1642, the year that Galileo died, provided that description. The force, he said, was gravity, and the equation he formulated—the *universal law of gravitation*—described how it worked.

Newton didn't actually know what gravity *was*. It took Einstein to figure that out. (A warp in the fabric of spacetime, caused by mass.) But in Newton's equation- and diagram-filled *Principia*, he proposed that anything with mass exerted this force. So the Sun attracted the Earth, and vice versa. And the same gravity that was said to have caused an apple to fall on Newton's head (exerted by Earth's mass) was responsible for locking the planets in their orbits (exerted by the Sun's mass).

To craft his law of gravity, Newton took his cues from observations of nature, and then he showed how it affected the motions of various kinds of matter:

*"... I derive from the celestial phenomena the forces of gravity with which bodies tend to the sun and the several planets. Then from these forces, by other propositions which are also mathematical, I deduce the motions of the planets, the comets, the moon, and the sea."*[14]

When applied to the Sun-centered solar system, Newton's equations supplied what was missing to fully physicalize the cosmos. In this way, Newton did much more than describe gravitation. He described *universal* gravitation, proposing "that everything attracts everything else," Gingerich says. So the same force that operates here on Earth operates celestially. "That was the core of Newton's achievement."

\*    \*    \*

With this contribution to the Copernican revolution, the medieval cosmic edifice finally lay in ruins. Risen in its place was the modern solar system, and the stars beyond. With the recognition that the planets revolved around the Sun, with Kepler's laws of planetary motion, and with Newton's universal law of gravitation, others could now contemplate how the physical bodies of the solar system actually came into being by *physical* processes. A little less than 100 years later, Immanuel Kant did just that, proposing that planets form within a solar nebula—a disk of primordial material swirling around the Sun. To this day, Kant's model remains the overarching blueprint for the construction of a solar system.

It was a visionary leap of imagination, but the blueprint was missing the important physical details. Filling in those details has been a major preoccupation of modern astronomy. And until recent years, scientists had only one solar system to worry about: our own. It simply was the only one known. Few astronomers imagined that when other solar systems were discovered they would be vastly different from our own. So they felt confident basing their theories on planet formation on a sample of one. In hindsight, it's easy to say they should have known better.

In 1994, Alex Wolszczan of Penn State University made the first widely accepted discovery of planets orbiting another star. But this ''solar system'' was fundamentally different from our own. The planets orbit a pulsar—the dead star that remains after a supernova explosion. Astronomers believe the pulsar planets formed out of the rubble left by the explosion. (There is no way they could have existed beforehand and survived the explosion intact.) To some astronomers, this was heartening news. After all, if planets could form in the immediate aftermath of such a violent event, maybe planets were common. On the other hand, they knew that they would have to look elsewhere for life. A pulsar planet is no place for life to get started, because a supernova's fires would drive off the relatively light elements need for biology, such as carbon, oxygen and nitrogen.

Wolszczan's discovery was an important milestone. But the question of whether *Sun-like* stars harbored planets was left up in the air. A definitive answer came in October of 1995, with the announcement by Michel Mayor, Didier Queloz and colleagues at the Geneva Observatory in Switzerland that they had indirectly detected the presence of a Jupiter-like planet circling a Sun-like star designated 51 Pegasi.

In discovering this solar system, Mayor, Queloz and colleagues didn't actually spy a giant, Jupiter-like orb of gas circling 51 Peg. Direct imaging of a so-called *extrasolar planet* (means *outside our solar system*) is difficult, to say the least. Planets do not emit their own light, at least not in the visible end of the spectrum. The light we see coming from Mars, Venus and the other planets in our own solar system is actually reflected from the Sun. These objects are visible because they're close to us. But an extrasolar planet like 51 Peg's is many light-years away. And this presents problems. To an observer here on Earth, a planet and its parent star are separated by a very tiny fraction of a degree on the sky.[15] Moreover, the star is extremely bright, while the reflected light from the planet is dismally dim. The result is that the star's light simply swamps the planet's, making it exceedingly difficult to distinguish between the two. Because of this, no one has yet imaged an extrasolar planet directly.

Mayor and Queloz's discovery of an extrasolar planet orbiting 51 Peg was made with another technique—one that has become the mainstay of planet detection efforts. Called precision

Doppler detection, it relies on the gravitational tug an extrasolar planet gives to its star. The star's gravity may be very much larger, but the planet still manages to pull on the star hard enough to make it move ever so slightly. (Jupiter causes the Sun to shift at a speed of about 10 yards per second.) The result is that star and planet both orbit a common center of mass. Given the huge size of the star, its orbital motion is nothing more than a subtle wobble.

Planet seekers like Mayor and Queloz, and the renowned team of Geoffrey Marcy of the University of California at Berkeley and Paul Butler of the Carnegie Institution of Washington, try to detect this wobble by measuring the Doppler shifts it causes in the star's light. As the wobble brings the star slightly closer to us, the light waves shift toward the blue end of the spectrum. As the wobble takes it away from us, the light reddens. And the time it takes for the star to make one complete cycle — one full wobble, with a period of blueshift and a period of redshift — is equal to the time it takes the planet to complete one orbit around the star. Thus, by measuring the wobble, astronomers can not only detect the presence of the planet, but they can also deduce the planet's orbital period and how close to the star it orbits.

The data also permit another calculation. The bigger the planet, the greater is its tug on the star. So a more massive planet will pull the star into a bigger wobble. This is reflected in bigger redshifts and blueshifts. So by measuring the *amplitude* of the spectral change, astronomers can estimate the mass of the planet. (But for technical reasons, they can be absolutely sure only of a lower limit on the mass.)

Since Mayor and Queloz's discovery of 51 Peg's extrasolar planet, they and other planet hunters have fundamentally altered our view of the cosmos by showing that planets are not all that uncommon. Between 1995 and late September of 2002, astronomers had used the Doppler technique to monitor about 1200 Sun-like stars. These efforts turned up 102 extrasolar planets. And the rate of new discoveries seems to be increasing.

But it's one thing to discover a host of extrasolar planets and another thing entirely to discover solar systems that look something like our own. ''We don't really know whether our solar system is typical or not,'' notes Ray Jayawardhana, an astronomer

282

at the University of Michigan. "A lot hinges on the answer. As a human being, I want to know whether we are rare or not."[16]

In our solar system, rocky terrestrial planets—Venus, Earth and Mars—move in almost circular orbits around the Sun. Our planet resides in a so-called *habitable zone*, where liquid water can exist on the surface and climatic conditions are conducive to life. We also share the solar system with a giant gaseous planet—Jupiter—that resides in a nearly circular orbit at about 5.2 astronomical units (about 512 million miles).

Before 1995, there wasn't much reason to think that solar systems elsewhere in the galaxy would look dramatically different from this. But that changed with Mayor and Queloz's discovery. The 51 Peg solar system was like nothing any astronomer had imagined. It's planet turned out to be 150 times more massive than Earth, yet it orbited 20 times more closely to its parent star than Earth circles the Sun.

"I expected there to be a diversity of solar systems," comments Jack Lissauer, a planet formation expert at the NASA Ames Research Center in California. "But I never suspected anything like 51 Peg."[17]

This was just the first surprise. Based on the relationships between stellar parents and planetary siblings, the vast majority of solar systems discovered since then look more like dysfunctional families than nurturing environments for life. As is the case with 51 Peg, many giant gaseous planets in these systems cling disturbingly close to home—even closer than Mercury orbits the Sun. Such planets have been dubbed *hot Jupiters*. Meanwhile, other extrasolar planets have adopted wildly eccentric orbits around their parents.

Based on our own solar system, astronomers had thought that giant planets would form only between 4 and 5 astronomical units from their Sun-like parent stars. So the close-orbiting hot Jupiters have been a bit of a puzzle. Astronomers don't believe it is possible for giant gaseous planets to form close in, so one possibility is that they form farther out and migrate in—movement that would spell doom for any Earthlike planets in their way.

Similarly, giant planets orbiting in eccentric orbits would not be healthy for plants, children and other living things on alien terrestrial planets. If our Jupiter were in an eccentric orbit, it would gravitationally pitch the Earth and Moon out of the solar system. Conversely, because Jupiter is in a nearly circular orbit, we stay in a nice, stable, nearly circular orbit that is conducive to complex life. "Jupiter enforces circularity on everybody else," says Paul Butler of the Carnegie Institution of Washington. "He's the big boy on the block. If a planet gets out of a circular orbit, Jupiter will grab it and toss it out of the solar system."[18]

The bottom line is that a planet of Jupiter's mass orbiting at a safe distance and with a low eccentricity seems to be a major plus for life. So astronomers have been working hard to find solar systems with Jupiters that look like ours. And in June of 2002, Butler and Geoff Marcy announced that they had come closer than ever before to accomplishing this goal with the discovery of "a first cousin" of Jupiter's. It orbits a Sun-like star designated 55 Cancri at a distance of about 6 AU, which is similar to Jupiter's distance to the Sun. And the orbit is not that far out of round.[19]

"All other extrasolar planets discovered up to now orbit closer to the parent star, and most of them have had elongated, eccentric orbits," Marcy said in a press release about the discovery. "This new planet orbits as far from its star as our own Jupiter orbits the Sun."

Unfortunately, there was a catch. Two of them, actually — *two hot Jupiters* that also orbit 55 Cancri, both within 0.2 AU of the star. So even though this solar system has one reasonably well adjusted child, by our standards it is still a dysfunctional family.

But as reported by Richard Kerr in the journal *Science*, literally an hour before Butler and Marcy made their announcement at a televised press conference at NASA headquarters, Michel Mayor and his colleagues were telling journalists they had discovered a solar system even more similar to our own. As Mayor described it at a scientific conference later that week, it consists of a Sun-like star (designated HD 190360), and a planet 1.1 times the mass of Jupiter. The planet orbits at a distance quite similar to Jupiter's, and with a very low eccentricity. The main difference between this system and 55 Cancri is that it harbors no hot Jupiters that could have ejected Earth-like planets. As Kerr reported in *Science*, this system "looks nearly identical to what alien observers would see if they looked at ours."[20]

Although this discovery brings hope that many more ordinary Jupiter-like planets will be discovered, the extrasolar planetary zoo still looks astoundingly diverse. In retrospect, Paul Butler says astronomers should have expected this. "In the galaxy, there are 100 billion to 200 billion stars, and each one is a laboratory where an experiment in planet formation could have been run," he says. "So it really shouldn't be a surprise that there are many, many, many ways to put a solar system together."[21]

Based on what astronomers have seen so far, there appear to be three solar system "archetypes," Butler says. The first is the 51 Pegasi kind of system, with a gargantuan planet orbiting at a tiny distance from its star. The second is a system consisting of a massive planet (or more than one) moving around its star in an eccentric orbit. And the third archetype, according to Butler, is the "classical" kind of solar system—our kind.

At the moment, it's much too soon to reach any conclusions about which archetype may be "normal." That's because the Doppler technique has some limitations. For example, detecting a planet with a long orbital period, and thus a big, Jupiter-like orbit, can require a significant portion of an astronomer's career. Imagine an alien astronomer on a planet 20 light-years from us using the technique to detect planets in our solar system. She'd have to take periodic measurements of our Sun for almost 12 years before feeling confident enough to announce that she *may* have discovered a Jupiter. That's because Jupiter, by far the most massive planet in our solar system, takes that long to make one orbit. (If she wanted to be more confident about her findings, she might want to observe *two* orbital cycles before declaring her detection a done deal.) Based on her detection of Jupiter, she'd probably manage to retain her grant funding. But it had better be a long-running contract, because she'd have to monitor the Sun for nearly *30 years* to be absolutely sure she had detected Saturn. (And Uranus? Eighty-four years.)

An even more fundamental limitation of the Doppler technique is that using it to detect planets much smaller than Saturn is virtually impossible with current technology. That's because the smaller the mass of the planet, the smaller the wobble of the star. So there may well be plenty of stars out there with planets the same size as Mars and Earth. But astronomers just can't detect them yet.

285

Despite this limitation, however, astronomers have detected quite a few gaseous planets smaller than Jupiter. And it turns out that relatively smaller planets actually dominate over the bigger ones.

To chart the distribution of extrasolar planet masses, Butler and his colleague Geoff Marcy have "placed" them into "bins" reserved for planets of different sizes. So, for example, one bin contains planets up to 1 Jupiter mass, the next from 1 to 2 and so on. This analysis reveals that there are far more planets in the small-mass bins than in the bigger mass bins.

"It's critical to understand that large mass planets are *easy* to find," Butler says. "They gravitationally tug the star much more easily. So that from five Jupiter masses on out to 80 Jupiter masses, those things are *trivial* to find. On the other hand, planets down at 1 Jupiter mass and less are incredibly *difficult* to find. They're barely tugging the star, so they're really pushing our measurement precision. And yet even though planets that would go in this bin are the hardest to find—and thus we're probably not detecting all of them—we're already seeing that this is where most of the planets are."[22]

As for finding extrasolar planets like Earth, astronomers will have to wait for technology to improve. NASA is planning a space-based instrument called Terrestrial Planet Finder that would be designed to measure the size, temperature, and placing of planets as small as the Earth in the habitable zones of far away solar systems. And with spectroscopy, TPF would help astrobiologists determine whether these planets have life-sustaining gases such as carbon dioxide and water vapor present. The European Space Agency is planning a similar mission, called Darwin. Like TPF, it would be designed to detect and study Earthlike worlds. No firm launch dates have been set for either TPF or Darwin, but it's unlikely that they would get off the ground before 2015.

\*   \*   \*

Extrasolar planets comprise just one piece of the puzzle of solar system origins. By fitting it together with the insights offered by planet formation theory, as well as the ground truth of chemicals and chondrules embedded in meteorites, scientists are trying to complete the puzzle. When they succeed, we should have answers to a fundamental question. Are the extra-solar planetary systems discovered so far only *seemingly* strange? Might they turn out to be the norm, and might our own solar system actually be the one that's strange? Answering that question definitively will take a long time. But already, the outlines of the puzzle picture are beginning to take shape.

# 11

# PEBBLES TO PLANETS

## The Origin of the Solar System

*It's a tough universe out there.*
— John S. Lewis[1]

*Planets are the apartment buildings that life moves into.*
— Anne Kinney[2]

When completing a jigsaw puzzle, it's often best to work on the border of the picture first to establish a framework that makes it easier to complete the puzzle. And in a sense, that's what the evidence from meteors has allowed astronomers to do. It hints at the conditions that prevailed during the formation of our solar system, and has revealed the actual materials that formed planet embryos.

So at a minimum, a successful theory explaining how planets form around young stars must account for those conditions and materials. In this way, the meteoritic evidence has established scientific boundaries that theoreticians must stay within. And just as piecing together the edge of a puzzle begins to build the overall picture, so has the meteoritic evidence given astronomers some sense of what the ultimate picture of planet origins will look like.

What do meteorites reveal? Specifically, what have scientists learned from *chondrites*? — the kind of primitive meteorite that rained down in fragments over Portales, New Mexico. Recall that chondrites are believed to be chips off the old block, fragments of planetary embryos called planetesimals that later grew up into full-fledged terrestrial planets. The chondrites consist of the chondrules — millimeter-size spheres of rock — embedded in a matrix rock, in much the same way that little stones are embedded within the cement matrix of concrete. Sometimes associated with the chondrules are calcium–aluminum

288

inclusions, or CAIs, which contain a specific mix of radioactive isotopes.

According to Frank Shu, water found within the matrix rock of chondrites provides a key insight. "Water gets driven out of rocks when we heat them to more than a few hundred degrees above room temperature," he says. "This tells us that these rocks have *never* experienced very high temperatures. The matrix material never was close to the Sun."[3] Based on the chemical composition of the matrix rock, scientists estimate that it formed no closer to the Sun than about 2.5 astronomical units, Shu says. So here is one key boundary condition that the ultimate theory of solar system origins must satisfy. Somehow, the matrix rock has to form relatively far out in the disk.

How about the chondrules? Their appearance and composition indicates that they were melted, Shu says. And the CAIs found with chondrules seem to have been heated to even higher temperature. Somehow, then, the chondrules and CAIs had to have been subjected to flash heating by the forming Sun. Moreover, the evidence shows that they were heated on a timescale of minutes and hours, and then cooled in a matter of days.

Another puzzle piece: chondrules come in a very narrow range of sizes. "This is very unusual in astronomy," Shu says. "When we see large populations of rocks, we usually find them with a large range of sizes. But in meteorites, there's a certain median size for chondrules; very few are twice as big or half as big." So any successful theory must account for this unusually narrow range of chondrule sizes.

The conventional theory for the origin of chondrules holds that they formed relatively far out in the disk, at about 2.5 astronomical units. Here, according to the theory, tiny dust balls were melted by periodic flares from the forming Sun, fashioning spherical droplets. As the flares subsided and the droplets cooled, they solidified into chondrules. As for the CAIs, the theory holds that their specific mix of radioactive isotopes was created when the solar flare activity irradiated precursor molecules.*

But Shu argues that the conventional theory does not match the meteoritic evidence. For example, he says the theory does

---

* An alternative theory, discussed in Chapter 9, is that some of these isotopes were created by irradiation from the explosion of a nearby supernova—an event theorized to have triggered the collapse of the original cloud core.

not readily explain why chondrules all come in a narrow range of sizes. Moreover, the kind of solar flares that could melt dust balls at 2.5 AU would not necessarily do the job in the required time of minutes to hours. And the specific mix of radioactive isotopes in CAIs cannot be explained by the conventional theory.

Shu and his colleagues formulated the X-wind theory to tie the disparate phenomena of star formation and planet formation into one framework, while also conforming more closely to the meteoritic ground truth than the conventional theory of planet formation. So let's journey once again to the now fully-fledged star we explored in Chapter 9 so we can see how the X-wind accomplishes all that.

The Sun is now satiated — it has finished gobbling gas. And the remaining disk, now called a *protoplanetary* disk, because this is where planets will form, is where the action has shifted. So we swoop low into the disk to have a look and collect some dust. As expected, hydrogen and helium gas is abundant, as well as other, heavier gases. We also find a haze of fine dust — particles of interstellar material as well as solids that have condensed into grains from the disk gas itself. Farther out in the disk we detect the presence of water ice, but not here. We're too close to the Sun. We're within the snow-line of the disk — the radius beyond which temperatures are cool enough for water to freeze. Even so, the mutual shade offered by all the fine dust particles keeps temperatures below the melting point of the dust particles.

We extend a collector to harvest dust grains from the disk and bring them on board for chemical analysis. They clearly are not like the stuff that covers my long-unattended furniture back at home. Household dust consists of bits of lint, dirt, dead skin cells, bug excrement, and the like. The disk grains we have collected turn out to be made mostly of carbon, oxygen and silicon in a variety of forms. Silicates and graphite are common. We also find some microscopic flecks of diamond. Organic compounds are abundant too, including polycyclic aromatic hydrocarbons, or PAHs, which are just like the soot produced by fossil fuel burning. (Maybe if all goes well, these organic molecules will some day be incorporated into lint, skin cells, bug excrement, fossil fuels...)

An individual particle is too tiny to be seen with the unaided eye, so we put a sample of the dust under the microscope to see what the grains look like.[4] Each one seems to be built up of smaller pieces. Some grains are compact and rounded with little projections and other irregularities; others are spiky and filamentous. In both cases, the structures are somewhat jagged and fractal in character, which may help the grains stick together when they collide—the first stage in the construction of a planet. (Alternatively, particles may also shatter, producing jagged fragments that may then collide and stick the next time around.)

One other thing: the grains are *really* tiny, measuring just a fraction of a micron across. By way of comparison, an *E. coli* bacterium is about 3 microns long. So once again, nature is confronted with a dynamic range problem. To form a planet the size of the Earth, these tiny particles, each smaller than some bacteria, must somehow agglomerate into a dense ball thousands of miles in diameter. (Earth's diameter is 7928 miles, or 12 758 kilometers.)

Actually, just building a planetesimal from this stuff is enough to make an astronomer like Shu pause. "To build a planetesimal, you have to get from a tenth of a micron to a kilometer in size, and that spans 10 orders of magnitude. Now that's a *big* problem."

Starting the tiny particles on the growth path toward a planet is not the complicated part. Static cling of the ordinary laundry-day variety probably does the trick, taking the microscopic raw ingredients of planet formation and fashioning them into basic building blocks that are about a millimeter across, which is about 10 times thicker than the width of an average strand of human hair. This happens as the attractive force of static electricity sticks dust particles together when they collide. First to form are tiny dust-bunny agglomerations. These rain down into the mid-plane of the disk, where they form a thin sub-layer of denser material sandwiched by less dense stuff, mostly gas, above and below. In this denser sub-layer, collisions are even more frequent. So growth of the dust-bunny agglomerations accelerates until the sub-layer is dominated by millimeter-size grains—basically, very small bits of "rock," as Shu puts it.

That much seems simple enough. But the next stage in building a planet is not: sticking enough rock bits together to form a gravel of pebble-sized material; packing enough gravel together to form boulders; and cementing enough boulders together to build a planetesimal on the order of a kilometer in diameter. Like the original microscopic dust grains, the millimeter-size rock bits are suspended in the mid-plane of the nebula, where they collide frequently. So you'd think that they would stick. "But do rocks really stick when they collide?," Shu asks. "You can go to the beach and throw millimeter-size sand grains together. Do they stick? Come on. Of course they don't."

In this case, static cling doesn't work because the rock bits are just too big. Static electricity operates at the surface of an object. So the greater the surface area of nearby objects, the greater the likelihood that electrostatic force can pull them together. But the other part of the equation is mass. And the mass of a millimeter-size rock grain is high compared to its surface area. As a result, electrostatic force cannot easily overcome the grain's inertia. At the same time, the grain is not massive enough for its gravitational field to attract and hold on to others. So the grains are in limbo: too massive for one force to get them to stick, yet not massive enough for another to work either.

"Now you can say, 'Well, if you have ice on the grains, the situation is better.' And that's probably true," Shu says, commenting on one idea for how rock bits might stick together. "We all know—those of us who have lived in cold climates— that you can take snow flakes and pack them into snowballs. You can take snowballs and pack them into snow men. And nature can take lots of snow men and pack them into glaciers. A glacier is essentially the size of a planetesimal. But that whole scenario depends on the snow being near the melting point. If you have very dry, powdery snow, it's not so easy to make a good snowball. And in the conventional picture of the forming solar system, the snow line is at Jupiter or somewhat interior to Jupiter. Inside the snow line, there is no ice. We just have bare rock," Shu says. So in the outer reaches of the disk, icy dust grains may form comets (and in our solar system, Pluto). But how do you build planetesimals from millimeter-size rock grains in the warm, inner part of the solar system, where the terrestrial planets eventually grew?

Over the years, astronomers have proposed a number of theories to pull the rock grains out of limbo. In 1973, Bill Ward, now of the Southwest Research Institute in Boulder, Colorado, and Peter Goldreich of the California Institute of Technology, showed that gravitational instabilities — enhanced regions of gravity — within the mid-plane of the disk might be capable of growing a planetesimal from millimeter-sized material. "Gravitational instability is the same kind of thing we see in Saturn's rings and in spiral galaxies," Shu says. "So it is well understood." (He should know, since he helped pioneer the theories explaining Saturn's rings and the arms of spiral galaxies.)

But in 1993, Stuart Weidenschilling of the University of Arizona and Jeff Cuzzi of the NASA Ames Research Center argued that this mechanism just wouldn't work. They pointed out that the sub-layer of millimeter-size grains in the mid-plane of the disk would move at a slightly different speed than the gas just above it. The rock grains orbit freely, their motion dictated by the gravitational pull of the Sun. The gas, however, is under pressure, which pushes on the gas, counteracting some of the Sun's gravitational pull. Because of this, they orbit somewhat more slowly than the free grains. The difference in speed is tiny, according to Shu. "It may be just one part in a thousand difference," he says. "That doesn't sound like much. But it amounts to a difference in speed of tens of meters per second. So think of the gas as a wind that blows at tens of meters per second. That's comparable to a pretty strong gust of wind."

Such a gust can easily kick up sand in the desert. And in the disk, it can keep solid grains stirred up, preventing them from settling easily toward the mid-plane of the disk. This, Weidenschilling and Cuzzi pointed out, was a problem because gravitational instability depends on there being a dense enough sub-layer. The reason? The self-gravity of any embryonic clump of rock grains must be strong enough to overcome the gravity of the Sun, which tries to pull any such clump apart.

According to Shu, detailed calculations support Weidenschilling and Cuzzi's claim that the gas wind would tend to keep the sub-layer too turbulent. For gravitational instability to work, the sub-layer would have to be more quiescent.

In 1993, Cuzzi and several colleagues also proposed another mechanism. The gas wind, they noted, would impose drag on the rock grains, naturally slowing them down, but not all at the same rate. Smaller grains would drift more quickly inward than larger ones. Cuzzi and his colleagues proposed that as the smaller grains zipped by, some would collide with the bigger grains and stick. And in less than a million years, objects up to 100 kilometers in size could grow. But how would the grains actually stick? That remains something of a problem.

In the late 1990s, Ward modified his gravitational instability model so that turbulence could actually be turned to advantage. He proposed that rock grains of similar size might become packed together at *stagnant points* that would naturally arise in the turbulent flow of material within the disk. In these small quiescent pockets, enough stuff might get together for their self-gravity to begin pulling more stuff together. In this way, the gravitational instability might still work, Ward proposed. But when presenting his idea to scientists at a NASA-sponsored conference on cosmic origins in 1997, he also acknowledged modeling results showing that the process could build small boulder-sized objects, not planetesimals. His conclusion? "The mystery is still as cloudy as ever."[5]

As grains come together to build bigger and bigger bodies, somehow the chondrules found in meteorites must form. According to the conventional view, they formed as a result of stellar temper tantrums—sudden increases in flare activity on the infant Sun—that melted rock grains through flash heating. When the solar activity died down, they cooled and fused into the solid, spherical chondrules. And these, in turn, later were incorporated into planetesimals. But Frank Shu believes his X-wind theory is a better idea, one that not only explains how the chondrules formed but also the central mystery of how they could agglomerate into kilometer size planetesimals. He admits his proposal is speculative. But it is part and parcel with his X-wind theory, and any idea that can explain so much, he says, warrants serious consideration.

Remember those Frank Shu designer sunglasses we donned in Chapter 9? They enabled us to watch as the X-wind sprayed droplets of molten rock out over the disk. *This* is the origin of chondrules, Shu argued. But at the time, we didn't examine what happened when these chondrules actually hit the disk. In Shu's view, the arrival of chondrules is what triggers the agglomeration of millimeter-size grains into planetesimals. As the chondrules rain down, they add enough mass into the disk's mid-plane sub-layer to allow the self-gravity of material there to begin pulling clumps together. How is this able to happen?

Once again, imagine a wind blowing across desert sands. If it's strong enough, it will kick up sand into a dust storm. "But even in a sandstorm," Shu says, "*not all of the sand in the desert gets lifted up*, *right*? Similarly, whether you can keep the entire sub-layer of the disk stirred up depends on how much rock you have there compared to the gas."

Up until now, astronomers have assumed that the ratio of rock grains to gas was the cosmic ratio, meaning 0.4% of the stuff in the disk is in rock. "However, I told you that in our picture there is a recycling of rock back into the solar nebula through the X-wind," Shu says. "The 0.4% rock—the stuff that was there to begin with—is in our view what the *matrix* of chondrites was made from. And then, if you increase that by a factor of five or ten, which the X-wind can do, you don't have the cosmic ratio of rocks to gas anymore. You've *enhanced* it. And it turns out that a factor of five to ten enhancement is just enough to make the gravitational instability work."

So the X-wind sprays the chondrules out over the disk. The small ones are thrown clear of the solar system. The big ones fall down close to the Sun and get re-melted. But ones with just the right size—in a *narrow range of sizes* centered on about one millimeter—rain down on the disk, adding their bulk to the 0.4% stuff, the matrix stuff, already there. And this is just enough to make gravity work. Rock bits and chondrules stick together, forming pebble-sized chunks of matrix with embedded chondrules. With gravity working now, the pebbles stick together to form boulders, and the boulders to form a planetesimal.

About 4.5 billion years later, chunks of this planetesimal rain down on Earth as meteorites, and we find that they are made of chondrules in a narrow range of sizes embedded within a matrix rock.

"So in our picture, the appearance of chondrules may have actually triggered—this is pretty speculative, okay?—the formation of planetesimals," Shu says.

In one, unifying grand leap, then, the X-wind theory solves the angular momentum problem for a forming star *and* the stubbornly uncrackable mystery of how planetesimals form. And it does this by tying together these seemingly disparate phenomena with chondritic string. Of course, explanatory power alone is not enough to enshrine a theory in textbooks. It must also obey the available physical evidence and make predictions that future observations can falsify or corroborate. Shu's theory satisfies the first requirement, staying within the boundary conditions established by chondritic meteorites. And it makes specific predictions about what astronomers will find when they have the technology to look directly at the point of origin of a jet emanating from a forming star. Since that technology doesn't exist yet, Shu will have to wait to see whether some of his theory's predictions are corroborated.

In the meantime, other predictions of the theory already have been checked out, he says. For example, the prediction that chondrules should come in a very narrow range of sizes. Another prediction concerns the composition of calcium–aluminum inclusions. In all meteorites containing CAIs that have been discovered, the *abundance* of each isotope is the same. In other words, some isotopes are always found in very tiny amounts whereas others are also found to be many times more abundant. Any theory purporting to explain how the CAIs formed must also explain this very specific pattern. In fact, the X-wind theory actually *predicts* the pattern, whereas the conventional theory, which invokes irradiation by solar flares at 2.5 AU, does not, Shu says. Nor does the proposed supernova explosion.[6]

Shu's theory makes still another prediction that can be tested: "It's not just any old disk that can build planetesimals," he says. "You've gotta have a disk that is *augmented* by these chondrules." To have that augmentation, a disk has to be rich enough in dust, which consists of elements heavier than hydrogen and helium — the elements astronomers call "metals" (even though chemically speaking, not all of them are metals). If a disk's "metallicity," as astronomers call it, is not rich enough, Shu says his X-wind will not throw out enough chondrules to trigger the formation of planetesimals.

Observations of extrasolar planets is now providing a hint that only disks relatively enriched in metals can build planets, Shu says. Paul Butler and Geoff Marcy, as well as other planet hunters, have found that the parent stars to extrasolar planets actually have a mysterious spectral signature indicating a *higher metallicity* than stars without planets. One explanation is that giant gaseous planets form far out in disks and then migrate inward. Some stop before being gobbled by the star, becoming close-orbiting "hot Jupiters." Others stop even farther out, becoming merely "warm Jupiters." But others plunge full speed ahead and are swallowed up, adding their complement of heavy elements to the star and thereby enriching its metallicity.

But Shu says there's a problem with this conventional explanation. Unlike fully grown stars like our Sun, the hot plasma inside young T-Tauri stars (the stage in stellar evolution when planets form) mixes well from the center to the surface. So if a planet dissolves in such a star, the planetary material becomes fully mixed — "and the metals are just a drop in the bucket compared to the entire star," Shu says. When a star grows up, however, mixing becomes stratified. A relatively thin upper layer mixes on its own, separate from the star's interior. If a planet is digested in *this* layer, it can raise the metallicity enough for astronomers to notice. So for the conventional explanation to work, one must assume that migrating planets are swallowed by their stars *just as they make the transition to stratified mixing.* (It can't happen much later because the era of planet formation and migration would be over — for reasons that will become clearer a little later in the chapter.)

Why would so many stars with extrasolar planets just happen to cannibalize some of their children at precisely the same moment? Shu considers this to be an unlikely coincidence.

In his alternative model, a disk must inherit enough heavy elements from its parent molecular cloud core if it's going to build planets. Otherwise, the rain of chondrules produced by the X-wind would not be sufficient to trigger the gravitational instability that leads to the formation of planets. Since a star and its retinue of planets is made from the same stuff, it will naturally have an excess of heavy elements if it has planetary children. Otherwise, the star would be childless.

Beyond explaining the higher metallicity of stars with extrasolar planets, Shu's theory has implications for the big question astronomers are trying to answer. How common are planets in general, and Earth-like planets in particular? "If this picture is correct," Shu concludes, "then planet formation is not quite as common as people think. It means that just because you have a disk it doesn't mean that you will automatically form planets."

297

Some scientists who specialize in gleaning clues from meteorites also are not so sure whether Shu's theory is right. John A. Wood of the Smithsonian Astrophysical Observatory is one of them. ''Though I admire Frank's approach to the problem, I have found it difficult to embrace the X-wind model,'' he says. ''I cannot give you clearly thought out arguments; it's a pit-of-the-stomach feeling I have about making chondrules and CAIs near the sun and then delivering them 3 AU away for incorporation in planetary material.''[7]

The bottom line is that astronomers just don't know yet how millimeter-size rock grains grow into kilometer-size planetesimals. Scientists like Shu and Ward have devised several compelling theories. But until the technology makes it possible to zoom in more closely on planetary construction sites, astronomers probably will not be able to agree which is correct.

''So somehow, maybe better living through chemistry, plane-tesimals form,'' jokes Jack Lissauer of the NASA Ames Research Center. And once they do, larger planetesimals easily have the gravitational capital to transform themselves into planets through a process of mergers and acquisitions. That's because larger planetesimals in the evolving disk are massive enough for their gravity to hold onto almost all objects they collide with, and so they grow larger. Meanwhile, smaller planetesimals are either sucked up by the bigger ones or are stomped to dust in collisions. In this way, Lissauer says, ''the rich get richer and the poor get poorer.''[8] And calculations show that the rich accumulate their wealth faster than the likes of Bill Gates, growing much faster and fatter than the rest of the swarm in a runaway accretion process.

In this way, a planetesimal can grow to the size of the Moon on the order of 100 000 years, Lissauer says. At this point, the object has sucked up most of the material within its ''feeding zone,'' so the runaway ends and the lunar-sized ball of rock grows by accretion at a slower pace.

Slower, but hardly sedate. Lunar-size objects collide cataclys-mically, adding their bulk together until planets with masses on the order of Earth's are formed. For rocky terrestrials like Mercury, Venus, Mars and Earth, construction is nearly complete at this point. All that remains to be added are the peculiar characteristics that define an individual planet. If a terrestrial planet is lucky, it will reside in the habitable zone around its star and also secure an atmosphere with just the right constituents in just the right amounts to insure the cosmic equivalent of a Sun Belt climate — one that can allow liquid water, a prerequisite for life, to exist.

But to build giant, gaseous planets like Jupiter and Saturn, much more growth is required. To finish the job, the conventional theory holds that such planets first build up rock and ice cores through the same runaway accretion process that builds the terrestrial planets. Then, in a second phase of runaway growth, they acquire a huge envelope of gas. But just how that happens is something of an open scientific question.

"We still don't have an acceptable model for giant planet formation," observes Alan Boss. "And without that, it's hard to say that we understand the planet-formation process in its entirety."[9]

Theorists have certain constraints that they must work within. Chief among these is a constraint of time. Runaway accretion of a core and then the gathering of a gas envelope must happen quickly enough so that the planet is formed before all the gas in the disk dissipates. Otherwise, it will run out of raw material. So how much time does that take?

According to Ray Jayawardhana of the University of Michigan, astronomers have made remarkable progress in observations of disks as they evolve over the course of 1 million, 10 million and 100 million years. "Now we have direct images of 10-million-year-old disks that have inner regions about the size of Pluto's orbit cleared out. This is presumably a result of planet formation," he says. "These observations provide a timescale for planet formation. Basically, the bulk of it needs to happen within the first 10 million years of a star's lifetime, because by then, most of the dust and gas out to a distance equal to 30 or 50 times Earth's distance from the Sun appears to be depleted."[10]

One way to beat the clock, Lissauer says, is through a second episode of runaway accretion. At first, the gravitational pull of the rock and ice core would gather gas around it slowly. But modeling shows that as the planet becomes steadily more massive, its increasing gravitational field draws in gas at an ever-increasing rate, leading to runaway accretion of gas. The entire process — from the first dust grains that stick together to the completion of the giant gas atmosphere—can take something like 8 million years. And this has an important implication. If it really takes that long, disks with relatively small amounts of gas and dust may run out of raw materials before they can build an Earth-like solar system, with its complement of giant gaseous planets. Only relatively massive disks will be capable of completing the job in time.

Working with Peter Bodenheimer of the University of California, Santa Cruz, Lissauer has been exploring a way for nature to go about its business more swiftly.

Like a glutton who eats too quickly, a forming gaseous planet that's gobbling gas from the disk can find it increasingly difficult to swallow additional gas. As the planet feeds on gas, it expands. At the same time, though, its gravity hugs the gaseous envelope, tending to contract it. When the planet feeds quickly, expansion outpaces contraction, and the surface extends farther and farther from the center of gravity. Since the planet feeds by gravitationally drawing gas onto its surface, the more it expands, the less able it is to feed quickly. At a certain point, the edge of the envelope extends so much that the gravity is too weak to suck in gas very quickly.

But if the growing planet can avoid becoming hot and bothered during its gluttonous feast, it can continue to gobble gas quickly. The more effective it is at ridding itself of heat from its gaseous envelope, the faster the gas will contract, thereby leaving more room within the planet's zone of effective gravitational attraction to pull in yet more gas.

How fast the envelope can cool off, and thus how fast it can gobble more gas, depends on the amount of dust it's got. That's because dust tends to make the envelope opaque, which blocks heat from escaping. In their modeling, Lissauer and Bodenheimer have made what they believe is a realistic assumption: there's less dust than previously thought, making the envelope less opaque and therefore more effective at cooling off. This model can build a giant gaseous planet like Jupiter in as little as 2 million years. So if nature worked the same way, a wider range of disks could build Jupiter-like planets.[11] And that would be a good thing if your concern is building an Earth-like solar system.

As is so often the case in this field, astronomers don't know for sure whether nature works like Bodenheimer and Lissauer's model, or any other model for that matter. To see whether it does will require observations of forming Jupiter-like planets—a challenging task, but one that astronomers are coming closer to achieving. "With adaptive optics on large telescopes, it is now technically possible to image a newborn Jupiter-mass planet, provided it's separated enough from its star," says Jayawardhana. "Several groups are conducting searches."[12] Moreover, observations of disks at different stages in their evolution are already helping to constrain the models of theorists.

Another puzzle theorists are working on is how planets like Uranus and Neptune form. These gaseous giants are considerably smaller than Jupiter and Saturn, and any model of planet formation must explain this size dichotomy. Lissauer and other scientists suspect that Jupiter and Saturn experienced the full gas runaway effect, whereas Uranus and Neptune did not. This is plausible because these smaller planets formed farther out in the disk, where the density of material available to build planets was lower. Consequently, it's possible that they simply ran out of gas before they could experience a full runaway. At still more distant and colder reaches of the disk, Pluto might have had only dust and ice as raw material, possibly explaining why it lacks a hefty atmosphere like its gaseous neighbors.

And why didn't the terrestrial planets develop their own giant gas envelopes? The answer is that they were so close to the Sun, high temperatures drove away hydrogen and helium, the light gases required for a Jupiter-like atmosphere.

In summary, then, this is how planets are believed to have formed in our own solar system:

- Static cling formed dust bunnies from microscopic dust particles, and dust bunnies grew and consolidated into millimeter-sized rock grains.
- Some of these grains melted and re-solidified, forming chondrules.
- Millimeter-size chondrules and other grains somehow built up into kilometer-sized planetesimals.
- These collided in a runaway process, building Earth and the other terrestrial planets, as well as the rocky cores of Jupiter and the other giant gaseous bodies.

- In a second runaway process, the rocky cores gathered hydrogen and helium from what was remaining in the disk, finishing the construction of the giant gaseous planets. By the luck of the draw, this happened to occur just as the disk was running out of gas. The density of gas closer in was sufficient to make planets as big as Jupiter and Saturn big, but not sufficient farther out, explaining why Uranus and Neptune are smaller.

As the canonical wisdom on the origin of planets, this scenario is widely accepted by astronomers. But not everyone abides scrupulously by the canon. One such person is Alan Boss. Driven by concerns that growing Jupiter-like planets could run out of gas before they managed to accrete their gas, he has followed a different path.

Unlike many others in the field, Boss does not cite a child-hood astronomical epiphany as the source of his fascination with planets and stars. As a child growing up in Clearwater, Florida, a town on an island in the Gulf of Mexico, his imagination was captured by the sea. Accordingly, he thought he would pursue a career in marine engineering.

But as an undergraduate in college, Boss discovered a love for physics. Enrolling in a graduate physics program at the University of California at Santa Barbara in 1973, he worked under the guidance of a professor named Stan Peel—known among the students as "flunk 'em Peel," Boss says. "It was pretty brutal."[13] Which was fine by him, because he did quite well. So well, in fact, that Peel asked Boss to be his graduate research assistant. It was then that Boss discovered the physics of solar system formation. He was inspired both by Peel and a Russian scientist named Victor Safronov, whose 1972 book on the subject, *Evolution of the Protoplanetary Cloud and the Formation of the Earth and the Planets*, has held up well even after several decades of astronomical discovery. It was Safronov's *theoretical* insights that excited Boss in particular, pushing him toward exploration of the galaxy using the lens of his mind.

Today, Boss is a widely respected theorist who spends most of his time building solar systems—not real ones, of course, but solar systems constructed of mathematical algorithms in a computer. To accomplish these modeling efforts, he has built his own multi-processor computing facility at the Carnegie Institution of Washington's Department of Terrestrial Magnetism.

Located in a leafy corner of Washington northwest of the White House, the now misnamed unit of the renowned Carnegie Institution is a sanctuary where Boss and his colleagues can devote themselves to pure research. It's the kind of place where astrophysical theorists will drop what they're doing and rush downstairs for a spot of tea and a talk by a colleague about ancient lava flows.

Tucked away in a windowless room in the department is Boss's computing facility. When I expressed an interest in taking a look at it, Boss reacted like a kid eager to show off a new bike. He built his "supercomputer" by linking eight personal computers together. (Each one is equipped with two fast processing chips.) This computing arrangement may seem jury-rigged and unequal to the task of building a solar system in a box. But the system has more computing speed than a $1 million Cray supercomputer—and for just $90 000 worth of processing chips and computers, plus the cost of the rack to hold them. The machine's power derives in part from its ability to divide a computing task into chunks, with each processor handling a different piece.

But hardware alone will not build you a solar system. A machine has no more idea of how to do that than it knows how to open a web page and display its contents on a monitor. So Boss's machine must be endowed with a set of instructions based on the details of what is to be simulated, and on the laws of physics, which determine how things will come out. To do this, Boss has written his own software for simulated solar-system construction—a Sim City for the astrophysically minded. Except the output is not nearly as slick—at least not visually. For example, in simulations of planet formation, the model produces a series of diagrams mapping the distribution of dust and gas within an evolving disk. As the simulated disk evolves forward in time, clumps may appear. On the map, such clumps are portrayed graphically in much the same way that a hill or mountain is portrayed on a topographic map. If all goes well, the clumps will then transform themselves into something more substantial, something more planet-like in shape and density.

When I visited Boss, he had 16 different simulations running on his system as part of an effort to understand the origin of planets, the formation of a disk around a protostar, and the collapse of a molecular cloud. Except when they crash or need upgrading, the computers run continually.

Computer simulations enable theorists like Boss to play what-if games. For example, what if a disk starts off with some random clumps of dust that are denser than the average distribution of dust all around? Such clumps could naturally be inherited from the clumpy molecular cloud itself, Boss says. Can they quickly transform themselves into giant gaseous planets like Jupiter, bypassing the need to build a rocky core first? Or will competing forces in the disk tear them apart before they can become a planet?

That's precisely one of the scenarios Boss is testing. ''If a clump of gas falling onto the disk from the parent cloud is large enough,'' he explains, ''it can induce a gravitational instability''—a region with a slight excess of gravity. This enhanced gravity could in turn start pulling a bigger clump together, triggering further infall of gas toward the clump. In this way, an initial gravitational instability could theoretically lead to the formation of a Jupiter-like planet before the disk runs out of gas—or so the theory goes.

He is well aware of the objections to this idea. ''The conventional wisdom is that gravitational instabilities in a disk would make spiral arms, not long-lived clumps,'' he says. ''This would transport angular momentum around the disk and thereby pull clumps apart before they could become big enough to be long-lasting. So it's basically a race between how fast the arms pull things apart and how fast the clumps contract down.''

Can the clumps win? At the beginning of the simulations Boss has run to answer the question, he has manually planted the seeds of a gravitational instability into his model. So the simulation doesn't reveal *whether* the instability can occur. Instead, it shows what would happen if you assumed that an instability *could* occur.

In his best model run, a clump formed and grew from an initial gravitational instability. And this was progress, because it showed that something could hold together in the face of spiral arms in the disk. Unfortunately, though, the clump lasted for just two orbits and then was pulled apart. "That was unnerving," Boss admits. But he also notes that with an earlier, less realistic version of his model, the clump lasted for just one orbit. And in still earlier versions, the clump was even more short-lived. So the trend is encouraging: as the realism of the model improves, the clump lasts longer. Boss feels confident that with enough work and processing time on the computer, he'll make some Jupiter-like planets.

But as a respected colleague at Carnegie, George Wetherill, points out, "Alan has made this thing shaped like a half moon. He hasn't made a planet."[14] And that seems to be the cautious consensus among many astronomers. "Talk to us when you've managed to make a real planet," has been the basic reaction.

Whether he pulls this off is not just an academic exercise, because much may depend on the outcome. John Bally and others argue that if giant gaseous planets like Jupiter cannot form quickly, it's quite possible that intelligent life is rarer than many of us would like to believe.

Bally's argument goes like this: As Wetherill was first to suggest, Jupiters are vital gravitational guardians that protect against a high rate of comet impacts on terrestrial planets like Earth. So to avoid extinction by comet impact, species with the potential to evolve into intelligent creatures may need to live in solar systems with just such a guardian. But in environments such as Orion—in which the overwhelming majority of stars are born—massive O stars rapidly strip the gas away from disks around Sun-like stars. So in such an environment, a potential Jupiter around a forming Sun-like star must somehow manage to pull itself together before the fusion fires ignite in nearby O stars. Otherwise, its birth will be aborted.

"John is exactly right," notes Boss. Unless Jupiters can form more quickly than conventional theories seem to indicate, Earth-like solar systems and maybe intelligent life will be rarer than previously suspected.

And what about rocky terrestrial planets? As gas is stripped away from a disk by a nearby O star's ultraviolet radiation, dust is carried away with it, depleting the reservoir of raw material needed to build them. On the other hand, if enough dust grains manage to reach a large enough size quickly, they will resist being carried off and can theoretically go on to form rocky

305

planets. Here then is another race—between the forces trying to tear the dust away from the disk and the growth processes that are trying to build it up into bigger and bigger bodies. Once again, can growth win?

Henry Throop, an astronomer with the Southwest Research Institute in Boulder, Colorado, believes it can. Throop led a team of researchers, which included John Bally, that produced new insights into this issue. The astronomers set out to determine the degree to which dust grains have grown in disks within Orion even as they have been assaulted by nearby O stars.

The researchers focused the Hubble Space Telescope on a disk called 114-426. Despite its undistinguished name, it is the largest circumstellar disk known in the nebula. Using the Wide Field Planetary Camera, or WFPC, and the Near Infra-red Camera and Multi-Object Spectrometer, or NICMOS, Throop and his colleagues observed how dust particles in the disk scatter background light from the nebula. Small particles on the order of a tenth of a micron or so—the kind that have not yet managed to grow appreciably and thus are vulnerable to erosion—scatter light selectively. The shorter the wavelength, the more of the background light the particles scatter. Conversely, the longer the wavelength, the less the light is scattered. Because of this, wavelengths toward the blue, or shorter, end of the spectrum are scattered away, whereas those toward the red, or longer, end pass through to the telescope cameras unmolested. The result is that light filtering through a disk with very tiny particles is reddened. But there is no such selective effect when particles have grown appreciably. These grains scatter all wavelengths equally, producing a drab gray light in Hubble images.

When Throop and his colleagues carried out their observations in 1998 and 1999, the results were unmistakable: "There was no measurable reddening," Throop reports. "The disk was gray. This is very difficult to explain if the disk contains small particles."[15]

To check their results, the astronomers made related observations of other disks in Orion, this time using ground-based telescopes. The results supported the initial findings.

In a paper published in 2001 in the journal *Science*, the astronomers concluded that micron-sized dust particles had clumped together in disk 114-426, forming rock bits about the size of sand grains.[16] "These grains are gigantic compared to what we expected to see," Throop says.

To gain insight into how the growth process plays out in Orion's nasty environment, Throop and his colleagues then used a computer model to simulate the growth of dust grains. Their model disk was analogous to 114-426, and in the simulations it was placed close enough to the core of the Orion nebula to be blasted by intense ultraviolet radiation. "Many people have modeled the formation of solar systems," Throop notes. "But in these previous simulations, they essentially put the forming solar system in a box, isolating it from the outside environment. I became interested in this problem because in Orion, the outside environment really affects these disks. They are being destroyed rapidly, yet we have this evidence for a lot of grain growth. So what's going on?"

To answer that question, Throop designed his model to account not only for the processes that promote grain growth, but also for the destructive forces that would retard it. And he ran a number of simulations based on different assumptions, such as how well dust grains might stick, how turbulence might affect growth, etc. The results, published in the same paper in *Science*, complemented the Hubble observations. Within 10 astronomical units, which is equivalent to the orbit of Saturn in our solar system, dust particles agglomerated into meter-sized boulders within just 100 000 years of simulated time. There's no way that objects this large can be carried off by the gas as it's stripped away from a disk. So these results support the idea that rocky planets can form in the inner reaches of solar systems in Orion-like environments. Meanwhile, in the outermost reaches of the simulated disk, dust particles smaller than a millimeter in size were easily entrained by the gas and carried out of the system. This makes sense because the density of dust in a disk decreases with distance from the central, forming star. With a higher density in the innermost part of a disk, there are more collisions and more opportunity for sticking.

Because the simulated disk was eaten away from the outside in, it shrank considerably over time. "On the order of 100 000 simulated years, we boiled down a 1000 AU disk to one with a diameter of just 40 AU," Throop says. (By comparison, our solar system is about 110 AU across.[17]) After a million years of simulated time, all ice and gas were stripped from the disk, leaving behind just the rocky stuff.

These simulations, combined with the observations of disk 114-426 in Orion, suggest that rocky terrestrial planets can theoretically form in Orion-like environments despite the erosive power of nearby O stars. But unless giant gaseous planets can form in less than a million years—the time it takes for the O stars to strip away all the gas from the disk—these solar systems will lack planetary guardians like Jupiter.

"In Orion, we might have just as many terrestrial planets as anywhere else," Throop concludes. "But Jupiter-size planets? That depends on how fast you can make them." If Boss is wrong and it's not possible to build a Jupiter quickly, it would be "a sad day for planet enthusiasts like myself," he says. Based on the evidence available right now, "planets are probably rarer than we thought."

But there is one silver lining to Throop's simulations. They suggest that solar systems in Orion would lack a large population of icy bodies such as comets in their outermost reaches, because ultraviolet radiation seems to strip away all the ice from a disk before these bodies would have a chance to form. That might be good news for potentially intelligent species on rocky planets in Orion, because then the need for a planetary guardian like Jupiter would be reduced. There just wouldn't be as many potential impactors to protect against.

But can life get a toe-hold on such planets? In Bally's view, the chances are not good. And the reason is that Orion-like environments have many ways to snuff it out.

For example, Bally and University of Hawaii astronomer Bo Reipurth have shown how the tight clustering of stars in some parts of the Orion Nebula could easily lead to the destruction of planets. In one region of Orion, there may be as many as 100 stars crammed together in a space smaller than our own solar system. Within that volume, these stars would emit 10 000 times more light than the Sun does.

"Picture a very small volume with a lot of stars in it," Bally says. "In this situation, we found that the most massive object will swallow less massive objects." And there's evidence for this in Orion: massive, exploding outflows from a star that are visible in some images as glowing gas spraying outward in giant fingers from the heart of the nebula. In this case, Bally and Reipurth believe one star has swallowed another, releasing a massive amount of energy in a short period of time.

What would become of any embryonic planets circling around stars in densely clustered environments like this? Cosmic abortion.

But what about other kinds of star-forming environments? Might they pose less of a hazard? In some ways, yes. In a molecular cloud complex in the constellation of Taurus, 450 light-years from us, star formation is much more placid. The reason? There are no O stars. But just because circumstellar disks aren't being shredded by hurricane winds and torched by ultraviolet radiation doesn't mean there are no obstacles to forming planets, Bally says. He cites research by Reipurth, who detailed how violent interactions between young stars in multiple systems could make planet formation difficult.

In these systems, one, two or even three stars are close companions to a primary star. According to Reipurth, such systems are incredibly common throughout the cosmos, not just in violent, Orion-like environments. Nearly 90% of all young stars are thought to have companion siblings at birth, he says.

Reipurth's research involved Hubble observations of a triple-star system and computer modeling of interactions between the stars. This work, published in the year 2000, suggests that interactions between companions in multiple-star systems can eject one or more from the system and literally tear up circumstellar disks. This would send any planets careering into their parent stars or into interstellar space.

On the other hand, recent observations suggest that multiple-star systems don't necessarily have to be a hazard to forming planets. In the year 2000, two teams, one that included Ray Jayawardhana, announced that they had independently spotted a disk around a star in a binary system. The star, designated HR 4796A, is in the southern constellation Centaurus. (What was most exciting for Jayawardhana is that he found the disk when he was making observations for his Ph.D. thesis.) Then, in January of 2002, a team co-led by Jayawardhana announced that they had discovered a dusty protoplanetary disk orbiting one of the stars in a newborn *quadruple* star system. The disk is about three times the size of Pluto's orbit. The astronomers imaged it edge-on using the

Gemini North telescope atop Mauna Kea equipped with adaptive optics. In both cases, Jayawardhana believes the disks might last long enough to form planets. In fact, in the quadruple system, the astronomers found evidence for the growth of dust grains.

Are these systems the exception, not the rule? And can the disks in these systems really last long enough to make planets? No one can say just yet. Getting the answer will require more observations. So John Bally is sticking by his claim that planets—and therefore life—are rarer than many people would like to believe.

"I admit, my claim is heretical," Bally says. "And what I'm saying may be dead wrong. But part of the problem I see right now is the Star Trek mentality. People *expect* life to be common. I worry that much of our enthusiasm for this idea is driven by the popular culture, and less by scientific fact. And I think the scientists themselves are blinded by the *desire* to find planets." That desire has created a popular misconception, he says, that most stars have planets, that solar systems abound in the universe, and that therefore ET is a reality, not just a cinematic invention. "The reality seems to be that Earth's solar system is a special place, and that we are lucky to be here."

Of course, scientists actually have found extrasolar planets. So it's obvious that despite all the violence now known to characterize star-forming regions, planets somehow manage to win out around some fraction of Sun-like stars. But even these discoveries have given scientists like Bally reason to wonder how common systems like our own may be. The reason is that the theories invoked to explain the weird orbits of hot Jupiters and warm Jupiters make it difficult to envision life flourishing in these solar systems.

For theorists trying to explain hot and warm Jupiters, the problem is simple. Geometry dictates that there's not enough material within the inner part of a disk to build a massive planet before the disk dissipates. Also, the inner regions are "as hot as a furnace," Alan Boss notes—way too hot for hydrogen and oxygen to form water ice, thought to be a key ingredient in building the core of a massive planet.

"We still have a very hard time imagining forming a planet half the mass of Jupiter out of ice and gas at just 0.05 AU— close to the hottest portion of the protoplanetary disk," Boss says.

Long before the discovery of hot Jupiters, Douglas Lin of the University of Santa Cruz proposed that giant gaseous planets could form farther out in the disk and then naturally migrate inward as a result of a complex interplay of tidal forces. But at the time, migration wasn't needed to explain anything in our solar system—the only one known at the time. So his theory didn't garner much support.

Says Bill Ward of the Southwest Research Institute, "At first it was a *heresy* to think that a full-grown massive planet could migrate vast distances within a disk. But the discovery of large extrasolar planets close to stars is circumstantial evidence that giant planets *have* moved vast distances."[18]

If they have, it may be a matter of push and pull. One migration hypothesis holds that a giant planet's gravitational field tugs strongly on the disk. But the disk doesn't just sit there; it tugs back. The resulting gravitational drag reduces the planet's orbital angular momentum, forcing it to spiral in toward the star.

The interplay of gravitational forces may also have a different effect, Jack Lissauer notes, but with the same outcome. The planet's gravitational field may be so strong that it torques open a groove in the disk, partitioning the disk into inner and outer regions. According to this hypothesis, the inner region and the planet lose angular momentum to the outer region, causing the planet to spiral inward.

What applies the brakes to keep the planet from cruising to its death in the star? One possibility is that the disk simply is truncated by magnetic fields, as in Shu's X-wind model. In this case, the planet would stop at the inner edge of the disk. Another hypothesis invokes tidal forces.

But the bottom line, once again, is that scientists really aren't sure. And neither theory explains why many extrasolar planets seemed to have stopped their migrations far short of the star, thereby becoming merely warm Jupiters. Or, even more critically, why our own, cool Jupiter did not spiral in close to the Sun.

One explanation is that some protoplanetary disks are more short-lived than others. According to this line of thinking, right after our Jupiter and Saturn formed, there was just enough gas left over to make Uranus and Neptune, but not enough to induce the weird gravitational interplay that otherwise would have forced Jupiter to move. Astronomical evidence showing that protoplanetary disks do exist for varying amounts of time supports this view, Boss says.

If that's what happened in our forming solar system, we're darn lucky to be here—just as Bally argues. Because if the disk had been somewhat longer-lived, Jupiter would have migrated and acted like the Death Star in Star Wars. It would have spelled doom for any planets in its way—including Earth. Basically, Jupiter would have shoved us into interstellar space or simply swallowed us up. And since the hot and warm Jupiters in other solar systems did seem to migrate, they may have killed off quite a few terrestrial planets, including some that might have given life a toe-hold.

''The discovery of hot Jupiters means that there will be some fraction of planetary systems that are inhospitable to life,'' Boss observes. ''In the process of migrating inward, a future hot Jupiter will kick any Earth-like planets it encounters out of its way, ejecting the Earths altogether to freeze in interstellar space, or vaporizing them through a collision with the system's star.''

A prospect that would no doubt make Darth Vader envious.

Astronomers looking at all this evidence reach different conclusions about whether Earth-like planets capable of sustaining intelligent life, or even life overall, are common.

Paul Butler, co-discoverer with Geoff Marcy of many extra-solar planets, is adamant that it's way too early to draw any conclusions. ''The critical issue for us right now is this: What fraction of planetary systems are going to end up looking like our own?,'' he says. ''And that question is *completely* up in the air right now. If anybody tells you they know or even have a clue, there's something *wrong*, because we just don't have any clue at all.''[19]

For astronomers like Butler, the answer will have to come from better observations. Alan Boss, a theorist, agrees. ''This is the question we all want to answer. What is the frequency of higher life?'' he says. ''Step one is to find some rock that could serve as a home base for such life. If we could just find a terrestrial-mass planet more or less in the habitable zone where water could be liquid, that would be a tremendous advance, even if that thing isn't beaming us their version of 'I Love Gork,' or whatever the popular television show is there. It would be great if we could find such a planet.''

NASA's Jack Lissauer also emphasizes that astronomers are just beginning a long quest for answers. "Today we know that there are 5 to 10% of stars which have planetary systems quite unlike our own, and they don't seem to be suitable for life," he says. "There are big planets in nasty orbits, and in some cases models suggest that big planets migrated through and therefore caused problems. So these solar systems are not good abodes for life. But we can't say anything about the other 90 or 95% because we just don't have good information about them."[20]

Given current plans by NASA and European nations, it will take years for astronomers to acquire the technology capable of finding Earth-like planets around other stars. In the meantime, all we have to go on is fragmentary information. And while that information may not allow firm conclusions, some astronomers are willing to venture a bit further than Paul Butler's "we have no clue" position.

"I believe that there are probably a lot of *habitable* planets out there," Lissauer speculates. "But I have no idea whether they're *inhabited*." He admits that he's worried about something called Fermi's paradox—a question first asked by the physicist Enrico Fermi in the 1940s.[21] It goes something like this: "If there is intelligent life out there, *where is it?*" Maybe it hasn't made its presence known by now because it's just not there.

Well, maybe it *is* there but we haven't detected it because communicating across vast distances is extremely difficult, even for an advanced civilization. So we're left in the same position of just not knowing.

But astronomers can agree on one thing. While the universe of popular culture is inhabited with all manner of complex life, science suggests that quite the opposite may well be true. We may be more special in the cosmic scheme of things than the Copernican revolution seemed to have taught us.

# 12

## TO LIFE

### *The Organic Soup Kitchen*

*"We are the dust of long dead stars. Or, if you want to be less romantic, we are nuclear waste."*
— Sir Martin Rees[1]

*For the poet, the eagle soars; the equations and DNA do not.*
— Frank Shu[2]

*The physical species Homo may count for nothing, but the existence of mind in some organism on some planet in the universe is surely a fact of fundamental significance. Through conscious beings the universe has generated self-awareness.*
— Paul Davies[3]

Scott Sandford's "kitchen," as he calls it, is a curious place to prepare a meal.

Mind you, the empty bottles of champagne on a shelf (momentos of particularly successful culinary creations) might seem at home in an ordinary kitchen. But while the bearded and bearish middle-aged "chef" may slave over a hot stove at home, here at the NASA Ames Research Center near San José, California, he labors over a high-tech piece of equipment that should, by all rights, carry the Sub Zero brand name. That's because his device *chills* his raw ingredients to just a tad above absolute zero. Only then does it cook them, and not with a flame but with ultraviolet radiation.

As one of the chefs in NASA's Astrochemistry Laboratory, Sandford uses this high-tech device to transform some interstellar medium, prepared from ingredients in what he calls his "spice rack," into a nutritious organic gelato. Something similar to this rich frozen goo seems to form within the cold molecular clouds that spawn stars and planets. And it may be just the stuff needed to tempt life into getting started on a young planet.

Based on tell-tale infra-red emissions from star-forming regions, astronomers have identified sand grains mantled by ices of methanol, hydrocarbon and water within molecular clouds. Theory has it that the compounds in these ices are slowly cooked by ultraviolet light from nearby stars into the organic goo. To be more precise, ultraviolet irradiation transforms the ices into life-critical molecules, such as quinones, which play a key role in the ability of cells to transfer energy, and amino acids, which are the raw materials for proteins. When a solar system forms within the cloud, these compounds are then bound up in comets and asteroids. And as these objects rain down on a young terrestrial planet like Earth, the chemicals, along with organics produced by other processes, provide the raw materials needed for the origin of life.

Or so the theory goes. To test it, and to tease out some of the details about how the process unfolds, Sandford uses his device to simulate the conditions within a molecular cloud. In these experiments, a pump and refrigerator suck the air and heat out of a metal chamber about 8 inches on a side, creating an icy cold vacuum. "That's space," Sandford says. Compounds in the spice rack, such as methanol and hydrocarbons, are then sprayed onto a lollipop-size disk, where they freeze. "That's a dust grain," he notes. And a lamp bathes the ices with ultraviolet light. "That's a local star."[4,5]

The results? "We're using molecules that you can buy in a grocery store, as well as others that you can get from the tailpipe of your car, and we're finding that if we expose them to ultraviolet light, we get all these amazing organic compounds," Sandford says. "This stuff looks a whole lot like the stuff in meteorites, which looks a whole lot like the stuff being used in living things."

The stuff in meteorites and living things ultimately was built up from star-dust—carbon, oxygen, nitrogen and other elements manufactured in stars, as well as hydrogen made originally in the big bang. Later these materials came together within molecular clouds to make planets, meteorites and the first living thing, not to mention astrochemists who use the same basic ingredients to make cosmic gelato. But animating star-dust—transforming these simple ingredients into single-celled microorganisms capable of reproduction and evolution (let alone intelligent creatures)—is another one of nature's dynamic range problems.

Scientists believe nature spanned this huge gulf with a long progression of chemical transformations, each one representing a step up in complexity. No one knows the exact nature and sequence of these chemical events. But starting with simple organic compounds in an early ocean, they are believed to include the development of self-replicating chemicals, the formation of cell-like vesicles encapsulating complex organics, and later, after many other intermediary steps, the origin of the first entities we would consider to be alive.[6]

Sandford and his colleagues are working on an earlier stage in the sequence: the source of the organic compounds that scientists believe must have been available for life to get started. Of course, they are not the first scientists to tackle this question. In the 1920s, the British geneticist and popularizer of science John B. S. Haldane, as well as Soviet biochemist, Aleksandr Ivanovich Oparin, proposed that the organics could have come from the sky. Earth's early atmosphere, they hypothesized, was chemically *reducing*, meaning the reactions that took place within it were the opposite of the oxidizing reactions that produce rust. And in a reducing atmosphere, they pointed out, plentiful hydrogen would have combined readily with oxidizing chemicals such as carbon dioxide and nitrogen to forms such compounds as ammonia ($NH_3$) and methane ($CH_4$). Most important, these compounds could then undergo chemical reactions to produce organic molecules useful to early life.

But once these organic molecules were produced, *where* did the critical jump from inanimate compounds to replicating, evolving living things take place? Charles Darwin had speculated that a "shallow, sun-warmed pond" was life's incubator.[7] Haldane and Oparin took this idea further, proposing that organic molecules synthesized in a reducing atmosphere rained down to the ocean, where they produced a rich "primordial soup." Evaporation in shallow lakes and lagoons concentrated the soup, setting the stage for the origin of life.

In the 1950s, Harold Urey and Stanley Miller put experimental flesh on these theoretical bones. In their famous pioneering experiments, Miller simulated a reducing atmosphere and an early ocean in a simple laboratory flask by filling it with water, hydrogen, ammonia and methane gases. Then he delivered simulated lightning strikes with electrical discharges in the gas. With repeated discharges over the course of days, liquid condensing in the apparatus began to turn color as new chemicals were produced. Later analysis of the liquid revealed a rich soup of organic molecules, including amino acids.[8]

These results were stunning. None other than Darwin himself had actually discounted the idea of exploring the origin of life scientifically. "It is mere rubbish thinking at present of the origin of life; one might as well think of the origin of matter," he said. But in their experiments (and ones to follow substituting ultraviolet light for electricity), Miller and Urey showed that it wasn't rubbish at all. At the least, the steps leading up to life could be explored with basic chemicals and laboratory equipment so simple it could be used in a high school chemistry class.

With Miller and Urey's experiments, a science-based view of the origin of life took root in the popular imagination. From the early Earth's stormy, lightning-filled skies, amino acids and other compounds rained down, collecting little by little in the oceans. And at the edge of the sea, in a warm, stagnant tide pool, evaporation left behind an increasingly concentrated organic scum that somehow gave rise to something living.

Today, there are alternatives to this classic view, theories that have gained some credence in part because of potential problems with the Miller–Urey scenario. One problem is that the early Earth might not have had a reducing atmosphere. It turns out that ultraviolet light could well have broken apart methane and ammonia molecules, and the hydrogen released could have escaped into space. If that had happened, the now not-so-reducing atmosphere would have produced far less in the way of organic compounds.[9] So Sandford's astrochemical kitchen is a way of testing an alternative to the classic lightning-bolts and life-chemicals theory. Much of the primordial soup that was whipped into life on the early Earth could have come from space, he and his colleagues believe.

And there's an even bigger problem with the theory that life formed from tide pool scum. The evidence available today suggests that the early Earth was a perfectly hellish place, beset by huge impacts with rocky and icy bodies left over from the origin of the solar system. While it lasted — and there is some question about the timing of the impacts — the violence would have evaporated entire oceans and even melted the planet through and through. Once the ocean-evaporating impacts died down, smaller cataclysms still would have evaporated what's known as the *photic zone*—the upper layer of the oceans where light penetrates and thus allows living things to make a living through photosynthesis. So until the rate of both ocean-evaporating and photic-zone-destroying impacts dropped off, tide pools would have been a nasty place for organic molecules struggling to become animate.

On the other hand, recent research by astrobiologists shows that the earliest common ancestor of all living things on Earth today—think of it as the trunk of the tree of life—was a *heat-loving* organism, a type of *extremophile*. That doesn't necessarily mean it was the *first* living thing. But it suggests that, at the least, very early life evolved in a very hot place. So even while impacts were stripping away the photic zone of an early ocean, the first forms of life might have been thriving at hot springs on the deep, protected sea floor. Today, heat-loving microorganisms thrive in the boiling hot waters of such hydrothermal vents. They gain their living from chemical energy and a rich stew of compounds spewing from the vents. And they form the bottom of the food chain of a richly tangled web of life thriving in total blackness, and in utter isolation from whatever may be going on closer to the surface.

Clearly, impacts posed challenges to the origin and sustenance of early life. But without them, the Earth would itself have evolved into a very different place, a planet that might not have been so nurturing of life. The impacts brought more than the raw materials for the construction of the rocky portions of the planet, and perhaps for the origin of life. They also brought the water and gases for oceans and an atmosphere. And they delivered the energy and radioactive materials that fuel Earth's internal heat, which drives the mobile, ever shifting segments of crust that continually build mountains and recycle important geochemicals. This process of plate tectonics may actually help keep Earth's climate stable enough for complex life forms to thrive.

In short, the violence of Earth's youth had a lasting, constructive legacy. When the impacts subsided, Earth was left with life support systems that have nurtured a stunning diversity of living things—from single-celled microorganisms to animals with the intelligence to ponder how their planetary oasis came to be. In this regard, Earth stands alone in the solar system. Venus, our closest planetary neighbor, has an extreme, hothouse climate that has left it sterile. And while there is some hope that microorganisms may hide somewhere on frigid Mars, no convincing signs of biological activity have yet been found. Earth alone appears to have acquired a climate temperate and stable enough to sustain life over the long haul, but also varied enough geographically and over time to stimulate bio-diversity.

So in the pages that remain, let's examine the connections between Earth's early history and life. How did the materials and energy imported by violent impacts help establish our planetary oasis, and what lasting legacy have those events had for living things, including us?

As we saw in the previous chapter, the actual birth of our planet was the result of violence on a scale we can scarcely imagine. Earth was formed through cataclysmic collisions and mergers of baby planets called planetesimals. Simply by chance, the rocky body that would grow through these mergers into our planet developed into one of the bigger guys in its neigh-borhood. Because it was more massive, it swept up everything within its gravitational feeding zone. The resulting mega-collisions are believed to have knocked our planet's spin axis off kilter. By causing variations in the amount of solar energy reaching different parts of the globe at different times of the year, Earth's tilt endowed us with seasons, and it may also contri-bute to the comings and goings of ice-age glaciers. Both aspects of Earth's climate system — seasonal variation and long-term climate cycles like ice-ages — have encouraged biodiversity by creating a wide variety of habitats and putting pressure on species to adapt to new conditions.

But if Earth's tilt, or *obliquity*, were to get *too* far out of whack, life-threatening climate swings could result. Luckily, the Moon's gravity appears to prevent that from happening. According to Lynn Rothschild, an astrobiologist at the NASA Ames Research Center, without the Moon's stabilizing grip, Earth's obliquity could vary enough to cause large variations in the amount of solar radiation reaching us. To survive, ''you'd better be very small, very versatile, and live in the water," she says. And what about humans? "Earth would not be a good place for us."[10]

Where did the Moon come from? It appears to be another legacy of the Earth's violent upbringing. Scientists believe it formed when an object the size of Mars crashed into the Earth about 4.51 billion years ago. According to this theory, the impact splashed a huge volume of melted debris into orbit — some from the Earth and some from the impactor itself — which eventually cooled and coalesced to form the Moon.[11]

The pummeling from planetesimals, comets, asteroids and other bodies while Earth was accreting had other important effects as well. The impacts delivered cargoes of water and gases from the solar nebula, leaving the Earth a steaming and fuming mess. Once the rate of impacts trailed off enough, and the Earth cooled sufficiently, the planet's gravity clutched gases such as nitrogen and carbon dioxide close to the surface, forming the atmosphere. Later, water vapor began to condense and rain out, forming a hydrosphere — a salient event in the pre-history of life. Liquid water is considered a prerequisite for life because nutrients and waste products can dissolve in it, essential chemicals can be moved within it, and biochemical reactions occur with it.

The construction of planet Earth and the Moon was essentially complete by about 4.47 billion years ago.[12] Unfortunately for geologists, detailed evidence of what happened on Earth over the next 500 million years has been obliterated by erosion. Not so on the Moon, however, where wind and water have never abraded and scraped the surface. As a look through a pair of modest binoculars will reveal, the Moon's face bears the scars of continued bombardment by solar system debris. Analysis of those scars and rocks brought back by Apollo astronauts have allowed scientists to calculate the rate of impacts on both the Moon and Earth.

There is some question as to when the bulk of the big, early impacts occurred. According to the conventional view, a continuous but slowly declining rain of solar system debris fell down on the Moon and Earth after the lunar-formation event. These were not planet-sized bodies like the impactor responsible for the formation of the Moon. The bodies included comets and very large asteroids, as well as smaller chunks of solar system flotsam and jetsam. The very largest of these impacts occurred earlier and trailed off in frequency as time passed. But based on another interpretation of the lunar impact record, some scientists believe the rate of impacts dropped steeply right after the Moon formed, and then spiked catastrophically about 3.9 billion years ago in an event known as the late-heavy bombardment.

Whether they were concentrated in a spike or spread out over some 600 million years, some of the impacts that occurred even after the Moon and Earth were complete were huge. The largest, telling impact scars on the Moon are dark basins on the front side that vaguely suggest a human visage. The biggest seems to be the South-Pole/Aitken Basin, which is about as wide as the distance between Chicago and the southern tip of Florida, or London and Athens. To gouge out that much rock, the colliding object (which hit about 4.1 billion years ago) had to have been about 200 kilometers across, or roughly as wide as the state of Massachusetts. Just a little smaller is the Imbrium Basin, 1500 kilometers across and gouged out by an object at least 100 kilometers in size. According to Kevin Zahnle, a scientist at the NASA Ames Research Center, the Moon suffered about a dozen such impacts in its early history.[13]

The Moon clearly suffered terribly as it took one hit after another on the chin. In the region of the big impact basins, its rocky crust thinned so much that molten rock welled up to the surface, filling the basins with oceans of magma that subsequently hardened into darker, lowland rocks. Meanwhile, the Earth, being a heavyweight in comparison to the Moon, had a more powerful gravitational grasp. "Earth, being bigger, picks out the largest available impacts," Zahnle says. As a result, our planet suffered hundreds of impacts of the kind that dug out the big basins on the Moon.

Zahnle uses an unconventional method for categorizing the various sizes of impacts that afflicted Earth: the U.S. Department of Agriculture's ratings for olives. So a *Super Colossal* impact is big enough to splash out enough material to form the Moon. A *Colossal* melts the entire crust of the Earth. A *Jumbo* vaporizes the oceans. An *Extra Large* vaporizes the photic zone of the oceans. And a *Large* cauterizes the continents.

Just a single jumbo impact would have created a hell on Earth. As Zahnle describes such an event, the energy of the impact evaporates large amounts of rock. The result? A noxious, rock vapor atmosphere "a hundred times thicker than the atmosphere we know and love," he says. Toward the top, cooling causes clouds to form—not the ordinary kind made of water but *rock clouds* made of *silica*. Lower down, temperatures exceed 2000 Kelvin, causing the sea surface to boil off. "It takes a few months at most to evaporate the oceans," Zahnle says. Then the rains come. But not water. As the rock vapor condenses, rock drops pelt Earth's surface. Once the rock rains out, a steam atmosphere is left behind. With this insulating blanket thrown over the planet, heat escapes slowly. According to Zahnle, it takes about 3000 years for conditions to cool enough for the steam to condense. Then, in perhaps just a few years, torrential downpours "give you your oceans back," he says.[14]

But pacific conditions would not have lasted for long. Based on the lunar impact record, scientists have estimated that during its early years, about five ocean evaporators hit Earth. The last of these cataclysms could have occurred as late as 3.8 billion years ago.[15] So woe to any early life forms that may have managed to get it together in a fetid tide pool during this period.

At some point early in Earth's history (probably before 4.4 billion years ago), conditions actually were hot enough for iron and nickel to melt and fall to the center of the planet, where it collected to form a metallic core.[16] Today, an outer shell of the core remains molten. But over time, liquid metal in this liquid shell has been freezing onto the surface of the solid inner core. As it freezes, the metal releases heat, which flows outward, roiling the molten iron and nickel like a pot of water on a stove. This movement of liquid metal is believed responsible for Earth's magnetic field, which is useful not just for people trying to navigate with compasses. The magnetic field also shields life on Earth from hazardous radiation coming from space.

The intense heating resulting from impacts, as well as the decay of radioactive elements, and the friction generated by iron and nickel sinking to the planet's center also caused other materials to fractionate into layers inside the Earth. These layers, which were in place about 4.47 billion years ago,[17] were the precursors of today's 2900 kilometer-thick mantle and the thin, overlying crust of brittle rock that makes up the continents and seafloors. In the billions of years that have elapsed since,

Earth's interior heat, today supplied mostly by radioactive decay, has been escaping little by little. It does so through movement of rock within the mantle (which has the consistency of stringy taffy). Hot buoyant rock rises in the mantle and releases heat near the surface through volcanic eruptions and other processes. Cooler and denser now, the rock sinks—only to be replaced by fresh batches of hot rock from below. With these movements, gargantuan rollers of material called convection cells become established within the mantle.

As they roll on (in place, like the rollers in a conveyor belt), they push and pull on the crust and top-most portion of the mantle, causing these layers to segment into distinct tectonic plates that drift atop the viscous mantle at about the speed that fingernails grow. In some places, plates grind laterally past each other along faults like California's San Andreas, triggering earthquakes. Along the mid-ocean ridge, which girdles the globe on the seafloor like the seams on a baseball, the plates spread apart, allowing magma to well up continuously to make new seafloor crust. In places where plate movements bring continental crust into collision, giant mountain ranges like the Himalayas form. And in places where seafloor crust converges (or where seafloor converges with a continent), one plate dives beneath the other in a process known as subduction. In the back of the trench that marks the subduction zone, volcanoes rise as subducting rock melts, mixes with mantle materials, and punches back up to the surface as magma. These volcanoes form island arcs, like the Lesser Antilles in the Caribbean, or chains of volcanic mountains, like the Andes in South America and the Cascades in the United States.

Among the solar system's three terrestrial planets, Venus, Earth and Mars, ours is the only one with plate tectonics and the long, massive mountain ranges that it builds. Earth also is the only one with liquid water at the surface. And these factors— plate tectonics, mountain ranges, and liquid water—all seem connected in a way that may just solve a mystery known as the Goldilocks problem. When it comes to the existence of complex animal life like us, Mars is too cold, Venus is too hot, but Earth is just right.

Two University of Washington scientists, paleontologist Peter Ward and astronomer Donald Brownlee, champion this argument in their recent book, *Rare Earth*.[18] Plate tectonics, they say, is essential for the existence of complex, animal life on a planet. It is made possible on ours by a number of factors, including the presence of liquid water.

Based on a theory first advanced in the early 1980s, Ward and Brownlee argue that plate tectonics is a critical component in a feedback system that prevents Earth's climate from getting too cold or too hot, like Mars or Venus. How many planets in the universe have just the right conditions to favor plate tectonics? Perhaps not many, Ward and Brownlee say. And they use this argument to advance their main theme: while microbial life may be exceedingly common in the universe, complex animal life may be very rare.

Whether they are right about this or not, a case can be made that we owe our existence as a species—as *Homo sapiens*, modern humans—to one of the most significant tectonic events of the last 100 million years: India's collision with the underbelly of Eurasia. By turning down Earth's thermostat, this event may have been the cause of a well documented slide in global temperatures. In the latter part of this cooling trend, according to one theory of human origins, habitat changes in the hominid homeland of Africa prodded our primate ancestors to step out from the forest and into the savanna—upright and on two legs.

Since climate and plate tectonics seem to be such important aspects of Earth's character—and since they seem to be connected—let's examine some of these issues now.

Overall, a few things push the climate in one direction or another: how much radiation the planet receives from the Sun; how much is reflected back into space; and how much of the heat resulting from absorption and re-radiation of solar energy is trapped within the atmosphere by the greenhouse effect. The climate depends on the balance of these things, as is illustrated by what would happen if Earth had no heat-retaining greenhouse gases in the atmosphere. The most important of these gases are water vapor, carbon dioxide and methane. Without the natural insulation they provide today, Earth would have a global average temperature of minus $18°$ C, equal to that of the airless Moon. In such a deep freeze, Earth could not harbor animal life. But thanks to Earth's greenhouse gases, the planet's global average temperature sits at a comfy $15°$ C.[19]

324

If scientists' understanding of solar physics is correct, the balance of factors that force the climate was quite a bit different 3.5 billion years ago. At this point, the output of solar energy from the Sun should have been about 30% lower than it is today. This should have left Earth in a deep freeze. But there is good evidence for copious liquid water on the planet during this time. So with the Sun so weak, why wasn't the Earth a planetary snowball at this early time in its life? The answer, many scientists believe, is that the planet's insulating blanket of carbon dioxide was thicker than it is today, compensating for the reduced sunlight by retaining more heat.[20]

The importance of the balance between incoming solar radiation and insulating greenhouse gases is also illustrated by our ongoing experiment with heat-retaining greenhouse gases and the global climate. Since the dawn of the industrial era, we humans have been pumping additional carbon dioxide into the atmosphere by burning fossil fuels, and the concentration of $CO_2$ has risen accordingly. Between the years 1750 and 2000, the concentration increased from about 280 parts per million to 370 parts per million, or by about a third.

The thicker insulating blanket of greenhouse gases seems to be having an effect. During the twentieth century, the global mean surface temperature increased by about a $0.5°$ C. Over the land surface of the Northern Hemisphere, the increase in temperature has likely been greater than during any other century in the past 1000 years. And in the Northern Hemisphere, the 1990s were likely the warmest decade of the millennium. According to the U.N.-sponsored Intergovernmental Panel on Climate Change, the most recent analysis of these trends provides stronger evidence than before that most of the warming observed over the past 50 years is attributable to human activities.[21]

Projections of what might happen in the future as we continue to pump $CO_2$ into the atmosphere are uncertain. But given current trends, the IPCC predicts that atmospheric carbon dioxide concentrations could reach as high as 1200 parts per million by the year 2100, a 330% increase over 1750 levels. And the rate of warming that could theoretically produce "would very likely be without precedent during at least the last 10 000 years," according to the IPCC.[22]

On geologic time scales, however, Earth has a hedge against things getting too far out of whack. Although the planet has been much warmer in the past than it is today, it has never suffered a runaway greenhouse effect like Venus seems to have experienced.

In some ways, Venus is our planetary twin. It's about the same size and mass as the Earth. It also has about the same inventory of carbon dioxide as our planet, and there are good reasons to believe that it started out with about the same amount of water.[23] Finally, Venus is a close planetary neighbor, at least by the standards of the solar system overall. So one might reasonably guess that the climate of the two planets would be fairly similar. Yet temperatures on the surface of Venus hover around 900° F (or about 470° C), which is high enough to melt lead, not to mention evaporate any liquid water that might otherwise be present. That's because most of Venus's carbon dioxide is in its atmosphere. By contrast, most of ours is locked up in carbonate deposits on the seafloor. In fact, carbonates on Earth store close to 170 000 times the amount of carbon dioxide as exists in the atmosphere today.[24]

Venus may not have started out this way. In its early years, it might have been cool enough to harbor an ocean. But it is closer to the Sun than we are, so early Venus was warmer than Earth. This caused heightened evaporation of water from the surface—extra water the atmosphere easily held because it was warmer. Remember now that water vapor is a potent greenhouse gas, so with more of this insulator in the atmosphere, warming increased. That caused even more evaporation, more atmospheric retention of water vapor, still higher temperatures, even more evaporation, and so on, until all of the water that might have been present on the surface of Venus had evaporated. The result? No ocean, a runaway greenhouse effect, and a hellish steam-bath climate.[25]

Now you may be wondering about the carbon dioxide. Isn't it responsible for the heat on Venus? Today it certainly is. The water that originally made things hot has been chipped away by solar energy. (Solar photons break water molecules apart, allowing the hydrogen to escape to space.) So almost all of the water vapor is now gone, leaving carbon dioxide behind. But why is almost all of Venus's $CO_2$ in the atmosphere today, where it maintains the hellish climate, whereas most of Earth's is locked away in carbonates on the seafloor? These differences are a bit puzzling, because both planets appear to have a mechanism for removing carbon dioxide from the atmosphere.

On Venus, carbon dioxide may react chemically with minerals on the surface, removing the gas from the atmosphere and depositing it in the form of carbonates. If something didn't counteract this removal, carbon dioxide would be drawn down so low that the planet would cool. Similarly, $CO_2$ in Earth's atmosphere dissolves into the ocean, where it precipitates to form carbonate compounds that make up limestone. Moreover, atmospheric $CO_2$ also reacts chemically with minerals in rocks over land. As the byproducts of this *chemical weathering* are carried by streams and rivers to the sea, they too wind up as carbonates in limestone on the seafloor. In fact, this removal system is so effective that without any resupply of carbon dioxide to the atmosphere, it would remove all $CO_2$ within 400 000 years.[26]

The resupply comes from volcanoes and vents. And a good thing too. Without it, $CO_2$ would drop so low we'd freeze. On the other hand, if limestone formation didn't remove $CO_2$ added to the atmosphere by volcanoes, it would build up unabated, and Earth's climate would be much hotter. Luckily, the life-support system on our planet maintains a nice balance. Huge amounts of $CO_2$ are stored safely on the seafloor, and just the right amount of $CO_2$ is left in the atmosphere to keep conditions temperate over the long haul. Meanwhile, on Venus, huge amounts of atmospheric $CO_2$ cannot be drawn out of the atmosphere and locked safely away largely because temperatures are just too high for this. The extreme heat breaks down any carbonates within rocks near the surface, releasing the carbon dioxide back into the atmosphere.[27]

And now we come to a key question. What is it about the carbon recycling system on Earth that keeps most of it locked away in limestone on the seafloor? Part of the answer is that Earth, being a little farther away from the Sun than Venus, never had a runaway greenhouse effect. So it never got so hot. But one could imagine that the climate system could get seriously out of balance if, say, volcanoes pumped out more of the gas than was removed by chemical processes. And that would cause temperatures to rise. If nothing intervened to reverse the situation,

$CO_2$ and temperatures would go higher and higher until Earth would cease to be such a homey place. Of course, something *does* intervene—a beautiful feedback system that works like a global thermostat to keep carbon dioxide levels and temperatures in line over the long haul. And the system seems to depend on plate tectonics. Here's how it works.

By building large mountain ranges, plate tectonics subject large amount of rock to chemical weathering. According to a theory proposed in the 1980s, when too much $CO_2$ builds up, the climate warms, precipitation increases, and both factors accelerate chemical weathering. With more chemical weathering and more carbonate formation in the ocean, carbon dioxide is drawn out of the atmosphere at an increasing rate, eventually cooling the climate.

Now if the $CO_2$ drawdown continued unabated, temperatures would be driven *too low*. This doesn't happen, because cooling *slows* chemical weathering and formation of carbonates in the ocean. Meanwhile, volcanoes continue to pump out carbon dioxide, so atmospheric stocks are replenished and the planet warms up again.

Peter Ward and Donald Brownlee argue that without the massive mountain ranges built by plate tectonics, the efficiency of chemical weathering might be much lower. Also, plate tectonics insures that carbon dioxide can continue to be drawn out of the atmosphere by chemical weathering, and added by volcanoes, as needed. Ward and Brownlee point out that chemical weathering depends on a reaction between $CO_2$ and calcium minerals in rocks. In the process, the calcium becomes sequestered on the seafloor as limestone. If weathering continued without a way to resupply calcium to the system, over long periods of time, calcium would become depleted, removal of $CO_2$ would slow, and the planet would heat up. Plate tectonics insures that this doesn't happen because it recycles calcium sequestered in limestone back into surface rocks. This can occur because plate movements continually shove seafloor crust, along with its load of carbonates, into subduction zones. As the limestone plunges into the Earth's hot interior, the carbonates break down. Volcanic eruptions behind the subduction zone bring the resulting calcium and carbon dioxide back to the surface in magmas. The $CO_2$ goes on up into the atmosphere. And the calcium winds up in igneous rocks that are once again available for chemical weathering.[28]

The bottom line is that plate tectonics promotes chemical weathering by building big mountain ranges. And it insures that weathered materials can be recycled back into the system to keep it going. Without it, Earth would lose its climatic thermostat, and our planet might then follow a story line uncomfortably similar to that of Venus.

Or things could go the opposite way and we could wind up looking more like Mars. Unlike Venus, the red planet has a very tenuous atmosphere — not much carbon dioxide to keep things warm. As a result, the average surface temperature is an icy cold $-81°$ F ($-63°$ C). Any water that made its way to the surface from some underground source would either freeze or instantly evaporate because of the low atmospheric pressure. There is, however, evidence that water once flowed copiously on the surface of Mars in the planet's early days: valleys forming dendritic patterns that look like watersheds on Earth; gullies on the walls of craters that look like they were carved by run off; and channels that appear to have been dug by catastrophic floods. Scientists believe Mars once had a thicker, more $CO_2$-rich atmosphere, and therefore temperatures warm enough to allow water to exist at the surface in liquid form. But then, somehow, Mars lost much of its atmospheric carbon dioxide to space.[29]

If Mars had been more Earthlike in the past — with an atmosphere richer in carbon dioxide and a warmer and wetter climate — then carbonates may have formed on the planet. This would have drawn $CO_2$ out of the atmosphere. But Mars, like Venus, has no plate tectonics. So there would have been no way to recycle carbon dioxide from carbonates back into the atmosphere. And according to some estimates, the extra carbon dioxide needed for a warmer, wetter climate could have been removed from the atmosphere in as little as 100 million years.[30]

Although it has avoided heating up like Venus, or cooling down like Mars, our planet certainly has experienced dramatic swings in temperature, including long stretches of tropical heat in which the poles were quite temperate. But the most intense climatic swings have been toward extreme cooling, including global freeze-ups in which almost all of the oceans may have been covered in ice. There is evidence for one such *snowball Earth* episode, as Cal Tech's Joseph Kirschvink has dubbed it, about 2.5 billion years ago; there is further evidence for a series of several more between 800 million and 600 million years ago.

Major milestones in the evolution of life on the planet seem to have occurred in the aftermath of these episodes. The earlier one was the appearance in the fossil record of the first *eukaryotic* life. Eukaryotic cells have a higher degree of organization than the cells of the more primitive prokaryotes. For example, eukaryotes have membrane-enclosed structures within their cells, including a nucleus containing DNA, and various organelles such as energy-producing mitochondria.

It took quite some time for eukaryotic life to take off on an evolutionary path of diversification that ultimately led to plants, fungi and animals. But the first fossils of a eukaryotic life form, a creature named *Grypania* consisting of chains of nucleated cells, are 2.1 billion years old. Of course, these are just the first such creatures to be discovered; eukaryotic life forms surely evolved even earlier.

According to the conventional explanation for the evolution of eukaryotes, photosynthesizing prokaryotic bacteria paved the way by bringing about a radical change in Earth's atmosphere dubbed the *oxygen revolution*. The prokaryotes lived on solar energy, water and carbon dioxide, releasing oxygen into the atmosphere as a result of their metabolic activities. Over hundreds of millions of years, oxygen built up.

For organisms adapted to living off carbon dioxide, oxygen was a toxin. And when oxygen concentrations reached a certain point, some could no longer cope. Those that could migrate to low-oxygen environments, such as hot springs deep on the ocean floor, survived. Those that couldn't became extinct. Meanwhile, those capable of adapting by actually *using* the oxygen in life processes—such as eukaryotes—gained a big advantage, because aerobic metabolism produces much more energy than anaerobic metabolism. And according to this theory, eukaryotes were chief among these organisms.

In a newer theory explaining the rise of eukaryotes, some scientists argue that the planet's emergence from the first snowball Earth episode was what played the decisive role. That emergence occurred thanks to the planetary thermostat. Here's how the process could have unfolded. With the oceans capped by ice, they were effectively sealed off from the atmosphere. Volcanoes on the seafloor continually added $CO_2$ to seawater. And volcanoes on the surface kept pumping $CO_2$ to the atmosphere. But there was no exchange — no $CO_2$ from the atmosphere could dissolve into the oceans. As a result, atmospheric concentrations of the greenhouse gas eventually grew high enough to warm the climate and melt the global snowball. And with the sudden disappearance of the icy lid that had previously isolated the oceans, photosynthetic organisms that had managed to survive, perhaps in a few ice-free areas around the equator, suddenly had a field day. Spurred on by the huge amount of $CO_2$ available in seawater (as well as nutrients that had built up too), and with the sudden flooding of the ocean basins by sunlight, they bloomed in a frenzy of renewed photosynthetic activity. "For a brief time, the Earth's oceans would have been as green as Irish clover," Kirschvink says.[31] So in addition to the climate warming dramatically, oxygen exploded into the atmosphere. And the evolutionary pressure this caused favoured the evolution of eukaryotes.

Kirshvink and his colleagues also argue that Earth's emergence from the freeze-ups between 800 million and 600 million years ago once again brought evolutionary pressure to bear on surviving species. And this, they argue, ultimately led to the biological big bang known as the Cambrian explosion — a dramatic increase in the diversity of animals that occurred about 550 million years ago.

Starting about 55 million years ago, the plate tectonics–$CO_2$ connection may have had another dramatic impact on climate and life. At about this time, one of the most monumental tectonic events of the past 100 million years occurred. The subcontinent of India, which had been detached from other landmasses, was shoved northward by plate movements into the underbelly of the Eurasian continent. Continental rock is too buoyant to subduct into the mantle. Instead, a collision like this causes the crust to crumple like the hood of a car in a head-on collision. So the result of India's collision with Eurasia was the Himalayan range and the Tibetan Plateau, the highest regions on Earth.

This event may have sent Earth's climate sliding toward progressively cooler temperatures. Evidence of the long-term cooling is readily visible near where I live at the base of the Rockies. Not far from Denver, scientists and volunteers from the Denver Museum of Nature and Science have been busy excavating remnants of a tropical forest that blanketed this part of Colorado 64 million years ago. Meanwhile, just a dozen or so miles away, windwhipped granite spires, serrations and pyramids poke up from the spine of the Continental Divide at elevations that exceed 14 000 feet. Glaciers that hung from mountain walls, flowed down valleys, and then retreated during on-again, off-again glaciations over the past 2 million years carved these rocky features.

The tropical forest grew here at the end of a long period of balminess that affected most of the globe. Up until 52 million years ago, for example, trees actually grew in the Antarctic, and alligators lolled about in the Canadian North. But then temperatures began the long slide, which occurred in a series of steps. The ride culminated about 3 million years ago with a big drop in temperature marking the onset of the modern glacial epoch that has gripped the Earth ever since. The epoch has been characterized by long periods of intense glaciation punctuated by short, relatively warm *interglacial* periods—like the one we live in today.

The last step in the long-term cooling trend happened at about at the same time that an ancient, pre-human lineage of primates in East Africa, the *australopithecines* split into different branches. One became the genus *Homo*, which eventually led to *Homo sapiens*. Was this just a coincidence? In the 1990s, Yale University paleontologist Elizabeth Vrba and other scientists championed the idea that it was not. According to the theory they developed, the cooling brought about environmental changes that put pressure on the primates to adapt or perish.

Peter deMenocal, a paleo-oceanographer at the Lamont–Doherty Earth Observatory at Columbia University, has conducted groundbreaking research on the climate changes in East Africa. In the mid 1990s, his analysis of sediments blown off the African continent and preserved in rocks on the ocean floor revealed a dramatic climate shift at about 2.8 million years ago. As a result of cooler sea surface temperatures characteristic of glacial conditions, the East African climate became cooler and periodically and progressively more arid, deMenocal found.[32]

332

To determine how the vegetation changed as a result, scientists have analyzed East African pollen grains and carbon isotopes from soils at sites where hominid remains have been discovered. According to Steven Stanley, a Johns Hopkins University researcher who studies evolution and ecosystem responses to environmental change, this work reveals that by 2.5 million years ago, tropical forests were giving way to wooded grasslands.

"The cooling and drying and the accompanying conversion of tropical forests to grasslands is exactly the kind of mechanism you would expect to force *Australopithecines* out of the trees," Stanley argues.[33] Those that managed this adaptation to the new conditions, members of the new genus *Homo*, went one way and survived. Those that did not manage it went another way — down an evolutionary dead end.

There is some uncertainty about exactly how this divergence happened. It likely depended on the details of how the environment changed. In the 1990s, scientists believed that forests gave way almost completely to grasslands. But more recent evidence suggests that forests fragmented rather than disappearing. This left forested islands separated by grassland seas. If this is correct, perhaps some *australopithecines* ventured out of their island homes, crossing the grasslands in search of food, and then retreated at night to the cover of the trees for protection. Those comfortable standing up and walking on two legs may have had an advantage. After all, bipedality might have allowed them to move about more easily, to peer above the tall grasses to watch out for predators, and to keep their hands free so they could carry food back to their compatriots in the forest. Moreover, by moving out into the grasslands in search of foods, the primates would have been challenged to make use of new kinds of food resources, putting evolutionary pressure on them to become more intelligent.

With these advantages — a bipedal posture and greater intelligence — the primates could have stood a better chance of survival. If so, they would have been more likely to reproduce successfully and therefore pass on their genes to succeeding generations. And in this way, according to theory, the genus *Homo* developed.

"Our genus was born of an ecological crisis," Stanley concludes.[34]

But what triggered the entire sequence of climate events? What sent the temperatures sliding 52 million years ago? No one really knows the answer. But Earth scientists Maureen Raymo and William Ruddiman have proposed that an intensification of chemical weathering brought on by the uplift of the huge Himalayan range and the vast Tibetan Plateau was the cause. The theory remains untested and controversial. But if it's right, it would establish a direct connection between plate tectonics (which caused the Himalayan uplift), climate change, and our own origins.

Why were we so lucky? Why does Earth, alone among the terrestrial planets, have plate tectonics? This is another question at the frontier of knowledge. But part of the answer may involve *liquid water*.

The *plates* in plate tectonics are chunks of what's known as the lithosphere, which consists of the brittle crust and the very upper part of the mantle. For plate movements to occur, lithosphere must be capable of subduction, of plunging down into the mantle to be recycled. On its own, chunks of lithosphere might resist making the necessary downward bend to angle into the mantle. But research cited by Ward and Brownlee suggests that liquid water—oceans of it—might weaken the rocks enough to allow this to happen. On Venus and Mars, there is no liquid water, and that may be one of the reasons why there is no subduction and plate tectonics on those planets.[35]

So here on Earth, we seem to be the lucky beneficiaries of chance and an exquisite system of feedbacks between water, rock and air. By the luck of the draw, Earth formed with the right mass to hold onto an atmosphere, and at just the right distance from the Sun—not too close to lead to a runaway greenhouse effect like Venus, and not too far to suffer perpetual iciness like Mars. Just by chance, Earth was struck by a large enough object to form a moon that stabilizes the tilt of our spin axis, perhaps preventing epic climate changes. With these benefits as a foundation, our greenhouse atmosphere has kept things clement enough for oceans to exist, which allows plates to drift across the face of the globe, which provides thermostatic control over the climate that prevents life-threatening swings, which insures that oceans of liquid water can exist. Meanwhile, the connection between plate tectonics and climate may have played a salient role in the evolution of complex life, including our own origins.

Now we will come full circle, back to Earth's very early days, when liquid water—a hydrosphere to sustain a biosphere—is first becoming permanent. Let's work our way backward, from the first fossil evidence for life, to the first chemical traces of life, and finally to the first evidence of a hydrosphere, which will mark the time that life could have gotten started.

The earliest fossil evidence of life comes from Australia, including the remains of *stromatolites*, large layered structures built by colonies of bacteria. Stromatolites consisting of living bacteria still exist today. Suspected fossils of stromatolite structures from Western Australia date to as old as 3.45 billion years ago—although a biological origin for these structures is sometimes disputed.

More controversial are discoveries by William Schopf, a paleontologist at the University of California, Los Angeles. In 1993, he announced that he had found fossilized remains of photosynthesizing cyanobacteria in chert, a flint-like sedimentary rock, from a site near the Western Australian town of Marble Bar. The findings were dated to 3.465 billion years ago, making them the oldest fossils of living organisms ever found—a claim that became widely accepted in the scientific community.[36] Perhaps the most arresting aspect of the discovery was Schopf's interpretation that the microorganisms were photosynthesizing. The ability to convert light into energy usable by a cell is a fairly advanced biological trait, so if these really were cyanobacteria, they must have evolved from even more primitive forms of life. And that would mean that life began considerably earlier.

As I was completing work on this book, Schopf's discoveries came under fire. In March of 2002, Martin Brasier of the University of Oxford argued in a paper in the journal *Nature* that Schopf's "fossils" were actually blobs of non-biological material formed at ancient seafloor hot springs.[37] And one of Schopf's former graduate students surfaced to say that the paleontologist had withheld from his paper evidence that might have cast doubt on his claims.[38]

Schopf contests Brasier's arguments, saying he is unskilled in interpreting ancient microfossils. But he does concede now that

335

the microorganisms he says he found were not photosynthesizing bacteria. Meanwhile, some paleontologists have come to Schopf's defense, whereas others are no longer so sure that the nearly 3.5 billion year old remains actually do constitute the earliest evidence for life on Earth.

As I'm writing this, the debate between Schopf and Brasier continues, and it's unclear how it's going to come out. If Brasier's arguments win the day, the 3.45 billion year old stromatolite fossils may remain as the earliest, direct marker of life's emergence on Earth.

Even earlier evidence of life has been discovered, but it's indirect. (This evidence too has recently been questioned, as we'll see in just a bit.) One of the pioneers in the search for this evidence is University of Colorado geochemist Stephen Mojzsis. (His last name is pronounced *Moyzish*.) As a graduate student at UCLA in 1996, he found a chemical signature suggestive of biological processes in 3.85 billion-year-old rocks from the southwest coast of Greenland. Follow up work using new samples he and his colleagues collected from the site on Akilia Island in 1999 has supported the original findings.

The chemical signature came from a kind of rock known as a banded-iron formation. Within them, Mojzsis found faint residues of the two stable isotopes of carbon: carbon-12 and the slightly heavier (because it has an extra neutron) carbon-13. These forms of carbon can come from biological and non-biological sources, and each source leaves a distinctive signature. Because carbon-12 is lighter, the chemical reactions in living things tend to use more of this isotope than the heavier carbon-13. So fossilized plants and even primitive bacteria show an enrichment in carbon-12. When Mojzsis analyzed the ratio of the two forms of carbon found in the Greenland banded-iron formation, he found a similar enrichment in carbon-12. So while he found no actual fossils, the isotopic residues pointed to the past presence of living things within materials preserved in the sediments.

And how old are these residues? The actual carbon can't be dated itself, so Mojzsis had to date the rock. He used a kind of natural clock: the known decay rate of certain radioactive elements. Mojzsis found the appropriate elements in tiny crystals called zircons that formed in gneiss, an igneous rock found in association with the banded-iron formation. The gneiss had originally erupted as magma that cut through the older banded-iron formation, forming a kind of seam. This means the zircon-containing gneiss is *younger* than the banded-iron formation. And the date returned by the radioactive clock? Stunningly ancient: 3.85 billion years old. Since the gneiss erupted after the banded-iron rocks were deposited, the biological activity that left its mark there should have occurred even earlier.

Mojzsis says the precise signature of carbon isotopes in the banded-iron formation points to fairly advanced life processes. So the organisms that may have left these residues could have evolved from still earlier forms. In other words, it's very possible that life could have gotten started even earlier than 3.85 billion years ago.

"These signatures are unique to life," Mojzsis says. "No process known on the Earth or on any planetary surface can mimic them. So this is not the interpretation of shapes. This is chemistry. It says that at about 3.9 billion years ago, there was a biosphere — *and this was not the origin of life*. The isotopic signatures indicate complex metabolism. But what kind of life was it? Was it a cyanobacterium or a hyperthermophile or a methanogen? I don't know. I can tell you, though, that this thing was living, a biosphere was in place, and it was far from the origin of life."[39]

In May of 2002, however, geologist Christopher Fedo of George Washington University and geochronologist Martin Whitehouse of the Swedish Museum of Natural History argued that Mojzsis had got it all wrong. The rocks in which Mojzsis had claimed to find signatures of life do not really comprise a banded-iron formation. They are instead volcanic rocks in which life could not have gotten a start.[40] Mojzsis vehemently disagrees, saying his interpretation of how these rocks came to look the way they do and attain their composition is simpler, and therefore more plausible, than the interpretation by Fedo and Whitehouse.[41]

Once again, it's not clear how the debate will turn out, so the date for the earliest indirect evidence of life is not yet firm.

Regardless of how the debate turns out, what is the earliest date that life *could have* gotten started?

"You need three things to get life going," Mojzsis says. "One is energy resources. That could be sunlight, that could be heat, or that could be chemical energy. Point the Hubble Space Telescope anywhere in space and you see energy *everywhere*, so the first criterion is easily met. The second is the availability of organic raw materials. Gook, gunk, carbonaceous organic material, stuff that's made from *SPONCH*—sulfur, phosphorous, oxygen, nitrogen, carbon and hydrogen—some of the most common elements in the universe,'' he says. SPONCH, and perhaps more complex organic compounds made from it, would have been in the materials Earth accreted from. SPONCH-containing molecules also were delivered to Earth by meteorites. So the second criterion was met.

''The third criterion for life is liquid water that's stable for long periods of time,'' Mojzsis continues. ''Now that's tough to fulfill, because there are four states of matter in the universe: solid, liquid, gas and plasma. The most common state is plasma, and it's not very good for living things.''

Liquid water could remain stable for long periods on the early Earth only when the rate of big impacts died down and conditions cooled sufficiently. So when did this occur? In addition to his sleuthing of early life, Mojzsis has tackled this question too. And his answer was a shocker.

''From our research, we think you have liquid water soon after the planet forms—and not just a little bit of it,'' he says. ''The kind of signals we've seen indicate a huge quantity of liquid water, a hydrosphere. I'm not saying there were seas or oceans. I'm not even saying that there were lakes. But there was a lot of liquid water coursing through the crust about 4.3 billion years ago. And two independent groups, ours and another, came to this same conclusion. Is it the truth? More work needs to be done to be sure. But I think we can feel confident that the stage was set for life *400 million years before any one thought*.''

How did Mojzsis and the other research group come to this conclusion? This time, the chemical signatures came from oxygen isotopes in rocks from Australia.

The oldest known continental rocks ever found on Earth come from Australia. They date to 3.8 billion to 4.0 billion years old. No older rocks have yet been found, so geologists like Mojzsis are in a bit of a bind if they want to study what the Earth was like at earlier stages in its evolution. But some evidence is available in the form of zircons, those crystal grains that helped Mojzsis date the ancient Greenland rocks. They are very tiny and very tough, managing to survive intact through billions of years of geologic activity.

Near Mt. Narryer and the Jack Hills in Western Australia, zircons can be found in rocks originally deposited in a fanlike delta as long as 3.7 billion years ago. But the zircons they contain did not actually form within these deltaic deposits. Using radioactive dating techniques, Australian researchers have shown that some of the zircons are 4.3 billion years old — much older than the host rocks. Mojzsis and his colleagues, Mark Harrison of UCLA and Robert Pidgeon in Perth, Australia, believe the zircons originally crystallized within molten rocks that were cooling and hardening underground. Later these rocks were exposed at the surface and eroded away; no trace is left of them. But the hardy zircons survived and found their way into the ancient Australian delta, where they became encapsulated in the sedimentary rocks. Here they have been preserved — ripe for the geologic picking by the likes of Stephen Mojzsis.

To recover the zircons, Mojzsis crushed, powdered and sieved host rocks he and his colleagues collected in Australia in June of 1999. He then used a device called an ion microprobe to measure the abundances of relevant radioactive elements in some of the zircons recovered in this way. With this technique he identified 17 zircons that were more than 3.9 billion years old. Two of these were 4.3 billion years old.

Next, Mojzsis analyzed the signature of an oxygen isotope within the zircons. Zircons form from molten rock as it is cooling, and the amount of oxygen-18 found within is a signature of the particular kind of rock. When zircons form within melted *crustal* rocks — for example, sedimentary rocks originally formed in water — they will have lower amounts of oxygen-18 than zircons that formed within primitive melted rocks derived from the mantle. The oxygen isotope signature of the most ancient zircons from Western Australia indicated that they had formed within melted, recycled crustal rocks that had originally interacted with liquid water under surface or near-surface conditions.

As Mojzsis and his colleagues described them in a paper in the journal *Nature* in January of 2001, the findings are consistent with the idea that a hydrosphere was in place at least 4.3 billion years ago.[42] The second research group, led by Simon Wilde, came to similar conclusions in a paper in the same issue of *Nature*. (They found evidence for water at 4.4 billion years ago.)

In an article accompanying the two research papers in the journal, Alex Halliday, an earth scientist at the Swiss research institute ETH, said the findings were an important step in understanding the early Earth. "Their results provide evidence that continents and liquid water were surface features of the earliest Earth," he wrote. "Part of one grain appears to have formed 4.4 billion years ago; this is the oldest terrestrial solid yet identified. Although it is unclear how general a picture a few tiny zircon grains can provide, the results represent a significant advance in reconstructing Earth's Dark Ages."

By the Dark Ages, Halliday means Earth's first geologic period, the *Hadean*, which lasted from the time of the planet's formation to 4 billion years ago.

"We are exploring the Hadean geological record," Mojzsis says. (He emphasizes the *de* as in *Hay-DEE-an*.) "This is really the first door that has opened for us onto this remarkable time." With the findings from Greenland and then from Australia, he and his colleagues say they have bracketed the period. "Now we've got two stakes, you might say, at either end of the Hadean. We have one where life is appearing at 3.9 billion years ago, and then another where an ocean is in place right after lunar formation. If there was life on Earth before the Moon formed, it's hard to imagine how it could have survived. So you can think of the lunar formation event as time zero for life."

Mojzsis's time zero is the point in Earth's evolution when the processes that led to the first living thing could have taken off. And according to Scott Sandford of the NASA Ames Research Center, those processes would have gotten a big boost if complex organic compounds were already available. "Life is too wacky, too complicated to simply fall together from simple molecules," he says. "But maybe it is the result of a whole chain of chemical events. And maybe life got started because what was needed was already available."

It's possible, for example, that once oceans formed, the necessary chemicals were created by chemical reactions at seafloor hydrothermal vents, and that the first life got started there. Today, these vents are common along the mid-ocean ridge, where tectonic plates spread apart. Magma from the upper mantle rises and solidifies, forming new crust. Seawater percolates down through cracks in the rock and is heated to temperatures as high as $1000°$ C. Then the hot water makes its way back into the ocean through the hydrothermal vents, emerging as hot as $450°$ C. As a result of this circulation in and out of the hot, mineral-rich environment of the vents, chemically-reduced fluids containing hydrogen sulfide mix with less reduced seawater. This mixing provides chemical energy that allows the conversion of carbon dioxide into organic compounds like those synthesized in the pioneering Miller–Urey experiments.[43]

Today, microorganisms thriving around hydrothermal vents make a living from this same chemical energy source. In 1985, John Baross and Sarah Hoffman proposed that the very first living things did likewise. They argued that hydrothermal vents would have been protected from large impacts, and that the chemicals and energy needed to get life going could have been available there. At first, many scientists didn't put much credence in the idea. The Miller–Urey tidepool scenario was just too well entrenched, and not enough evidence had been gathered to support the idea that life began at hydrothermal vents. But when later research showed that the earliest known common ancestor of all living things was a heat-loving extremo-phile—just like the microbes living in today's vent commu-nities—scientists began warming to the idea. Today, the hydrothermal vent idea is one of the leading theories being explored by astrobiologists seeking the origin of life.

Based on what he and his colleagues have cooked up in their lab, Sandford believes the raw ingredients for life probably did not come just from hydrothermal vents, even if life did get its start there. Through the solar system equivalent of FedEx—meteorite impacts—compounds that first formed in a molecular cloud were delivered to the early Earth, where they could have been available to participate in life's origins.

So we finish now where we began, in Scott Sandford's kitchen. Starting with ices consisting of frozen water, methanol and ammonia, Sandford's molecular-cloud machine has manu-factured a host of complex molecules, such as ketones, ethers and alcohols. The very same compounds have been spotted by astronomers in molecular clouds. And we know these compounds were delivered to the early Earth because they have also been discovered in meteorites.

Sandford and his colleagues have also produced a molecule called hexamethylenetetramine. Put HMT in warm acidic water and you'll wind up with amino acids. And put some of the other compounds produced in the kitchen in water, as David Deamer of the University of California, Santa Cruz did, and something very curious happens: they form little cell-like vesicles. These are very much like the ones Deamer produced more than 10 years ago using extracts from a meteorite found in Murchison, Australia. Analysis of the spherical structures by Jason Dworkin of Sandford's group reveals that they are built of complex organic molecules arranged in a particular way. The water-loving, or *hydrophilic*, ends of these molecules align facing the water; meanwhile, their *hydrophobic* ends face inside the membrane.[44]

These experiments suggest how simple chemical reactions using materials that should have been available on the early Earth could have formed the first things resembling cells. These certainly weren't alive by any stretch of the imagination. But early life would have needed to encapsulate its chemical machinery in a cell and mediate the passage of compounds in and out. In fact, Sandford and his colleagues were intrigued when the vesicles they produced in their molecular-cloud machine turned out to fluoresce, because it meant that they had trapped other organic substances inside.

"Living membranes need to be leaky to allow things in and out—but not *too* leaky," Sandford notes. "Our vesicles are like that. They have scavenged molecules to their insides. How do we know this? The molecules are fluorescent, so we can see them inside."

Some organic fluorescent compounds absorb ultraviolet radiation, which raises the possibility that within early cells these molecules may have functioned as a kind of sunscreen. "It doesn't do early life any good if amino acids being made inside cells are being broken down minute by minute by ultraviolet radiation," Sandford says.

Compounds known as quinones, common in living things today, are candidates for the role of sunscreen, because they absorb ultraviolet radiation. Sandford and his colleagues have prepared them in their kitchen starting with compounds known as polycyclic aromatic hydrocarbons, or PAHs. You and I know this stuff simply as soot. But PAHs are found not just in barbecue grills and automobile tailpipes. They are also plentiful in molecular clouds, and—you guessed it—certain types of meteorites.

Cells could have scavenged quinones delivered to Earth by impacts during one of the stages in the origin of life, Sandford speculates. Once within cells, they may have helped protect against damage before the Earth developed an ozone layer to screen out ultraviolet light from the Sun.

Quinones are also interesting because they help transport electrons across cell membranes, creating a separation of charges that drives subsequent cellular chemistry. "Why do we use quinones for electron transport?" Sandford asks. "Maybe it's because that's just what fell out of the sky."

The electron-transport ability of quinones also plays an important role in photosynthesis, the process by which plants convert light into chemical energy. "So quinones delivered by meteorites may have allowed some sort of primitive photosynthesis in early microorganisms," Sandford says.

Can anyone really tell for sure where the materials needed for the origin of life actually came from? Probably not. "But I'm sure that when life did get started, wherever it got started, it didn't care about the 'made-in' labels. It just used whatever was available."

And Sandford's research suggests that a lot was available from cosmic sources. "Through simple processes of astrochemistry going on everywhere that stars and planets are forming, you probably have this stuff around," he concludes. "We know that it can get to the surface of planets because we've seen it in meteorites. And this means the origin of life has a leg up. Any time you make a planet and it has the right conditions, in principle, on Day One, you have this stuff floating around in the water."

# AFTERWORD

## Cosmology and Environmentalism

*Also, despite appearances, life is a significant process on the largest scales of both time and space. The future behavior of life will determine the future behavior of stars and galaxies.*

— David Deutsch[1]

*The cosmic oasis on which man lives – this miracle of an exception, our own blue planet in the midst of the disappointing celestial desert – is no longer "also a star," but rather the only one that seems to deserve this name.*

— Hans Blumenberg[2]

*You must not be mistaken and think that the spheres and the angels were created for our sake.*

— Moses Maimonides[3]

As a member of the Society of Environmental Journalists, and an advisor to environmental journalism students, I'm often asked why I wrote this book. What connection could there possibly be between environmental issues and the search for our cosmic roots and place in the universe?

The short answer is that as a science journalist, I write about what interests me. And what interests me simply has expanded ever outward. Starting with geology and environmental science, my fascination has simply grown to encompass wider and wider environmental realms. And when I reached the upper edge of the atmosphere, I saw no reason to stop.

In this process of ever-expanding circles of fascination, I have come to see that our environment encompasses much more than prairies, forests, canyons, mountains, rivers, lakes, estuaries and oceans. So today, I'm just as eager to know how the universe came to be as I am to understand how the Grand Canyon formed. The motivation is the same: curiosity about the world in which we live and where we fit.

I first grew concerned about our impact on the environment when, as a teenager growing up in Brooklyn, I went on a backpacking trip to Harriman State Park north of New York City. As I walked by a massive jumble of rounded boulders, some as big as apartment houses, I found myself wondering how they got there. Later, I learned the answer. Glaciers that once advanced across the area (and stopped just miles from my house in Brooklyn), left the boulders there when they retreated. On subsequent backpacking trips in Shenandoah National Park, I wondered what caused the beautiful corrugations that comprise the Blue Ridge Mountains. Plate tectonics was the answer, of course — subduction, volcanism and continental collisions that crumpled the crust of the eastern United States between about 480 million and 220 million years ago.

This was the original source of my connection to the environment — a desire to understand *how* it came to be. (Although the horrendous air pollution of New York City at the time played a role too.) And as my interests have expanded, as I came to wonder how North America formed, and how the Earth and the solar system and the galaxy and finally the universe formed, this environmental sensibility has remained the foundation of it all.

I know that I'm not alone in this. As I did the reporting for this book, I came to appreciate that many of the scientists I interviewed shared my environmental perspective. In almost every case, their central motivation turned out to be simple fascination with nature. But another theme frequently emerged in conversations with my sources. While the Copernican revolution seemed to tell us that we are nothing special, the reality is quite different. As I've described them in this book, recent findings suggest that the complex animal life that adds such rich biodiversity to our planet may not be very common. Intelligent life may be rarer still. Indeed, in our galaxy — and possibly in the universe as a whole — we may be it. If that's true, our rarity would make us special. Of course, even if we are not rare in the cosmic scheme of things, intelligence is an incredible and very special outgrowth of our planet's biosphere.

So I'd like to conclude by sharing what some of the scientists whom I've interviewed said about these issues.

When I met with Michael Turner on the park-like campus of the Aspen Center for Physics, he focused on the philosophical ramifications of the multiverse.

"If the multiverse is right, it could be that we live in a little oasis," he said. "Most of the other bubbles [his description of other parts of the multiverse] are devoid of matter, or have two dimensions or eight dimensions—either way, not suitable for life… Martin Rees has speculated that we could well be the only intelligent beings in the observable universe. And that would be important to know. Just like that first picture of Earth-rise as seen from the Moon, this might encourage people to be more careful. There are 100 billion galaxies, 100 billion stars per galaxy, and we may be the only intelligent life. We'd better be really careful not extinguish it."

Sandra Faber and Joel Primack, of U.C. Santa Cruz, both were passionate about the connection between cosmology and environmental awareness.

"One lesson astronomy tells us is that we're a tiny mote in a hostile void, and the nearest help is too far away," Sandra said. "We're on our own, on spaceship Earth. So we have to solve our own problems. More important, we're fantastically lucky, because we've been given the gift of time. We have at least a billion years in which our home, if we continue to take care of it, will continue to suffice. What a chance, right?"

"And lastly," she said, "people like Joel and I are figuring it all out by sitting on this tiny little planet and collecting photons from space. So far from feeling dwarfed by the vast reaches and energy of the cosmos, what we really learn is that we are the most remarkable and complicated product of cosmic evolution, and our potential is unlimited. In little localized pockets, the universe is capable of building some beautiful complexity."

At first, I thought she was finished. But then she continued:

"I'd really like people to understand that fantastic ground-work has been laid—a cosmic experiment has been running 14 billion years, and it has gotten to an interesting point. How incredibly tragic it would be if we foolishly pulled our own plug."

346

Frank Shu had something similar to say. "The fact that we have no evidence for intelligent life elsewhere suggests that it's just not there," he said, echoing Fermi's paradox. "If intelligent life is indeed rare, then it is our duty and obligation to try to preserve it.... People glibly talk about extinctions. But the death of the human race—now that would be a tragedy beyond reckoning because it would mean the loss of any knowledge of Shakespeare, of Jane Austin, of Mozart. Suppose all that were erased. Wouldn't that be a tremendous tragedy? All the great achievements *just erased*. That's conceivable now because we are screwing things up."

In my conversations with Alan Dressler of the Carnegie Observatories, he concluded the session with a comment about our place in the universe. "The most amazing result of the Copernican revolution is that we're very special after all... Whether there are millions of civilizations in the galaxy or just one, it's still the case that every one, or just the one, can feel very special because life has come about as a result of these many, many, many steps. And the result of that life is to understand the steps."

Now that we understand the steps, we may be ready as a species for some truly extraordinary things. Looking close to home, we clearly have the ability to alleviate human suffering. And looking outward, we really are in the midst of what the late astrophysicist David Schramm called the "golden age" of cosmology. Sensitive instruments already launched or in the works, such as the Planck mission to map the cosmic microwave background in unprecedented detail, and the Next Generation Space Telescope, will soon be opening new windows on the early universe. Moreover, astrobiologists are moving closer to deciphering how life actually got started, and perhaps to answering one of the most important questions of all time. Are we alone or part of a vast cosmic community?

But we'll accomplish these things only if we take care of Earth. We need to make the next evolutionary leap in our development by finding a way to live sustainably on the planet. Otherwise, all the wonders we've learned about the universe and our place in it will have been for naught.

# NOTES

## Notes for Preface

[1] Erdman, D. V., Ed. (1982). "Auguries of innocence" in *The Complete Poetry and Prose of William Blake*. Garden City, Anchor Books.
[2] Greene, B. (1999). *The Elegant Universe*. New York, Vintage Books, p. 130.
[3] Wertheim, M. (2000). *The Pearly Gates of Cyberspace: A history of space from Dante to the Internet*. New York, W. W. Norton, p. 131.
[4] Faber, S. M. and J. Primack, interview with T.Y., Santa Cruz, California, Aug. 22, 2000.

## Notes for Prologue 1

[1] In his *Eureka: A Prose Poem*, Poe sketches out ideas that seem eerily similar to key concepts of the big bang—a century before the theory emerged. Poe proposed that a primordial, indivisible particle expanded to create the universe. In the conventional big bang theory, space and time and all matter and energy in the universe burst forth explosively from an infinitely small, infinitely dense primordial point—the singularity described here at the beginning of the prologue. (Dannielson, D. R., Ed. (2000). *The Book of the Cosmos*. Cambridge, Massachusetts, Helix Books.)
[2] In the course of researching this book, I conducted numerous interviews with Neil Turok—in person, by phone and by e-mail. All of the quotations from Turok in this chapter and those that follow, are from these interviews and from a talk ("Inflation and the Beginning of the Universe") at the Cosmic Questions conference in Washington D.C., on April 19, 1999.
[3] Guth, A. "Eternal Inflation," talk delivered at Cosmic Questions conference, National Museum of Natural History, Washington D.C., April 14, 1999.
{4} Ibid.
[5] Many years later, Turok's teacher saw Neil in a documentary film on television about Stephen Hawking. She had moved back to Scotland. As Turok tells the story, "She sent me a letter and asked, 'Are you the same Neil Turok I taught at age 7 in Tanzania?' Well, I nearly fell over." Turok was about to give an inaugural lecture to commemorate his being elevated to the mathematical

348

physics chair at Cambridge. "So I invited the sisters. They came, and at the opening of the lecture I told the story!"

[6] Linde, A. Telephone interview with T.Y., July 1998.

[7] Hawking, S. (2001). *The Universe in a Nutshell*. New York, Bantam Books.

[8] Guth, A. H. E-mail communication with T.Y., July 21, 1998.

[9] Wheeler, J. A. (1998) *Geons, Black Holes and Quantum Foam*. New York, W. W. Norton. The quotation comes immediately before the table of contents on an unnumbered page.

# Notes for Chapter 1

[1] In Chown, M. "Anything goes," *New Scientist*, June 6, 1998. In this story, Tegmark champions the idea that every set of mathematical equations that could possibly describe a universe is physically expressed as a real, physical universe. In this view, our universe is not the only one in existence. There are an ensemble of universes, each one based on its own, particular set of physical laws. These universes should differ from ours not just by having different values for, say, the strengths of the fundamental forces. "What I have in mind are universes which dance to the tune of entirely different sets of equations of physics," Tegmark says. In other words, these universes can't be described simply by substituting different numbers in the equations we use to describe our universe. They would require completely different equations.

Only those equations that lead to universes with observers (meaning any kind of life) will be perceived to exist. This, Tegmark says, explains a puzzling aspect of our universe. The laws of physics seem finely tuned to allow for life. Why should this be? It's not so puzzling if every conceivable kind of universe actually exists. With that kind of diversity, some universes are bound to have laws of physics that allow for biology. Turn this around, and Tegmark's quote at the heading of the chapter makes sense. Biology determines the laws of physics in the sense that it can exist only if those laws are just so.

[2] Cajori, F. (1934). *Newton's Principia, Motte's Translation Revised*. Berkeley, University of California Press, p. 6.

[3] Wheeler, J. A. (1998). *Geons, Black Holes & Quantum Foam*. New York, W. W. Norton, p. 235. Contrast the relativistic view with the Newtonian one, which has been described by astrophysicist Frank Shu like this: "Force tells mass how to accelerate, and mass tells gravity how to exert force."

[4] Kragh, H. (1996). *Cosmology and Controversy: The Historical Development of Two Theories of the Universe*. Princeton, Princeton University Press, p. 13.

[5] Feynman, R. P. (1985). *QED: The Strange Theory of Light and Matter*. Princeton, Princeton University Press, p. 10.

[6] Hawking, S. and R. Penrose (1996). *The Nature of Space and Time*. Princeton, Princeton University Press, p. 4.

[7] Kragh, H. (1996). *Cosmology and Controversy: The Historical Development of Two Theories of the Universe*. Princeton, Princeton University Press, p. 6.

[8] Ibid, p. 12.

[9] Ibid, p. 11.

[10] Ibid, p. 12.

[11] Guth, A. H. (1997). *The Inflationary Universe – The quest for a new theory of cosmic origins.* Reading, Mass., Perseus Books, p. 44.

[12] Kragh, H. (1996). *Cosmology and Controversy: The Historical Development of Two Theories of the Universe.* Princeton, Princeton University Press, p. 24.

[13] Ibid, p. 25.

[14] Ibid, p. 27.

[15] Ibid, p. 14.

[16] Encyclopedia of Astronomy and Astrophysics, http://www.ency-astro.com.

[17] Kragh, H. (1996). *Cosmology and Controversy: The Historical Development of Two Theories of the Universe.* Princeton, Princeton University Press, p. 30.

[18] Goldsmith, D. (2000). *The Runaway Universe: The Race to Find the Future of the Cosmos.* Cambridge, Perseus Books, p. 31.

[19] Freedman, W. L. (2001). "Final results from the Hubble space telescope key project to measure the Hubble constant." *Astrophysical Journal* **533**, 47–72.

[20] Kragh, H. (1996). *Cosmology and Controversy: The Historical Development of Two Theories of the Universe.* Princeton, Princeton University Press, p. 33.

# Notes for Chapter 2

[1] Feynman, R. P. (1985). *QED: The Strange Theory of Light and Matter.* Princeton, Princeton University Press, p. 77.

[2] In Kaku, M. "What happened before the big bang?," *Astronomy*, May 1996.

[3] Gribbin, J. (1998). *The Search for Superstrings, Symmetry, and the Theory of Everything.* Boston, Little, Brown and Company, p. 52.

[4] General relativity predicts that when mass is accelerated, disturbances in the gravitational field produce gravitational radiation, just as accelerating a charged particle produces electromagnetic radiation. These gravity waves have not yet been detected because gravity is extremely weak. This weakness means that only extremely massive objects undergoing strong acceleration would produce gravity waves of sufficient magnitude to be detectable.

[5] Einstein was led to the answer by a phenomenon known as the photoelectric effect. This phenomenon was discovered in 1887 by the German physicist Heinrich Hertz, who found that light shining on certain metals knocks electrons free. Strangely, experiments showed that brighter light did not cause the electrons to fly off the metal at higher speed, as you might expect. Instead, a greater *number* of electrons were pried loose. On the other hand, experiments also revealed that electrons did fly off with greater speed when light with a higher frequency was used. For visible light, a higher frequency means shorter wavelength and a shift in color from a redder part of the spectrum to a bluer one. So why should electrons care about the color of the light in this way?

Einstein solved the problem by proposing that Planck's energy lumps were actual particles — they were later dubbed photons — whose energy was determined by their frequency. In this view, unless a particular photon has enough energy, it will carom harmlessly off an atom without liberating an electron. But if it has a high enough frequency, and therefore enough energy, a photon will have enough oomph to pry the electron free. Once this energy threshold is passed, increasing the intensity of the light—which means increasing the total number of photons in the beam—will result in more photon-electron collisions. And that's why more electrons fly free.

[6] In Abrams, N. E. and Primack, J. [Internet]: "Einstein's View of God," cited Feb. 27, 2002. Available from: http://physics.ucsc.edu/cosmo/ primack_abrams/htmlformat/Einstein4.html.

[7] Wheeler, J. A. (1998). *Geons, Black Holes & Quantum Foam*. New York, W. W. Norton, p. 168.

[8] Abrams, N. E. and Primack, J. [Internet]: "Einstein's View of God," cited Feb. 27, 2002. Available from: http://physics.ucsc.edu/cosmo/ primack_abrams/ htmlformat/Einstein4.html.

[9] Feynman, R. P. (1985). Q*ED: The Strange Theory of Light and Matter*. Princeton, Princeton University Press, p. 78.

# Notes for Chapter 3

[1] Bahcall, J. (2000). "The Big Bang is bang on." *Nature* **408**, 916-917.

[2] Turok, N. "Inflation and the Beginning of the Universe," talk given at Cosmic Questions conference, National Museum of Natural History, Washington, D.C., April 14, 1999.

[3] Harris, J. Interview with T.Y. at Brookhaven, N.Y., June 2000. All quotations from John Harris in this chapter are from this interview.

[4] Videbeck, F. Interview with T.Y. at Brookhaven, N.Y., June 2000. Unless otherwise referenced, all quotations from Flemming Videbeck in this chapter are from this interview.

[5] Kharzeev, D. Interview with T.Y. at Brookhaven, N.Y., June 2000. All quotations from Dimitri Kharzeev in this chapter are from this interview.

[6] Wheeler, J. A. (1998). *Geons, Black Holes & Quantum Foam*. New York, W. W. Norton, p. 137–138.

[7] In Kragh, H. (1996). *Cosmology and Controversy: The Historical Development of Two Theories of the Universe*. Princeton, Princeton University Press, p. 90.

[8] Ibid, p. 99.

[9] Ibid, p. 105.

[10] Gamow, G. (1946). "Expanding Universe and the Origin of Elements." *Physical Review* **70**, 572–573.

[11] D'Agnese, J. "The Last Big Bang Man Left Standing," *Discover*, July, 1999.

[12] Kragh, H. (1996). *Cosmology and Controversy: The Historical Development of Two Theories of the Universe*. Princeton, Princeton University Press, p. 114.

[13] Gamow originally wanted to extend the mischief. He tried to convince Ralph Herman, a physicist who was working with Gamow and Alpher, to change his name to *Delter* so it could be the alpha-beta-gamma-delta, or $\alpha\beta\gamma\delta$ paper. Herman refused. Even as late as 1960, Gamow couldn't resist trying to get one more laugh out of the joke, saying that when problems with the

theory became known, Bethe considered changing his name to Zacharias!
Kragh, H. (1996). *Cosmology and Controversy: The Historical Development of Two Theories of the Universe*. Princeton, Princeton University Press.

[14] Alpher, R., H. Bethe *et al* (1948). "The origin of the chemical elements." *Physical Review* **73**, 803–804.

[15] Alpher, R. and R. Herman (1948). "Evolution of the Universe." *Nature* **162**, 774–775. Michael Turner of the Fermi National Laboratory points out that the physics in this paper wasn't really correct. "They got the right answer for the wrong reason," he says.

[16] It would be most accurate to say that their prediction checked out only in part. Alpher and Herman did not propose a specific, unique spectrum for the big bang's afterglow photons. That came in 1965, from Princeton University's Jim Peebles, who calculated that the big bang's remnant radiation should have a blackbody spectrum. After Penzias and Wilson had made their discovery, heralded by Walter Sullivan in the *New York Times* as a remnant of the origin of the universe, other scientists confirmed that the microwave background radiation does indeed have a blackbody spectrum, adding yet more evidence in favor of the big bang.

[17] John Archibald Wheeler, working with his graduate student Richard Feynman, came up with an alternative explanation for electron–positron annihilation. They proposed that a positron is actually one and the same particle as an electron, which accounts for why it has the same mass. It appears to have a positive charge because it is actually an electron moving backward in time! In this picture, an electron moves forward in time, releases radiation, then moves backward in time, then reverses with more radiation release, then backward and forward, and so on. Each reversal seems to be an annihilation of two particles when in fact it's just an electron zigzagging through time and releasing energy as it goes.

[18] Malamud, E. Fermi National Accelerator Laboratory [Internet]: "Fermilab's chain of accelerators: Why go to high energy?"; updated Aug. 30, 2000; cited June 13, 2002. Available from: http://www-bd.fnal.gov/public/why.html.

[19] Fields, B. D. (2001). "COSMOLOGY: a census of cosmic matter." *Science* **294** (5542), 529–530.

[20] Srianand, R., Petitjean, P. *et al* (2000). "The cosmic microwave background radiation temperature at a redshift of 2.34." *Nature* **408**, 931–935.

[21] Bahcall, J. (2000). "The Big Bang is bang on." *Nature* **408**, 916–917.

[22] Turner, M. S. (1999). "Cosmology solved? Quite possibly!" *Publications of the Astronomical Society of the Pacific* **111** (March), 264–273.

[23] Guth, A. H. E-mail interview with T.Y., July 21, 1998.

[24] Buzsa, W. Interview with T.Y. at Brookhaven, N.Y., June 2000. All quotations from Wit Buzsa in this chapter are from this interview.

[25] Young, G. Interview with T.Y. at Brookhaven, N.Y., June 2000. All quotations from Glen Young in this chapter are from this interview.

[26] The magnets are powered by electricity. To reduce the resistance in the electrical circuits to superconducting levels, the magnets are cooled by liquid helium to just slightly above absolute zero. With 1740 superconducting magnets, the RHIC accelerator requires an enormous amount of liquid helium: enough to fill every balloon in the Macy's Thanksgiving Day parade for 100 years.

# Notes for Chapter 4

[1] Wilczek, F. "The Simplicity of Matter and the Unity of Nature," talk given at AAAS annual meeting, San Francisco, February 16, 2001.

[2] Quigg, C. "A Decade of Discovery". Ibid.

[3] Greene, B. (1999). *The Elegant Universe*. New York, Vintage Books, p. 12.

[4] Between 1967 and 1970, Sheldon Glashow, Steven Weinberg and Abdus Salam had found a unified description of the weak force and electromagnetism, constructing a theory of an electroweak interaction. They theorized that this unification occurred at the higher temperatures of the early universe, and predictions of theory checked out. Their theory of the electroweak interaction was a major step in the construction of the standard model of particle physics.

[5] This is a useful but imperfect metaphor: The "directions" of the Higgs fields, as I describe them here, are not actual physical directions. Technically speaking, a Higgs field has no direction, which makes it different from an electromagnetic field. It is described instead with an abstract property labeled unhelpfully as "value". At each point in space, a Higgs field must have some value. And it is this property that actually allows it to have the effects described in this section. But the term value is so abstract that I've decided to use the concept of direction as part of an easy-to-picture metaphor of Higg's fields aligning in something akin to the axes of a crystal. In doing so, I've followed the lead of a number of physicists who've attempted to explain Higgs fields. For some examples, see http://hepwww.ph.qmw. ac.uk/epp/higgs.html.

[6] Guth, A. H. (1997). *The Inflationary Universe – The quest for a new theory of cosmic origins*. Reading, MA, Perseus Books, p. 31.

[7] Hamilton, A. E-mail communication with T.Y., July 23, 2002.

[8] Wheeler, J. A. (1998). *Geons, Black Holes & Quantum Foam*. New York, W. W. Norton, p. 247.

[9] Ibid.

[10] Hawking, S. (2001). *The Universe in a Nutshell*. New York, Bantam Books, p. 57.

[11] Guth, A. H. (1997). *The Inflationary Universe – The quest for a new theory of cosmic origins*. Reading, MA, Perseus Books, p. 151.

[12] The antigravity of the false vacuum sounds an awful lot like Einstein's cosmological constant, the term he entered into his equations to forestall cosmic collapse. And it certainly behaves that way—except for the fact that once the false vacuum decayed, the repulsive force went away, whereas Einstein's cosmological constant was present, well, constantly.

[13] Guth, A. H. (1997). *The Inflationary Universe – The quest for a new theory of cosmic origins*. Reading, MA, Perseus Books, p. 175.

[14] Greene, B. (1999). *The Elegant Universe*. New York, Vintage Books, p. 356.

[15] Guth, A. H. (1997). *The Inflationary Universe – The quest for a new theory of cosmic origins*. Reading, MA, Perseus Books, p. 184–186.

[16] Andrew Hamilton, a cosmologist at the University of Colorado at Boulder who very kindly reviewed portions of the Origins manuscript before publication, offered this thought experiment to describe what it would be like for observers in a universe that is inflating at greater than light speed: "Suppose you and I start together in an inflationary universe, sharing a cup of GUT tea. (Presumably we are sub-flea sized.) If we start a little distance apart, then the

repulsive vacuum will cause us to accelerate away from each other (assuming our own gravity is negligible). You accelerate away from me, and at some point you accelerate past the speed of light. This does not violate the special relativistic rule against exceeding light speed, because it is space itself that is expanding, carrying us with it. I never actually see you recede faster than the speed of light. Rather, as you approach my horizon, it takes longer and longer for light from you to reach me, and from my point of view, you seem to slow down and then freeze. If I am looking at the flea-sized watch on your flea-sized wrist, then I see the watch freeze to a halt. If your watch seems to freeze at half past one, then I deduce that you passed my horizon at half past one.''

[17] Guth, A. H. ''Eternal Inflation,'' talk given at Cosmic Questions conference, National Museum of Natural History, Washington D.C., April 14, 1999.

[18] Ibid.

[19] Greene, B. (1999). *The Elegant Universe*. New York, Vintage Books, p. 234.

[20] Guth, A. H. (1997). *The Inflationary Universe – The quest for a new theory of cosmic origins*. Reading, MA, Perseus Books, p. 25.

[21] Robert Dicke, the scientist who had brought the flatness problem to Guth's attention, made an anthropic argument back in 1961 in connection with another issue. It goes like this. For life to exist, there must be an appropriate chemistry, which probably must involve carbon. For carbon to exist, it has to be cooked up inside stars. This cooking process appears to take on the order of 10 billion years, give or take. Once the carbon is made during this long interval, it can be spewed out into interstellar space when certain stars explode as supernovae at the end of their lifetimes. The carbon, and other vital elements, can then participate in the origin and ongoing chemistry of life.

Now here's where the anthropic reasoning comes in. The universe is observed to be at least 10 billion years old, but not very much older. Why should this be? Anthropically speaking, it has to be at least 10 billion years old, otherwise life wouldn't have had enough time to get started. (This also is why the observable universe is as big as it is, since its size is equal to the distance light has been able to travel since the origin of the universe more than 10 billion years ago.) Also, the universe can't be very much older than 10 billion years because with enough time, carbon and other elements essential to life would be processed (as part of a giant, cosmic recycling system) into white dwarfs, neutron stars and black holes and therefore be unavailable to participate in the chemistry of life.

The anthropic bottom line, according to Dicke, is that the universe cannot be much younger or older (and therefore much smaller or bigger) than what we observe it to be because otherwise our existence would be impossible. (Carr, B. Encyclopedia of Astronomy and Astrophysics [Internet]: Anthropic Principle; cited May 4, 2002. Available from http://www.ency-astro.com.)

[22] Guth, A. H. (1997). *The Inflationary Universe*. Reading, MA, Perseus Books, p. 179.

# Notes for Chapter 5

[1] Plato (1959). *Plato's Timaeus*, translated by Francis M. Cornford. Indianapolis, Bobbs-Merrill, pp. 20–21.

[2] Dressler, A. (1994). *Voyage to the Great Attractor*. New York, Alfred A. Knopf, p. ix.

[3] Ibid, pp. 140–141.

[4] Peebles' map was made possible by the work of two astronomers, Shane and Wirtanen, who used magnifying glasses to pore over photographic plates from the Lick Observatory.

[5] Dressler, A. (1994). *Voyage to the Great Attractor*. New York, Alfred A. Knopf, p. 142.

[6] Ibid, p. 170.

[7] When imaging a single portion of the sky, the 2dF's 12-foot-wide Anglo-Australian Telescope uses fiber optics to collect light from hundreds of galaxies simultaneously. The optical fibers are positioned automatically within the telescope by computer. A spectrum is obtained for each of the galaxies, revealing its degree of redshift, and therefore a measure of its distance from Earth. When this is complete, the telescope is directed to the next portion of the sky, and the process is repeated. By plotting the distances and positions of thousands of galaxies using this method, 2dF scientists have been producing large-scale maps of the universe.

    The 2dF Redshift Survey is one of two major galaxy surveys underway. The other is the U.S.-based Sloan Digital Sky Survey.

[8] Verde, L., *et al* (2002). MNRAS, submitted, astro-ph/0112161 v3: The 2dF Galaxy Redshift Survey: The bias of galaxies and the density of the Universe.

[9] Wu, K. S., Lahav, O. *et al* (1999). "The large-scale smoothness of the universe." *Nature* **397**, 225–230.

[10] Glanz, J. "Robotic telescope affirms assumption on Universe's birth." *New York Times*. New York, June 7, 2000.

[11] Dodelson, S. "Cosmology Past the Crossroads," talk given at the Aspen Center for Physics, June 28, 2002, Aspen, Colorado.

[12] Lahav, O., *et al* (2002). MNRAS, submitted, astro-ph/0112162 v3: The 2dF Galaxy Redshift Survey: The amplitudes of fluctuations in the 2dFGRS and the CMB, and implications for galaxy biasing.

[13] Linde, A. Telephone interview with T.Y., July 10, 1998.

[14] Hamilton, A., e-mail communication, July 23, 2002.

[15] Turner, M. S., interview with T.Y. at Aspen Center for Physics, Aspen, Colorado, June 28, 2002. All quotations from Michael Turner in this chapter are from this interview.

[16] All quotations from Wayne Hu in this chapter and the next are from several interviews I conducted with him by telephone and e-mail in the spring of 2002.

[17] Dressler, A., interview with T.Y., Pasadena, California, July 17, 2000. Unless referenced otherwise, all quotations from Alan Dressler in this chapter and others are based on an interview I conducted with him at the Carnegie Observatories.

[18] Blumenthal, G. R., Faber, S. M. *et al* (1984). "Formation of galaxies and large scale structure with cold dark matter." *Nature* **311** 517.

[19] Dressler, A. (1994). *Voyage to the Great Attractor.* New York, Alfred A. Knopf, pp. 212–213.

[20] Rubin, V., telephone interview with T.Y., May 2002. All quotations from Vera Rubin in this chapter are based on this interview with her.

[21] Faber, S. M. and Primack, J., interview with T.Y., Santa Cruz, California, Aug. 22, 2000. All quotations from Sandra Faber and Joel Primack in this chapter and others are based on this interview conducted with the two scientists together at the University of California, Santa Cruz.

[22] Ibid.

[23] Axions, if they existed, would belong to the family of particles known as bosons—the force-carrying particles that include photons and gluons. Axions would be slow despite their low mass because they would have been created non-thermally in a phase transition. WIMPs, as well as neutrinos, on the other hand, would have been created thermally.

[24] MACHOs are a reality. Over the years, their presence has been detected in the way that their gravity bends light from other objects. (A phenomenon known as microlensing.) And in 2001, an international team of astronomers using the Hubble Space Telescope actually observed a MACHO directly for the first time. A small, dwarf star about 600 light-years away with a mass between 5 and 10% that of the Sun. Then, with the European Southern Observatory's Very Large Telescope, they succeeded in getting spectral information about the star, which supported the original identification. So MACHOs may well make up some percentage of the dark matter—perhaps even as much as 50% of the dark matter in galaxy haloes. But based on big bang nucleosynthesis, as well as other reasoning, scientists believe that baryonic material such as that in MACHOs cannot make up all of the dark matter in the universe. (Alcock, C. R. *et al* (2001). "Direct detection of a microlens in the Milky Way." *Nature* **441** 617–619.)

[25] When Faber, Primack and their colleagues wrote their paper, inflation was one of a number of processes they considered for generating spacetime wrinkles. As we'll see in the next chapter, improvements in the theory and, most recently, a host of observational tests, have given scientists greater confidence that the inflationary picture is the right one.

[26] To keep things moving along, I've left out one detail here. For a time in the early universe, the fluctuations in the dark matter actually are not too important. "That's because early on, the universe consists mostly of radiation—photons, neutrinos, things moving essentially at the speed of light," Primack says. (Neutrinos, unlike photons, do have mass, but it is so infinitesimal that they move at very close to the speed of light.) There are some fluctuations in this stuff—it does feel gravity. But according to Sandy Faber, "these never lead to collapse, partly because the radiation is moving so fast that gravity just can't affect it much. And not only that. Suppose you have a little region over here with higher than average radiation, and another over there with lower than average. Pretty soon, some radiation over here is going to stream over there, and vice versa. So it all just smears out—a process known as free streaming."

As time goes on, however, the universe cools and matter becomes increasingly important compared to radiation. At about 200 000 years, the universe reaches a milestone called matter–radiation equality: "That's when matter becomes the most important thing, gravitationally, in the universe," Primack says. "And suddenly," Faber adds, "the matter can 'see itself' via its gravity. Before then, gravity just didn't matter."

[27] Dey, A. "The Most Distant Galaxies," talk given at AAAS annual meeting, San Francisco, February 16, 2001.

[28] Astronomers use the degree of redshifting to express the age of a source of light in terms of a fraction of the time since the big bang. The actual fraction depends on the geometry of the universe. For a flat universe, it equals $1/(1+z)3/2$. In other words, one over three-halves power of $(1+z)$. For a $z$ of 1, that works out to $1/2.8$, or a smidgen over 35%.

[29] Steidel, C. "Galaxies at High Redshift," talk given at AAAS annual meeting, San Francisco, February 16, 2001.

[30] Bolte, M. "The Fossil Record — What stars in the Milky Way tell us," talk given at AAAS annual meeting, San Francisco, February 16, 2001.

[31] Guth, A. H. (1997). *The Inflationary Universe – The quest for a new theory of cosmic origins*. Reading, MA, Perseus Books, p. 206.

[32] Guth, A. H. "Eternal Inflation," talk given at Cosmic Questions conference, National Museum of Natural History, Washington, D.C., April 14, 1999.

[33] Leslie, J. (1989). *Universes*, London, New York, Routledge, p. 5.

# Notes for Chapter 6

[1] Weinberg, S. (1999) Comments at a press briefing during the "Cosmic Questions" conference at the National Museum of Natural History, organized by the Dialogue on Science, Ethics and Religion of the AAAS. April 15, 1999. Washington.

[2] Polkinghorne, J. (1999) Ibid.

[3] Perlmutter, S., interview with T.Y. July 19, 2002, at U.C. Berkeley, California. All quotations from Saul Perlmutter in this chapter are from this interview.

[4] Turner, M. S., interview with T.Y. at Aspen Center for Physics, Aspen, Colorado, June 28, 2002. All quotations from Michael Turner in this chapter are from this interview.

[5] Turner, M. S. (1990). "The Best Fit Universe." A talk given at *Nobel Symposium No. 79: The Birth and Early Evolution of Our Universe*, Graftavallen, Sweden, June 11–16, 1990. Available at: http://finalpubs/fnal.gov/archive/1990/conf/Conf-90-226-A.pdf

[6] Hamilton, A. E-mail communication with T.Y., August 5, 2002. All quotations from Hamilton in this chapter are from this communication.

[7] Those that do manage to penetrate to Earth's surface make their presence known. For example, about 10% of the "snow" seen between television channels in the days before cable was caused by microwave background photons. On the other hand, the microwave background photons that reach us don't pack much energy. The background microwave energy absorbed by a person is equal to about one ten millionth of the power from a 100 watt light bulb.

[8] According to Ian Morrison of the Jodrell Bank Observatory, technically speaking the map unveiled by COBE researchers in 1992 did not really reveal temperature variations. What it found, he argues, was noise. But these fluctuations in the signal were slightly higher than would have been expected if the cosmic microwave background radiation were totally smooth. This higher noise level proved that there were temperature variations in the radiation. But based on the initial data, the COBE researchers could not really say "which parts of the sky were hotter or colder than others," Morrison says. Later that year, astronomers at Jodrell Bank produced a map of the background radiation showing real temperature differences. And when four years of COBE data had been analyzed, the COBE team's final map did reveal temperature variations in the background radiation.

[9] Hu, W., telephone interview with T.Y., April 2002. All quotations in this chapter from Wayne Hu are from this interview.

[10] Sawyer, K. "New Findings Support Theory of Big Bang," *Washington Post*, April 24, 1992.

[11] Guth, A. H. (1997). *The Inflationary Universe – The quest for a new theory of cosmic origins.* Reading, MA, Perseus Books, p. 243.

[12] In Wilford, J. N. "Scientist Reports Profound Insight on How Time Began," *The New York Times*, April 24, 1992.

[13] Ibid.

[14] In Sawyer, K. "New Findings Support Theory of Big Bang," *Washington Post*, April 24, 1992.

[15] In Sawyer, K. "Big Bang Ripples Have Universal Impact," *Washington Post*, May 3, 1992.

[16] In Browne, M. E. "Despite New Data, Mysteries of Creation Persist," *The New York Times*, May 12, 1992.

[17] Guth, A. H., E-mail interview with T.Y., July 21, 1998.

[18] Krauss, L. M. and Turner, M. S. (1995). "The cosmological constant is back." *General Relativity and Gravitation* **27** 1137–1144.

[19] According to Turner, Perlmutter's group actually made the discovery first. But they wound up publishing the results in 1999, second to the High-$z$ Supernova Team. Turner says Perlmutter pulled him aside at a conference in 1998 to see a poster presentation he had prepared. "He calls me over and says, 'Mike, you may be interested in this.'" The data in Perlmutter's poster revealed that the expansion of the universe was accelerating. "I told him I thought it was great science. And he really was first with it. But the one who wins the race is not the one who gives a poster paper first; it's the one who publishes first." And the High-$z$ Supernova Team wound up beating Perlmutter's team to publication. "The other group caught that one lucky break," Turner says. But he believes history will judge the race to be a tie.

[20] Wilford, J. N. "Wary astronomers ponder an accelerating Universe." *New York Times*, March 3, 1998.

[21] Watson, A. (2002). "COSMOLOGY: cosmic ripples confirm Universe speeding up." *Science* **295** (5564), 2341.

[22] Seife, C. (2001). "Microwave telescope data ring true." *Science* **291** (5503), 441.

[23] In Seife, C. (2001). "ASTROPHYSICS: echoes of the Big Bang put theories in tune." *Science* **292** (5518), 823.

# Notes for Chapter 7

[1] Kolb, R. "A Recipe for Primordial Soup," talk given at Cosmic Questions Conference, National Museum of Natural History, Washington, D.C., April 14, 1999.

[2] Guth, A. H. "Eternal Inflation." Ibid.

[3] Hawking, S. and R. Penrose (1996). *The Nature of Space and Time*. Princeton, Princeton University Press, p. 76.

[4] In Guth, A. H. (1997). *The Inflationary Universe – The quest for a new theory of cosmic origins*. Reading, MA, Perseus Books, pp. 13–14.

[5] Guth, A. H., interviewed by T.Y. at "Cosmic Questions" conference, Washington, April 15, 1999.

[6] Guth, A. H. (1999) comments at "Cosmic Questions" press conference, Washington, April 15, 1999.

[7] Turok, N., interview with T.Y. Quotations from Turok in this and other chapters come from interviews I conducted with him in person, by phone and e-mail over the course of five years.

[8] Hawking, S. and Penrose, R. (1996). *The Nature of Space and Time*. Princeton, Princeton University Press, pp. 75–76.

[9] Hawking, S. (2001). *The Universe in a Nutshell*. New York, Bantam Books, p. 85.

[10] Hawking, S. (1988). *A Brief History of Time*. New York, Bantam Books.

[11] Encyclopaedia Britannica [Internet]: Creation Myth; cited June 23, 2002. Available from: http://search.eb.com/eb/article?eu=117208.

[12] Wheeler, J. A. (1998). *Geons, Black Holes & Quantum Foam*. New York, W. W. Norton, p. 168.

[13] Preskill, J., telephone interview with T.Y., July 14, 1998.

[14] Linde, A., telephone interview by T.Y., July 10, 1998.

[15] Ibid. All quotations in this chapter from Linde are from this interview.

[16] Hawking, S. and R. Penrose (1996). *The Nature of Space and Time*. Princeton, Princeton University Press, pp. 3–4.

[17] Ibid. p. 121.

[18] In Clarke, T. Nature Science Update [Internet]: Big Bangs Spark Row; updated April 26, 2002; cited June 24, 2002. Available from: http://www.nature.com/nsu/020422/020422-17.html.

[19] Encyclopaedia Britannica [Internet]: Paramenides; cited June 23, 2002. Available from: http://search.eb.com/eb/article?eu=59996.

# Notes for Prologue 2

[1] Pardue, L. Interviewed by T.Y., June 11, 1999. All information and quotations from Pardue are from this interview.

[2] Interviews with Wilson and other Portales residents were conducted in person May 22–25, 1999.

[3] A tiny handful of meteorites are thought to have come from the Moon and Mars, blasted free and flung toward Earth by asteroid impacts.

[4] "Meteor Man," *Portales News-Tribune*, June 28, 1998.

[5] University of New Mexico meteorite museum.

[6] Kring, D. *et al* (1999) "Portales Valley: A meteoritic sample of the brecciated and metal-veined floor of an impact crater on an H-chondrite asteroid," *Meteoritics and Planetary Science*, **34**, 663–669.

[7] Brearley, A. Interview with T.Y., May 26, 1999, University of New Mexico, Albuquerque. All quotations from Brearley in this chapter are from this interview.

[8] Kring, D. Telephone interview with T.Y., June 1999. All quotations from Kring are from this interview.

[9] When a polished surface of iron–nickel from certain kinds of meteorites is etched with a dilute acid, the pattern made by the metallic crystals is revealed. This is called the Widmanstätten pattern. From the width of individual crystals, it's possible to estimate how long it took for the metal to cool. The Widmanstätten pattern forms at rates of cooling of a few degrees per million years.

[10] Planetary bodies start their lives hot, in part because of the many impacts that occurred in the early solar system. As they age, they steadily cool. And the smaller the planet, the faster the cooling. So Mars, with a diameter slightly more than half that of Earth's, has pretty much given up its internal heat. Once exuberantly active with gargantuan volcanoes and lava flows—evidence of intense internal heat—Mars today seems not much more than a cadaver.

# Notes for Chapter 8

[1] Shakespeare, W. (1610). *The Tempest*. London, World Library, Inc. [Internet]: netLibrary. Available from http://emedia.netlibrary.com.

[2] Twain, M. (1981). *The Adventures of Huckleberry Finn*. New York, Bantam Books. The quotation comes from Chapter 19, "The Duke and the Dauphin Come Aboard."

[3] Bally, J. Interviews with T.Y., December 1999, at the CFHT and Hale Pohaku Lodge, Big Island, Hawaii. Information and quotations from Bally in this and succeeding chapters come from these interviews, as well as numerous discussions in person, by phone and by e-mail between 1999 and 2002.

[4] Maddalena, R. J. National Radio Astronomy Observatory [Internet]: "A Tour of Orion the Hunter," cited July 24, 2002. Available from: http://www.gb.nrao.edu/~rmaddale/Education/OrionTourCenter/index.htm.

[5] Ibid.

[6] Many astronomers prefer to say that there are two such clouds. But according to Bally, it's acceptable to think of the two as comprising a single cloud complex. For simplicity's sake, this is how I will refer to it here.

[7] For a map showing the patchy distribution of molecules within our galaxy, the Milky Way, see http://antwrp.gsfc.nasa.gov/apod/ap970430.html.

[8] Lada, C. J. and Kylafis, N. Eds. (1999). *The Origin of Stars and Planetary Systems*. NATO Science Series. Dordrecht, Netherlands, Kluwer Academic Publishers, p. 156.

[9] Blitz. Encyclopedia of Astronomy and Astrophysics [Internet]: "Interstellar Molecular Clouds;" updated November 2000; cited July 24, 2002. Available from: http://www.ency-astro.com/.

[10] In terms of mass, the Sun consists today of 73.5% hydrogen, 24.8% helium and 1.7% other elements. Grevesse, N. and Sauval, A. J. Encyclopedia of Astronomy and Astrophysics [Internet]: "Solar Abundances"; updated November 2000; cited July 25, 2002. Available from: http://www.ency-astro.com/.

[11] Here's a bit more detail about the nature of the Sun. At its center, neutral atoms and molecules are crushed up against each other so hard that the electrons can no longer orbit their nuclei normally. The result? Bare nuclei and electrons, otherwise known as a plasma. The density of these particles is astonishingly high: something like $10^{26}$ particles are crammed together in a cubic centimeter at the center of the Sun. But the particles are so tiny they still have plenty of room to move. As a result, the plasma behaves like a perfect gas. As with any gas, the higher the pressure, the higher the temperature. The pressure at the center of the Sun is 10 000 times greater than the pressure at the center of the Earth. And that's why the temperature is at least 15 000 000 million Kelvin.

In these unimaginably harsh conditions, electrons accelerate madly, emitting a great number of photons. What kind? The radiation is essentially that of a blackbody. You may recall from Chapter 2 that blackbody radiation has a peak in the spectrum that depends solely on the temperature. At the temperatures typical of the Sun's interior, that peak should be in the X-ray portion of the electromagnetic spectrum. So X-ray photons zing this way and that. But they never get very far — less than an inch — without crashing into matter particles. When this happens, they are either scattered or absorbed and re-emitted. The bottom line is that X-ray photons and matter particles race about in a frenzy, catapulting into one another with great violence.

In this maelstrom, X-ray photons zig and zag in what physicists call a "random walk." But it's not completely aimless because the photons do manage to make slow progress toward the surface. Along the way, they are degraded into lower energy photons, eventually emerging at the surface, where the temperature is about 5800 Kelvin, as visible light. The entire photon perambulation, from center to surface, takes on the order of a million years. And once photons fly from the Sun, it takes only about eight minutes for some to reach us here on Earth, providing the energy that makes our planet habitable.

Photons have been leaking away like this since the Sun was born some 4.5 billion years ago. Luckily, there has been a source of photon replenishment: the thermal energy residing in hot matter in the Sun's deep interior. But the thermal energy should have become depleted after just 50 million years. Yet the Sun still shines. So something keeps the Sun's tank from running dry.

In the 1920s, scientists figured out what it is: nuclear fusion. Hydrogen nuclei like the ones found in the Sun's interior ordinarily repel one another because of electrostatic repulsion. But the nuclei bang about so violently in the hot, dense conditions that the repulsion can be overcome, causing nuclei to fuse. In the process, some of the mass of the fusing nuclei is converted into energy, which is released in the form of neutrinos and photons. The neutrinos don't interact much with matter, so they go streaming off, carrying energy away from the Sun. But the photons do interact with the matter, smashing into it and heating it up, giving rise to new photons, which random walk to the surface.

One final thought. Where did the pressure and high temperature of matter at the Sun's center come from originally? The answer is gravity. As we'll see in coming chapters, when the solar system was forming, gravity pulled hydrogen, helium and other materials together into a dense ball. As that happened, gravitational energy was transformed into heat. (Sources: Stix, M. Encyclopedia of Astronomy and Astrophysics [Internet]: "Sun: Basic Properties;" updated November 2000; cited July 25, 2002. Available from: http://www.ency-astro.com/. Shu, F. H. (1982). *The Physical Universe: An Introduction to Astronomy*. Sausalito, CA, University Science Books.)

[12] European Southern Observatory. ESO [Internet]: "Secrets of a Dark Cloud;" updated July 2, 1999; cited July 25, 2002. Available from: http://www.eso.org/outreach/press-rel/pr-1999/phot-29-99.html.

[13] Ibid.

[14] Doug Johnstone has posted beautiful SCUBA images of the Orion cloud at http://www.cita.utoronto.ca/~johnston/jcmt.html.

[15] Frommert, H. and Kronberg, C. The Munich Astro Archive and Students for the Exploration and Development of Space [Internet]: "M 42;" updated April 9, 2002; cited July 26, 2002. Available from: http://www.seds.org/messier/m/m042.html.

[16] O'Dell, C. R. (1997). "Young stars and their surroundings." *Science* **276** 1355–1369.

[17] Britt, R. R. Space.com [Internet]: "Off the Charts: Hot Stars Surprise Astronomers;" updated Nov. 10, 2000. Available from: http://www.space.com.

[18] Scott, J. University of Colorado Office of News Services [Internet]: "CU Astrophysicists Use Chandra Satellite to Help Discover First Cries of Baby Stars;" updated Nov. 9, 2000. Available from: http://www.colorado.edu/NewsServices/NewsReleases/2000/936.html.

[19] Devitt, T. "Dissecting a Cocoon of Stardust, Scientists Begin to Tease Out a Hidden Star's Secrets," University of Wisconsin, Madison press release, July 24, 1998.

[20] Supernovae, of course, have other effects as well. To get a feel for one of those effects, hold your hand out for 10 seconds. Feel anything? No? That's funny, because a dozen electrons and muons just shot through your flesh and bone. These particles actually are by-products of collisions between cosmic rays and atoms in the atmosphere. The cosmic rays, in turn, are believed to be produced by supernova explosions. According to the theory, a supernova shock wave blasts interstellar atoms to bits — protons, helium nuclei and other particles — which are accelerated to energies as much as a billion times greater than those produced in the most powerful particle accelerators on Earth. It's a good thing that the atmosphere intervenes between us and these cosmic rays because the numerous electrons and muons that pass through us do much less damage.

[21] With some modern telescopes, such as the Kecks, the astronomers don't even have to be on the mountain. With video-conferencing, they can sit in a control room at the base of the mountain and tell the telescope operator up at the observatory what to do. And the data are piped down through high-bandwidth connections.

[22] Technically speaking, a Herbig–Haro object does not have to be powered by a jet. According to Bally, any shock-excited nebula powered in some way by a young star is classified as a Herbig–Haro object. In addition to jets, wide angle winds from a star can power the shocks too.

[23] Naeye, R. (1998). "The story of starbirth." *Astronomy* **26** 50–55.

[24] The momentum of an object moving in a straight line is equal to the mass of the object times its velocity. Angular momentum is a measure of the momentum of an orbiting or spinning body. The higher the mass or the greater the velocity, the greater the angular momentum. And if the mass or velocity of a protostar's rotation is high enough, the resulting forces would tend to tear the object apart.

[25] Kant, I. (1969). *Universal Natural History and the Theory of the Heavens*. Ann Arbor, The University of Michigan Press, p. 17.

[26] Cajori, F. (1934). *Newton's Principia, Motte's Translation Revised*. Berkeley, University of California Press, p. 544.

[27] Kant, I. (1969). *Universal Natural History and the Theory of the Heavens*. Ann Arbor, The University of Michigan Press, p. 74.

[28] Ibid, pp. 80–81.

[29] Wertheim, M. (2000). *The Pearly Gates of Cyberspace: A history of space from Dante to the Internet*. New York, W. W. Norton, pp. 157–158.

[30] Lada, C. J. and N. Kylafis, Eds. (1999). *The Origin of Stars and Planetary Systems*. NATO Science Series. Dordrecht, Netherlands, Kluwer Academic Publishers, p. 581.

[31] Ibid, pp. 583–584.

[32] Naeye, R. (1998). "The story of starbirth." *Astronomy* **26** 50–55.

[33] Boss, A. E-mail interview with T.Y., June 6, 1997.

[34] The heating of a cloud core as it collapses actually poses a problem: gas expands as its temperature rises. This expansion could counteract the force of gravity pulling the gas together, slowing or even stopping the collapse. As a result, for a cloud core to collapse, it must somehow get rid of heat. A widely accepted theory holds that the astronomical equivalent of a swamp cooler—water within the cloud core—does the trick. According to the theory, as the frenetically moving molecules within the core collide with each other, electrons within the atoms of water molecules are shoved into higher orbits, removing kinetic energy from the system. When the electrons drop back down into their former orbits, they give up potential energy, emitting photons that easily run off from the cloud, carrying energy away at the speed of light. In this way, energy is removed from the cloud core, allowing gravity to win out over heat-induced expansion.

[35] According to Frisch, the movement of the solar system itself combined with the movement of material from the Scorpius–Centaurus association produces a wind of interstellar material that's flowing toward the Sun at about 26 kilometers per second. It is coming from a direction close to the plane of the

ecliptic and within about 15° of the center of the galaxy. Because this material flows through the solar system, it has been dubbed the "local interstellar wind". (Frisch, P. C. (2000). "The Galactic Environment of the Sun." *American Scientist* **88**(1). Accessed online at http://www.sigmaxi.org/amsci/articles/00articles/frischintro.html#solar.)

[36] As noted earlier, many astronomers believe that supernovae can trigger the collapse of molecular clouds, triggering star formation. But whether such an explosion was the trigger for the formation of our own solar system is controversial—a topic that comes up in chapters 6 and 7.

[37] A collaboration led by Bally's former student, David Theil, lends support for this idea. In January of 2000, Theil presented new evidence that super-rings are the origin of giant molecular clouds. Working under Bally's supervision, Theil analyzed a survey of molecular clouds published in 1988 by Tom Dame of the Harvard-Smithsonian Center for Astrophysics. He found that the clouds, which include Orion's, are all expanding away from a central location in the Alpha Persei star cluster. This suggests a common origin for the clouds, Theil says. Moreover, the velocity of the clouds matches what one might expect for a super-ring about 40 million to 50 million years old, which is when the supernovae in Alpha Persei are believed to have gone off. (The clustered stars are what's left of the original star-formation event.) "One satisfying idea from this new research is that star formation can be cyclical and self-regulating. A young cluster releases energy that both destroys the cloud that spawned the cluster and starts new generations of star birth thousands of light-years away," Theil says.

[38] Wetherill, G. W. (1994). "Possible consequences of absence of 'Jupiters' in planetary systems." *Astrophysics and Space Science* **212** 23–32.

[39] Ibid.

[40] The cycling actually is not particularly efficient, Bally notes. A hydrogen atom may loop from an active star-forming region like the Orion Nebula into a new molecular cloud and then into a new star-forming region some 20 times before it finds itself becoming part of a star. That's because only 5% of the material in any given cloud actually winds up being incorporated into stars. The rest is flushed out by the kinds of processes Bally has been documenting in Orion. If the hydrogen atom is incorporated into a massive O-star, which has a short life, it may be forged into silicon or iron or some other heavy element in about 40 million years. If, on the contrary, it finds its way into a smaller and much more long-lived star like our Sun, it will be locked up there for billions of years. Eventually, though, even Sun-like planets die, and they spew their contents back out into interstellar space to be recycled again.

# Notes for Chapter 9

[1] Shostak, S. Talk given at "Intelligent Life in the Universe" symposium sponsored by the University of Colorado at Boulder's Center for Astrobiology, October 10, 2000, Boulder, Colorado.

[2] Rees, M. J. (1997). *Before the Beginning: our universe and others*, Reading, MA, Addison-Wesley, p. 2.

[3] Jayawardhana, R. (2000). "Meet the cosmic gambler." *Astronomy* **28** 42–47.

[4] Jayawardhana, R. Telephone interview with T.Y., January 29, 2002. All quotations from Jayawardhana in this chapter are from this interview.

[5] Shu, F. H., interview with T.Y., July 20, 2002. University of California, Berkeley.

[6] Boss, A. P., telephone interview with T.Y., July 10, 2001.

[7] Space Telescope Science Institute. STScI [Internet]: "Giant Molecular Clouds: Breeding Grounds for Star Birth;" updated October 21, 1997; cited July 19, 2001. Available from: http://oposite.stsci.edu/pubinfo/pr/1997/34/af2.html.

[8] Shu, F. H. "Mechanics of Star Formation," talk given at NASA/Michelson Fellowship Program Interferometry Summer School, August 21, 2000. Audio available from: http://sim.jpl.nasa.gov/michelson/iss2000/shu.html.

[9] These relationships are expressed formulaically as $J = Iw$, where $J$ is the angular momentum, $I$ is the moment of inertia (measured in $g \cdot cm^2$), and $w$ is the angular velocity (measured in radians per second). The skater's moment of inertia decreases as she draws in her arms because the diameter of the circle she inscribes as she spins contracts, which means her mass covers less area. (Since $I$ is measured in $g \cdot cm^2$, reducing the surface area of that circle reduces the moment of inertia.)

[10] Shu, F. H. "Mechanics of Star Formation," talk given at NASA/Michelson Fellowship Program Interferometry Summer School, August 21, 2000. Audio available from: http://sim.jpl.nasa.gov/michelson/iss2000/shu.html.

[11] Lada, C. J. and Kylafis, N. Eds. (1999). *The Origin of Stars and Planetary Systems*. NATO Science Series. Dordrecht, Netherlands, Kluwer Academic Publishers, p. 19.

[12] Ibid, p. 20.

[13] Ibid, p. 33.

[14] Shu, F. H. "Mechanics of Star Formation," talk given at NASA/Michelson Fellowship Program Interferometry Summer School, August 21, 2000. Audio available from: http://sim.jpl.nasa.gov/michelson/iss2000/shu.html.

[15] Lada, C. J. and N. Kylafis, Eds. (1999). *The Origin of Stars and Planetary Systems*. NATO Science Series. Dordrecht, Netherlands, Kluwer Academic Publishers, p. 209.

[16] Ibid, p. 210.

[17] Boss, A. P. E-mail interview with T.Y., July 20, 2001.

[18] Astronomy Picture of the Day [Internet]: "The Antennae Galaxies, October 22, 1997; cited July 19, 2001. Available from: http://antwrp.gsfc.nasa.gov/apod/ap971022.html.

[19] Space Telescope Science Institute. STScI [Internet]: "Star Clusters Born in the Wreckage of Cosmic Collisions"; updated July 19, 2001; cited July 19, 2001. Available from: http://oposite.stsci.edu/pubinfo/pr/2001/22/pr.html.

[20] Gallagher, S. E-mail interview with T.Y., July 20, 2001.

[21] Ibid.

[22] Shu, F. H. "Mechanics of Star Formation," talk given at NASA/Michelson Fellowship Program: Interferometry Summer School, August 21, 2000. Audio available from: http://sim.jpl.nasa.gov/michelson/iss2000/shu.html.

[23] I've included this virtual journey to a star-forming cloud core to bring the science to life. Needless to say, it is a flight of fancy. Such a trip is techno-logically infeasible—and not just because we don't have the technology to travel fast enough to traverse great swaths of the Milky Way as if we were flying coast to coast in an airplane. To observe the evolution of a molecular cloud core from the gravitational collapse phase to the point where an infant star is born would take thousands of years.

[24] Scientists believe there are actually two collapse phases. The first collapse phase ends when material in the central object becomes so dense, infra-red energy can no long escape. As a result, heat builds up, and thus, so does pressure. This prevents collapse for about 10 000 years. A second collapse begins when molecular hydrogen begins to dissociate, which absorbs thermal energy. This allows contraction to resume. This, at least, is what theory suggests. But observers have not actually confirmed it. "It would be a true theoretical prediction and a real triumph if this phase could be discovered to actually exist," says Alan Boss.

[25] Shu, F. H., interview with T.Y., July 20, 2000. University of California, Berkeley.

[26] Boss, A. P. E-mail interview with T.Y., July 20, 2001.

[27] Lada, C. J. and Kylafis, N. Eds. (1999). *The Origin of Stars and Planetary Systems.* NATO Science Series. Dordrecht, Netherlands, Kluwer Academic Publishers, p. 616.

[28] Ibid.

[29] Shu, F. H. "Mechanics of Star Formation," talk given at NASA/Michelson Fellowship Program Interferometry Summer School, August 21, 2000. Audio available from: http://sim.jpl.nasa.gov/michelson/iss2000/shu.html.

[30] Ibid.

[31] Ibid.

[32] The spectral energy distribution was considered odd because ordinary stars radiate essentially as a blackbody, and the objects being studied by Lynden-Bell and Pringle were far from looking like blackbody sources.

[33] Lada, C. J. and Kylafis, N. Eds. (1999). *The Origin of Stars and Planetary Systems.* NATO Science Series. Dordrecht, Netherlands, Kluwer Academic Publishers, p. 581.

[34] Shu, F. H. "Mechanics of Star Formation," talk given at NASA/Michelson Fellowship Program Interferometry Summer School, August 21, 2000. Audio available from: http://sim.jpl.nasa.gov/michelson/iss2000/shu.html.

[35] If the star tried to rotate faster, the magnetic cables connecting it to the disk would tend to tug it back into line. On the other hand, if the star tried to slow down, the attachment points of the cables on the disk would get ahead of the attachment points on the star. As a result, the cables would tighten and snap the star forward.

[36] Shu, F. H. "Mechanics of Star Formation," talk given at NASA/Michelson Fellowship Program Interferometry Summer School, August 21, 2000. Audio available from: http://sim.jpl.nasa.gov/michelson/iss2000/shu.html.

[37] There have actually been many steps in the development of the model. Shu and colleagues first published a paper proposing the X-wind in 1988. (Shu, F. H., Lizano, S. *et al* (1988). "Mass loss from rapidly rotating magnetic protostars." *The Astrophysical Journal* **328** L19–L23.) It was built on the

work of many other investigators. Then in 1994, Shu and fellow researchers published a more comprehensive version of the theory, which explained the production of jets and how they take care of the protostar's angular momentum problem. (Shu, F. H., Najita, J. *et al* (1994). "Magnetocentrifugally driven flows from young stars and disks. A generalized model." *The Astrophysical Journal* **429** 781–796.) Since then, Shu and his colleagues have published several other papers setting forth what they have seen as the consequences of X-wind, including the production of chondrules.

[38] Shu, F. H., interview with T.Y., July 20, 2000, University of California, Berkeley.

# Notes for Chapter 10

[1] This quote comes from Bruno's 1548 work, *On the Infinite Universe and Worlds*. In it, the Italian philosopher sets forth his controversial ideas in the form of a dialogue between three characters: Philotheo, who is Bruno's spokesperson, and his interlocutors, Fracastoro and Elpino. These particular words come from Elpino. And Philotheo answers, "That's right." (Dannielson, D. R., Ed. (2000). *The Book of the Cosmos*. Cambridge, Massachusetts, Helix Books, p. 144.)

[2] Ibid, p. 150.

[3] Ibid, p. 169.

[4] Wertheim, M. (2000). *The Pearly Gates of Cyberspace: A history of space from Dante to the Internet*. New York, W. W. Norton, p. 33. In her book, Wertheim beautifully describes how the medieval hierarchy of "soul-space" manifested in the art of the times. In The Last Judgement by Giotto, painted on a wall in the Arena Chapel, a giant image of Christ floats in the center of the image. Next in size come the angels and apostles. Further down in size and thus in the metaphysical hierarchy are humans who have been saved. And tiniest of all are the unfortunate souls who have been walled off in Hell.

[5] Ibid, p. 130.

[6] Gingerich, O., telephone interview with T.Y., March 7, 2002. All quotations from Gingerich in this chapter are from this interview.

[7] Dannielson, D. R., Ed. (2000). *The Book of the Cosmos*. Cambridge, Massachusetts, Helix Books, p. 142.

[8] Copernicus, N. (1995). *The Revolutions of Heavenly Spheres*. New York, Prometheus Books, p. 24.

[9] Ibid, p. 3.

[10] Ibid, p. 51.

[11] Ibid, p. 24.

[12] Wertheim, M. (2000). *The Pearly Gates of Cyberspace: A history of space from Dante to the Internet*. New York, W. W. Norton, p. 142.

[13] Ferris, T. (1997). *The Whole Shebang*. New York, Simon & Schuster, p. 26.

[14] Cajori, F. (1934). *Newton's Principia, Motte's Translation Revised*. Berkeley, University of California Press, p. xviii.

[15] There are $360°$ in the sky. Each degree consists of 60 arcminutes. And each arc minute consists of 60 arcseconds. The angular separation between an extrasolar planet and its parent star is on the order of a fraction of an arcsecond.

[16] Jayawardhana, R. Telephone interview with T.Y., January 29, 2002.

[17] Lissauer, J. Telephone interview with T.Y., August 2, 2001.

[18] Butler, P. Interview with T.Y. at the Department of Terrestrial Magnetism, Carnegie Institution of Washington, June 29, 2000.

[19] Kerr, R. A. (2002). "EXOPLANETS: Jupiter's brother joins the family." *Science* **296**(5576) 2124.

[20] Ibid.

[21] Butler, P. Interview with T.Y. at the Department of Terrestrial Magnetism, Carnegie Institution of Washington, June 29, 2000.

[22] Ibid.

# Notes for Chapter 11

[1] Dannielson, D. R., Ed. (2000). *The Book of the Cosmos*. Cambridge, Massachusetts, Helix Books, p. 515.

[2] Kinney, A. (2001). Talk given at National Astrobiology Institute meeting, Washington, D.C. April 12, 2001.

[3] Shu, F. H. Interview with T.Y. at University of California, Berkeley, July 20, 2000. All quotations from Shu in this chapter are from this interview.

[4] Scientists aren't actually sure exactly what interstellar dust grains look like. But based on the way interstellar dust clouds affect light coming from stars, on the fact that interstellar dust grains must somehow stick if they are to grow into bigger bodies that can be incorporated into planets, and on computer modeling of grain growth, researchers have a pretty good idea what to expect. The description I've included in this virtual journey to a protoplanetary disk is based on that work. For more information, see Li, A. and Greenberg, J. M. (1997). "A Unified Model of Interstellar Dust." *Astronomy and Astrophysics* **323** 566–584; Wright, E. L. NASA [Internet]: "Fractal Interstellar Dust Up-Close;" updated May 9, 1999; cited Aug. 10, 2001. Available from: http://antwrp.gsfc.nasa.gov/apod/ap990509.html; and Wright, E. L. [Internet]: "Fractal Dust Grains;" updated May 10, 1999; cited Aug. 10, 2001. Available from: http://www.astro.ucla.edu/~wright/dust/.

[5] Ward, W. R. "On Planet Formation and Migration," talk given at NASA Origins conference, May 22, 1997, Estes Park, Colorado.

[6] As you may recall from earlier in the book, some astrophysicists say the CAIs and their radioactive isotopes have their origin in a supernova. The explosion, they argue, irradiated materials in the original cloud core, producing the isotopes. At the same time, the blast triggered the collapse of the cloud core, leading ultimately to the formation of our solar system. But Shu argues that the isotopes within the calcium–aluminum inclusions are much more naturally a product of the X-wind. According to his theory, both the CAIs and chondrules were produced when little dust balls were melted by the high temperatures prevailing close to the Sun and then thrown outward by the X-wind. More specifically, based on their chemistry, the CAIs appear to have formed a little closer to the Sun than the chondrules. And

Shu proposes that the melting occurred as a result of impulsive flares triggered by fluctuations in the magnetic field lines threading both the Sun and the disk. This is all part and parcel of the X-wind theory.

If the theory is correct, those flares would blast the tiny rock bits in the inner part of the disk with a very specific flux of high energy particles. And that specific flux would produce a very specific mix of radioactive isotopes in the calcium–aluminum inclusions. In other words, this is another prediction of the X-wind theory: if it's correct, then the CAIs must have that specific mix of isotopes. And guess what? They do. Of course, a star exploding near the original cloud core might be capable of doing the same thing. "But supernovae are very unlikely to do that," Shu says. An exploding star must have very specific characteristics to transform tiny rock bits in the cloud core into the specific mix of isotopes found within CAIs. The likelihood of such a specific match is very low. Another way of thinking about this is to say that the exploding star would have to "know" the specific make-up of rock bits in the cloud core in order to produce exactly the right kind of irradiation to produce exactly the right pattern of isotopes. "What does a supernova know about rocks?" Shu asks jokingly.

[7] Wood, J. A. E-mail interview with T.Y., September 7, 2001.

[8] Lissauer, J. "Origin of Planets," talk given at NASA Origins conference, May 22, 1997, Estes Park, Colorado.

[9] Boss, A. P. E-mail interview with T.Y., June 6, 1997.

[10] Jayawardhana, R. E-mail interview with T.Y., July 29, 2002.

[11] Lissauer, J. Telephone interview with T.Y., August 2, 2001.

[12] Jayawardhana, R. E-mail interview with T.Y., July 29, 2002.

[13] Boss, A. P. Interview with T.Y. at the Department of Terrestrial Magnetism, Carnegie Institution of Washington, June 29, 2000. The remaining quotations from Boss in this chapter are from this interview.

[14] Wetherill, G. W. Interview with T.Y. at Department of Terrestrial Magnetism, Carnegie Institution of Washington, June 29, 2000.

[15] Throop, H. B. Interview with T.Y. at the Southwest Research Institute, Boulder, Colorado, July 27, 2001. All quotations from Throop in this chapter are from this interview.

[16] Throop, H. B., Bally, J. *et al* (2001). "Evidence for dust grain growth in young circumstellar disks." *Science* **292** (5522), 1686–1689.

[17] The "edge" of the solar system is actually a boundary where the solar wind—a flow of particles emanating from the Sun—meets the interstellar medium. This boundary is called the heliopause and is believed to be teardrop-shaped. The space inside the heliopause contains the planets, moons, asteroids, comets and other bodies of the solar system. For an artist's rendering of the solar system at this scale, go to the Astronomy Picture of the Day page at http://antwrp.gsfc.nasa.gov/apod/ap020624.html.

[18] Ward, W. R. "On Planet Formation and Migration," talk given at NASA Origins conference, Estes Park, Colorado, May 22, 1997.

[19] Butler, P. Interview with T.Y. at the Department of Terrestrial Magnetism, Carnegie Institution of Washington, June 29, 2000.

[20] Lissauer, J. Telephone interview with T.Y., August 2, 2001.

[21] Ibid.

# Notes for Chapter 12

[1] In Dreifus, C. "A Conversation with Sir Martin Rees: Tracing Evolution of Cosmos From its Simplest Elements," *The New York Times*, April 28, 1998.

[2] Shu, F. H. (1982). *The Physical Universe: An Introduction to Astronomy.* Sausalito, CA, University Science Books, p. 527.

[3] Davies, P. (1992). *The Mind of God.* London, Simon & Schuster.

[4] Bernstein, M. P., Sandford, S. A. *et al* (1999). "Life's Far-Flung Raw Materials." *Scientific American* (July 1999) 27–33.

[5] Sandford, S. A. Interview with T.Y. at NASA Ames Research Center, Mountain View, California, Aug. 24, 2000. All quotations in this chapter from Sanford are from this interview.

[6] Hazen, R. M. (2001). "Emergence and the Origin of Life." In: G. Palyi (Ed.) *Proceedings of the Workshop on Life*, Modena, Italy, September 2000.

[7] Ward, P. D. and Brownlee, D. E. (2000). *Rare Earth.* New York, Copernicus, p. 67.

[8] Jakosky, B. (1998). *The Search for Life on Other Planets.* Cambridge, Cambridge University Press, pp. 96–98.

[9] Ibid.

[10] Rothschild, L. J. "What Good is the Moon? A biologist's perspective." Talk given at Astronomical Society of the Pacific Astrobiology Symposium, Pasadena, California, July 18, 2000.

[11] Halliday, A. N. (2001). "Earth science: In the beginning..." *Nature* **409** 144–145.

[12] Ibid.

[13] Zahnle, K. (2000). "Refugia From Asteroid Impacts on Early Mars and Earth," talk given at Astronomical Society of the Pacific Astrobiology Symposium, Pasadena, California, July 18, 2000.

[14] Ibid.

[15] Jakosky, B. (1998). *The Search for Life on Other Planets.* Cambridge, Cambridge University Press, p. 31.

[16] Halliday, A. N. (2001). "Earth science: In the beginning..." *Nature* **409** 144–145.

[17] Ibid.

[18] Ward, P. D. and Brownlee, D. E. (2000). *Rare Earth*, New York, Copernicus.

[19] Ibid, p. 207.

[20] Jakosky, B. (1998). *The Search for Life on Other Planets.* Cambridge, Cambridge University Press, p. 65.

[21] Watson, R. T. *et al* (2001). "Climate Change 2001: Synthesis Report, Summary for Policymakers, Intergovernmental Panel on Climate Change." Available online at http://www.ipcc.ch/. An increase in the global mean temperature, or the average temperature over a hemisphere of the planet, is pretty much irrelevant to a dairy farmer in England, a native seal hunter in the Canadian North, or a home owner living in the floodplain of the Mississippi River. What matters for people is how the heat retained in Earth's atmosphere by increased carbon dioxide is translated into changes that affect our lives — changes in rainfall, snow cover and ice thickness, for example. According to the most recent IPCC report, summer drying and the associated incidence of drought has increased in some regions during the twentieth century. On

the other hand, heavy rainfall events have increased in mid- and high northern latitudes. Meanwhile, snow cover has decreased in area by 10% since global satellite observations became available in the 1960s, and permafrost has thawed in some polar and sub-polar areas. And in recent decades, the thickness of the Arctic Sea in late summer to early autumn has thinned by 40%.

[22] Ibid.
[23] Jakosky, B. (1998). *The Search for Life on Other Planets*. Cambridge, Cambridge University Press, p. 176.
[24] Ibid, p. 65.
[25] Ibid, p. 177.
[26] Ibid, p. 265.
[27] Ibid.
[28] Ward, P. D. and Browlee, D. E. (2000). *Rare Earth*. New York, Copernicus, p. 210.
[29] Jakosky, B. (1998). *The Search for Life on Other Planets*. Cambridge, Cambridge University Press, pp. 125–128.
[30] Ibid, p. 266.
[31] In Ward, P. D. and Brownlee, D. E. (2000). *Rare Earth*. New York, Copernicus, p. 117.
[32] deMenocal, P. B. (1996). "Climate and Evolution," talk given at Geological Society of America Annual Meeting, Denver, October 1996.
[33] Stanley, S. M. (1996). Ibid.
[34] Ibid.
[35] Ward, P. D. and Brownlee, D. E. (2000). *Rare Earth*. New York, Copernicus, p. 214.
[36] Kerr, R. A. (2002). "PALEONTOLOGY: Earliest signs of life just oddly shaped crud?" *Science* **295**(5561) 1812.
[37] Brasier, M. D. *et al* (2002). "Questioning the evidence for Earth's oldest fossils." *Nature* **416** 76–81.
[38] Dalton, R. "Microfossils: squaring up over ancient life." *Nature* **417** 782–784.
[39] Mojzsis, S. J., interview with T.Y. at University of Colorado, Boulder, January 22, 2001. Unless otherwise referenced, all quotations from Mojzsis in this chapter are from this interview.
[40] Fedo, C. M. and M. J. Whitehouse (2002). "Metasomatic origin of quartz-pyroxene rock, Akilia, Greenland, and implications for Earth's earliest life." *Science* **296**(5572) 1448–1452; Kerr, R. A. (2002). "PALEONTOLOGY: Reversals reveal pitfalls in spotting ancient and E.T. life." *Science* **296**(5572) 1384–1385.
[41] Mojzsis, S. J., telephone interview with T.Y., August 2, 2002.
[42] Mojzsis, S. J., Harrison, M. *et al* (2001). "Oxygen-isotope evidence from ancient zircons for liquid water at the Earth's surface 4300 Myr ago." *Nature* **409** 178–181.
[43] Ward, P. D. and Brownlee, D. E. (2000). *Rare Earth*. New York, Copernicus, pp. 77–78.
[44] Bernstein, M. P., Sandford, S. A. *et al.* "Life's far-flung raw materials." *Scientific American* (July 1999): 27–33; Sandford, S. A., interview with T.Y. at NASA Ames Research Center, Mountain View, California, Aug. 24, 2000.

# Notes for Afterword

[1] Deutsch, D. (1997). *The Fabric of Reality.* New York, Penguin Books, p. 193.
[2] Dannielson, D. R., Ed. (2000). *The Book of the Cosmos.* Cambridge, Massachusetts, Helix Books, p. 429.
[3] Ibid, p. 88.

# INDEX

2276